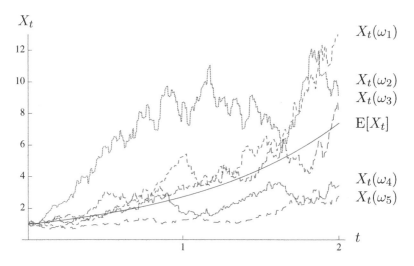

The cover figure shows five sample paths $X_t(\omega_1), X_t(\omega_2), X_t(\omega_3)$ and $X_t(\omega_4)$ of a geometric Brownian motion $X_t(\omega)$, i.e. of the solution of a (1-dimensional) stochastic differential equation of the form

$$\frac{dX_t}{dt} = (r + \alpha \cdot W_t)X_t \qquad t \geq 0 \; ; \;\; X_0 = x$$

where x, r and α are constants and $W_t = W_t(\omega)$ is white noise. This process is often used to model "exponential growth under uncertainty". See Chapters 5, 10, 11 and 12.

The figure is a computer simulation for the case $x = r = 1$, $\alpha = 0.6$. The mean value of X_t, $E[X_t] = \exp(t)$, is also drawn.

Courtesy of Jan Ubøe, Norwegian School of Economics and Business Administration, Bergen.

For other titles published in this series, go to
www.springer.com/series/223

Bernt Øksendal

Stochastic Differential Equations

An Introduction with Applications

Sixth Edition
With 14 Figures

 Springer

Bernt Øksendal
Department of Mathematics
University of Oslo
P.O. Box 1053 Blindern
0316 Oslo
Blindern
Norway
oksendal@math.uio.no

ISBN 978-3-540-04758-2   e-ISBN 978-3-642-14394-6
DOI 10.1007/978-3-642-14394-6
Springer Heidelberg Dordrecht London New York

Library of Congress Control Number: 2010930618

Mathematics Subject Classification (2000): 60H10, 60G35, 60G40, 60G44, 93E20, 60J45, 60J25

Cover design: WMXDesign GmbH, Heidelberg

Printed on acid-free paper

Springer is part of Springer Science+Business Media (www.springer.com)

To My Family
Eva, Elise, Anders and Karina

We have not succeeded in answering all our problems. The answers we have found only serve to raise a whole set of new questions. In some ways we feel we are as confused as ever, but we believe we are confused on a higher level and about more important things.

Posted outside the mathematics reading room,
Tromsø University

Preface to the Fifth Corrected Printing of the Sixth Edition

The biggest change in this printing is in Theorem 10.1.9, where the proof of part b) has been corrected and rewritten. I am indebted to *Yang Ge* for pointing out the gap in the previous proof and for other helpful suggestions. In addition some other corrections and improvements here and there have been carried out, mainly based on useful comments from (in alphabetical order), *Nikolai Chernov, Giulia Di Nunno, Ulrik Skre Fjordheim, Sindre Froyn, Helge Holden, Gautam Iyer, Torstein Nilssen, Steffen Sjursen, Jan Ubøe, Zhigus Yan* and *Yan Zhu*. I thank them all for their help.

I am also grateful to Dina Haraldsson for once again offering her valuable typing assistance.

Blindern, April 2010
Bernt Øksendal

Preface to the Fourth Corrected Printing of the Sixth Edition

Basically all the corrections in this printing are due to Ralf Forster. I am grateful for his valuable help to improve the book. I am also grateful to Jerome Stein for inspiring discussions, which led to the new Exercise 8.18.

I also want to thank Dina Haraldsson, who also this time helped me with her expert typing.

Blindern, February 2007
Bernt Øksendal

Preface to the Third Corrected Printing of the Sixth Edition

In this Second Corrected Printing some misprints have been corrected and other improvements in the presentation have been carried out. The exercises to which a (partial or complete) solution is provided (in the back of the book) are denoted by an asterix *. I wish to thank the following persons (in alphabetical order) for their helpful comments:

Holger van Bargen, Catriona Byrne, Mark Davis, Per-Ivar Faust, Samson Jinya, Paul Kettler, Alex Krouglov, Mauro Mariani, John O'Hara, Agnès Sulem, Bjørn Thunestvedt and Vegard Trondsen.

My special thanks go to Dina Haraldsson for her careful and skilled typing.

Blindern, August 2005
Bernt Øksendal

Preface to the Sixth Edition

This edition contains detailed solutions of selected exercises. Many readers have requested this, because it makes the book more suitable for self-study. At the same time new exercises (without solutions) have beed added. They have all been placed in the end of each chapter, in order to facilitate the use of this edition together with previous ones.

Several errors have been corrected and formulations have been improved. This has been made possible by the valuable comments from (in alphabetical order) Jon Bohlin, Mark Davis, Helge Holden, Patrick Jaillet, Chen Jing, Natalia Koroleva, Mario Lefebvre, Alexander Matasov, Thilo Meyer-Brandis, Keigo Osawa, Bjørn Thunestvedt, Jan Ubøe and Yngve Williassen. I thank them all for helping to improve the book.

My thanks also go to Dina Haraldsson, who once again has performed the typing and drawn the figures with great skill.

Blindern, September 2002
Bernt Øksendal

Preface to Corrected Printing, Fifth Edition

The main corrections and improvements in this corrected printing are from Chapter 12. I have benefitted from useful comments from a number of people, including (in alphabetical order) Fredrik Dahl, Simone Deparis, Ulrich Haussmann, Yaozhong Hu, Marianne Huebner, Carl Peter Kirkebø, Nikolay Kolev, Takashi Kumagai, Shlomo Levental, Geir Magnussen, Anders Øksendal, Jürgen Potthoff, Colin Rowat, Stig Sandnes, Lones Smith, Setsuo Taniguchi and Bjørn Thunestvedt.

I want to thank them all for helping me making the book better. I also want to thank Dina Haraldsson for proficient typing.

Blindern, May 2000
Bernt Øksendal

Preface to the Fifth Edition

The main new feature of the fifth edition is the addition of a new chapter, Chapter 12, on applications to mathematical finance. I found it natural to include this material as another major application of stochastic analysis, in view of the amazing development in this field during the last 10–20 years. Moreover, the close contact between the theoretical achievements and the applications in this area is striking. For example, today very few firms (if any) trade with options without consulting the Black & Scholes formula!

The first 11 chapters of the book are not much changed from the previous edition, but I have continued my efforts to improve the presentation throughout and correct errors and misprints. Some new exercises have been added. Moreover, to facilitate the use of the book each chapter has been divided into subsections. If one doesn't want (or doesn't have time) to cover all the chapters, then one can compose a course by choosing subsections from the chapters. The chart below indicates what material depends on which sections.

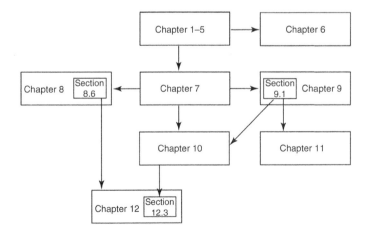

For example, to cover the first two sections of the new chapter 12 it is recommended that one (at least) covers Chapters 1–5, Chapter 7 and Section 8.6. Chapter 10, and hence Section 9.1, are necessary additional background for Section 12.3, in particular for the subsection on American options.

In my work on this edition I have benefitted from useful suggestions from many people, including (in alphabetical order) Knut Aase, Luis Alvarez, Peter Christensen, Kian Esteghamat, Nils Christian Framstad, Helge Holden, Christian Irgens, Saul Jacka, Naoto Kunitomo and his group, Sure Mataramvura, Trond Myhre, Anders Øksendal, Nils Øvrelid, Walter Schachermayer, Bjarne Schielderop, Atle Seierstad, Jan Ubøe, Gjermund Våge and Dan Zes. I thank them all for their contributions to the improvement of the book.

Again Dina Haraldsson demonstrated her impressive skills in typing the manuscript – and in finding her way in the L^{A}T$_E$X jungle! I am very grateful for her help and for her patience with me and all my revisions, new versions and revised revisions . . .

Blindern, January 1998

Bernt Øksendal

Preface to the Fourth Edition

In this edition I have added some material which is particularly useful for the applications, namely the martingale representation theorem (Chapter IV), the variational inequalities associated to optimal stopping problems (Chapter X) and stochastic control with terminal conditions (Chapter XI). In addition solutions and extra hints to some of the exercises are now included. Moreover, the proof and the discussion of the Girsanov theorem have been changed in order to make it more easy to apply, e.g. in economics. And the presentation in general has been corrected and revised throughout the text, in order to make the book better and more useful.

During this work I have benefitted from valuable comments from several persons, including Knut Aase, Sigmund Berntsen, Mark H. A. Davis, Helge Holden, Yaozhong Hu, Tom Lindstrøm, Trygve Nilsen, Paulo Ruffino, Isaac Saias, Clint Scovel, Jan Ubøe, Suleyman Ustunel, Qinghua Zhang, Tusheng Zhang and Victor Daniel Zurkowski. I am grateful to them all for their help.

My special thanks go to Håkon Nyhus, who carefully read large portions of the manuscript and gave me a long list of improvements, as well as many other useful suggestions.

Finally I wish to express my gratitude to Tove Møller and Dina Haraldsson, who typed the manuscript with impressive proficiency.

Oslo, June 1995   *Bernt Øksendal*

Preface to the Third Edition

The main new feature of the third edition is that exercises have been included to each of the chapters II–XI. The purpose of these exercises is to help the reader to get a better understanding of the text. Some of the exercises are quite routine, intended to illustrate the results, while other exercises are harder and more challenging and some serve to extend the theory.

I have also continued the effort to correct misprints and errors and to improve the presentation. I have benefitted from valuable comments and suggestions from Mark H. A. Davis, Håkon Gjessing, Torgny Lindvall and Håkon Nyhus, My best thanks to them all.

A quite noticeable non-mathematical improvement is that the book is now typed in T_EX. Tove Lieberg did a great typing job (as usual) and I am very grateful to her for her effort and infinite patience.

Oslo, June 1991   *Bernt Øksendal*

Preface to the Second Edition

In the second edition I have split the chapter on diffusion processes in two, the new Chapters VII and VIII: Chapter VII treats only those basic properties of diffusions that are needed for the applications in the last 3 chapters. The readers that are anxious to get to the applications as soon as possible can therefore jump directly from Chapter VII to Chapters IX, X and XI.

In Chapter VIII other important properties of diffusions are discussed. While not strictly necessary for the rest of the book, these properties are central in today's theory of stochastic analysis and crucial for many other applications.

Hopefully this change will make the book more flexible for the different purposes. I have also made an effort to improve the presentation at some points and I have corrected the misprints and errors that I knew about, hopefully without introducing new ones. I am grateful for the responses that I have received on the book and in particular I wish to thank Henrik Martens for his helpful comments.

Tove Lieberg has impressed me with her unique combination of typing accuracy and speed. I wish to thank her for her help and patience, together with Dina Haraldsson and Tone Rasmussen who sometimes assisted on the typing.

Oslo, August 1989   *Bernt Øksendal*

Preface to the First Edition

These notes are based on a postgraduate course I gave on stochastic differential equations at Edinburgh University in the spring 1982. No previous knowledge about the subject was assumed, but the presentation is based on some background in measure theory.

There are several reasons why one should learn more about stochastic differential equations: They have a wide range of applications outside mathematics, there are many fruitful connections to other mathematical disciplines and the subject has a rapidly developing life of its own as a fascinating research field with many interesting unanswered questions.

Unfortunately most of the literature about stochastic differential equations seems to place so much emphasis on rigor and completeness that it scares many nonexperts away. These notes are an attempt to approach the subject from the nonexpert point of view: Not knowing anything (except rumours, maybe) about a subject to start with, what would I like to know first of all? My answer would be:

    1) In what situations does the subject arise?
    2) What are its essential features?
    3) What are the applications and the connections to other fields?

I would not be so interested in the proof of the most general case, but rather in an easier proof of a special case, which may give just as much of the basic idea in the argument. And I would be willing to believe some basic results without proof (at first stage, anyway) in order to have time for some more basic applications.

These notes reflect this point of view. Such an approach enables us to reach the highlights of the theory quicker and easier. Thus it is hoped that these notes may contribute to fill a gap in the existing literature. The course is meant to be an appetizer. If it succeeds in awaking further interest, the reader will have a large selection of excellent literature available for the study of the whole story. Some of this literature is listed at the back.

In the introduction we state 6 problems where stochastic differential equations play an essential role in the solution. In Chapter II we introduce the basic mathematical notions needed for the mathematical model of some of these problems, leading to the concept of Ito integrals in Chapter III. In Chapter IV we develop the stochastic calculus (the Ito formula) and in Chapter V we use this to solve some stochastic differential equations, including the first two problems in the introduction. In Chapter VI we present a solution of *the linear filtering problem* (of which problem 3 is an example), using the stochastic calculus. Problem 4 is *the Dirichlet problem*. Although this is purely deterministic we outline in Chapters VII and VIII how the introduction of an associated Ito diffusion (i.e. solution of a stochastic differential equation) leads to a simple, intuitive and useful stochastic solution, which is the cornerstone of stochastic potential theory. Problem 5 is an *optimal stopping problem*. In Chapter IX we represent the state of a game at time t by an Ito diffusion and solve the corresponding optimal stopping problem. The solution involves potential theoretic notions, such as the generalized harmonic extension provided by the solution of the Dirichlet problem in Chapter VIII. Problem 6 is a stochastic version of F.P. Ramsey's classical control problem from 1928. In Chapter X we formulate the general *stochastic control problem* in terms of stochastic differential equations, and we apply the results of Chapters VII and VIII to show that the problem can be reduced to solving the (deterministic) Hamilton-Jacobi-Bellman equation. As an illustration we solve a problem about optimal portfolio selection.

After the course was first given in Edinburgh in 1982, revised and expanded versions were presented at Agder College, Kristiansand and University of Oslo. Every time about half of the audience have come from the applied section, the others being so-called "pure" mathematicians. This fruitful combination has created a broad variety of valuable comments, for which I am very grateful. I particularly wish to express my gratitude to K.K. Aase, L. Csink and A.M. Davie for many useful discussions.

I wish to thank the Science and Engineering Research Council, U.K. and Norges Almenvitenskapelige Forskningsråd (NAVF), Norway for their financial support. And I am greatly indebted to Ingrid Skram, Agder College and Inger Prestbakken, University of Oslo for their excellent typing – and their patience with the innumerable changes in the manuscript during these two years.

Oslo, June 1985   *Bernt Øksendal*

Note: Chapters VIII, IX, X of the First Edition have become Chapters IX, X, XI of the Second Edition.

Contents

Chapter 1
Introduction

To convince the reader that stochastic differential equations is an important subject let us mention some situations where such equations appear and can be used:

1.1 Stochastic Analogs of Classical Differential Equations

If we allow for some randomness in some of the coefficients of a differential equation we often obtain a more realistic mathematical model of the situation.

Problem 1. Consider the simple population growth model

$$\frac{dN}{dt} = a(t)N(t), \qquad N(0) = N_0 \text{ (constant)} \qquad (1.1.1)$$

where $N(t)$ is the size of the population at time t, and $a(t)$ is the relative rate of growth at time t. It might happen that $a(t)$ is not completely known, but subject to some random environmental effects, so that we have

$$a(t) = r(t) + \text{"noise"} ,$$

where we do not know the exact behaviour of the noise term, only its probability distribution. The function $r(t)$ is assumed to be nonrandom. How do we solve (1.1.1) in this case?

Problem 2. The charge $Q(t)$ at time t at a fixed point in an electric circuit satisfies the differential equation

$$L \cdot Q''(t) + R \cdot Q'(t) + \frac{1}{C} \cdot Q(t) = F(t), \ Q(0) = Q_0, \ Q'(0) = I_0 \qquad (1.1.2)$$

where L is inductance, R is resistance, C is capacitance and $F(t)$ the potential source at time t.

Again we may have a situation where some of the coefficients, say $F(t)$, are not deterministic but of the form

$$F(t) = G(t) + \text{``noise''} . \qquad (1.1.3)$$

How do we solve (1.1.2) in this case?

More generally, the equation we obtain by allowing randomness in the coefficients of a differential equation is called a *stochastic differential equation*. This will be made more precise later. It is clear that any solution of a stochastic differential equation must involve some randomness, i.e. we can only hope to be able to say something about the probability distributions of the solutions.

1.2 Filtering Problems

Problem 3.   Suppose that we, in order to improve our knowledge about the solution, say of Problem 2, perform observations $Z(s)$ of $Q(s)$ at times $s \leq t$. However, due to inaccuracies in our measurements we do not really measure $Q(s)$ but a disturbed version of it:

$$Z(s) = Q(s) + \text{``noise''} . \qquad (1.2.1)$$

So in this case there are two sources of noise, the second coming from the error of measurement.

The *filtering problem* is: What is the best estimate of $Q(t)$ satisfying (1.1.2), based on the observations $Z(s)$ in (1.2.1), where $s \leq t$? Intuitively, the problem is to "filter" the noise away from the observations in an optimal way.

In 1960 Kalman and in 1961 Kalman and Bucy proved what is now known as the Kalman-Bucy filter. Basically the filter gives a procedure for estimating the state of a system which satisfies a "noisy" linear differential equation, based on a series of "noisy" observations.

Almost immediately the discovery found applications in aerospace engineering (Ranger, Mariner, Apollo etc.) and it now has a broad range of applications.

Thus the Kalman-Bucy filter is an example of a recent mathematical discovery which has already proved to be useful – it is not just "potentially" useful.

It is also a counterexample to the assertion that "applied mathematics is bad mathematics" and to the assertion that "the only really useful mathematics is the elementary mathematics". For the Kalman-Bucy filter – as the whole subject of stochastic differential equations – involves advanced, interesting and first class mathematics.

1.3 Stochastic Approach to Deterministic Boundary Value Problems

Problem 4. The most celebrated example is the stochastic solution of the *Dirichlet problem*:

Given a (reasonable) domain U in $\mathbf{R}^n$ and a continuous function f on the boundary of U, ∂U. Find a function $\tilde{f}$ continuous on the closure $\overline{U}$ of U such that

(i) $\tilde{f} = f$ on ∂U

(ii) $\tilde{f}$ is harmonic in U, i.e.

$$\Delta \tilde{f} := \sum_{i=1}^{n} \frac{\partial^2 \tilde{f}}{\partial x_i^2} = 0 \quad \text{in } U \ .$$

In 1944 Kakutani proved that the solution could be expressed in terms of *Brownian motion* (which will be constructed in Chapter 2): $\tilde{f}(x)$ is the expected value of f at the first exit point from U of the Brownian motion starting at $x \in U$.

It turned out that this was just the tip of an iceberg: For a large class of semielliptic second order partial differential equations the corresponding Dirichlet boundary value problem can be solved using a stochastic process which is a solution of an associated stochastic differential equation.

1.4 Optimal Stopping

Problem 5. Suppose a person has an asset or resource (e.g. a house, stocks, oil...) that she is planning to sell. The price X_t at time t of her asset on the open market varies according to a stochastic differential equation of the same type as in Problem 1:

$$\frac{dX_t}{dt} = rX_t + \alpha X_t \cdot \text{"noise"}$$

where r, α are known constants. The discount rate is a known constant ρ. At what time should she decide to sell?

We assume that she knows the behaviour of X_s up to the present time t, but because of the noise in the system she can of course never be sure at the time of the sale if her choice of time will turn out to be the best. So what we are searching for is a stopping strategy that gives the best result in the long run, i.e. maximizes the *expected* profit when the inflation is taken into account.

This is an *optimal stopping problem*. It turns out that the solution can be expressed in terms of the solution of a corresponding boundary value problem (Problem 4), except that the boundary is unknown (free) as well and this is compensated by a double set of boundary conditions. It can also be expressed in terms of a set of *variational inequalities*.

1.5 Stochastic Control

Problem 6 (An optimal portfolio problem).
Suppose that a person has two investment possibilities:

(i)   A *safe* investment (e.g. a bond), where the price $X_0(t)$ per unit at time t grows exponentially:
$$\frac{dX_0}{dt} = \rho X_0 \tag{1.5.1}$$
where $\rho > 0$ is a constant.

(ii) A *risky* investment (e.g. a stock), where the price $X_1(t)$ per unit at time t satisfies a stochastic differential equation of the type discussed in Problem 1:
$$\frac{dX_1}{dt} = (\mu + \sigma \cdot \text{"noise"})X_1 \tag{1.5.2}$$
where $\mu > \rho$ and $\sigma \in \mathbf{R} \setminus \{0\}$ are constants.

At each instant t the person can choose how large portion (fraction) u_t of his fortune V_t he wants to place in the risky investment, thereby placing $(1 - u_t)V_t$ in the safe investment. Given a utility function U and a terminal time T the problem is to find the optimal *portfolio* $u_t \in [0, 1]$ i.e. find the investment distribution u_t; $0 \leq t \leq T$ which maximizes the expected utility of the corresponding terminal fortune $V_T^{(u)}$:

$$\max_{0 \leq u_t \leq 1} \left\{ E\left[U(V_T^{(u)}) \right] \right\} \tag{1.5.3}$$

1.6 Mathematical Finance

Problem 7 (Pricing of options).
Suppose that at time $t = 0$ the person in Problem 6 is offered the *right* (but

without obligation) to buy one unit of the risky asset at a specified price K and at a specified future time $t = T$. Such a right is called a *European call option*. How much should the person be willing to pay for such an option? This problem was solved when Fischer Black and Myron Scholes (1973) used stochastic analysis and an equlibrium argument to compute a theoretical value for the price, the now famous *Black-Scholes option price formula*. This theoretical value agreed well with the prices that had already been established as an equilibrium price on the free market. Thus it represented a triumph for mathematical modelling in finance. It has become an indispensable tool in the trading of options and other financial derivatives. In 1997 Myron Scholes and Robert Merton were awarded the Nobel Prize in Economics for their work related to this formula. (Fischer Black died in 1995.)

We will return to these problems in later chapters, after having developed the necessary mathematical machinery. We solve Problem 1 and Problem 2 in Chapter 5. Problems involving filtering (Problem 3) are treated in Chapter 6, the generalized Dirichlet problem (Problem 4) in Chapter 9. Problem 5 is solved in Chapter 10 while stochastic control problems (Problem 6) are discussed in Chapter 11. Finally we discuss applications to mathematical finance in Chapter 12.

Chapter 2
Some Mathematical Preliminaries

2.1 Probability Spaces, Random Variables and Stochastic Processes

Having stated the problems we would like to solve, we now proceed to find reasonable mathematical notions corresponding to the quantities mentioned and mathematical models for the problems. In short, here is a first list of the notions that need a mathematical interpretation:

(1) A random quantity
(2) Independence
(3) Parametrized (discrete or continuous) families of random quantities
(4) What is meant by a "best" estimate in the filtering problem (Problem 3)
(5) What is meant by an estimate "based on" some observations (Problem 3)?
(6) What is the mathematical interpretation of the "noise" terms?
(7) What is the mathematical interpretation of the stochastic differential equations?

In this chapter we will discuss (1)–(3) briefly. In the next chapter we will consider (6), which leads to the notion of an Itô stochastic integral (7). In Chapter 6 we will consider (4)–(5).

The mathematical model for a random quantity is a *random variable*. Before we define this, we recall some concepts from general probability theory. The reader is referred to e.g. Williams (1991) for more information.

Definition 2.1.1 *If Ω is a given set, then a σ-algebra $\mathcal{F}$ on Ω is a family $\mathcal{F}$ of subsets of Ω with the following properties:*

(i) $\emptyset \in \mathcal{F}$
(ii) $F \in \mathcal{F} \Rightarrow F^C \in \mathcal{F}$, *where* $F^C = \Omega \setminus F$ *is the complement of F in Ω*
(iii) $A_1, A_2, \ldots \in \mathcal{F} \Rightarrow A := \bigcup_{i=1}^{\infty} A_i \in \mathcal{F}$

The pair $(\Omega, \mathcal{F})$ is called a measurable space. *A probability measure P on a measurable space $(\Omega, \mathcal{F})$ is a function $P: \mathcal{F} \longrightarrow [0,1]$ such that*

(a) $P(\emptyset) = 0, \ P(\Omega) = 1$
(b)   *if $A_1, A_2, \ldots \in \mathcal{F}$ and $\{A_i\}_{i=1}^{\infty}$ is disjoint (i.e. $A_i \cap A_j = \emptyset$ if $i \neq j$) then*

$$P\left(\bigcup_{i=1}^{\infty} A_i\right) = \sum_{i=1}^{\infty} P(A_i) \ .$$

The triple $(\Omega, \mathcal{F}, P)$ is called a probability space. *It is called a* complete probability space *if $\mathcal{F}$ contains all subsets G of Ω with P-outer measure zero, i.e. with*

$$P^*(G) := \inf\{P(F); F \in \mathcal{F}, G \subset F\} = 0 \ .$$

Any probability space can be made complete simply by adding to $\mathcal{F}$ all sets of outer measure 0 and by extending P accordingly. From now on we will assume that all our probability spaces are complete.

The subsets F of Ω which belong to $\mathcal{F}$ are called $\mathcal{F}$-*measurable* sets. In a probability context these sets are called *events* and we use the interpretation

$$P(F) = \text{``the probability that the event } F \text{ occurs''} \ .$$

In particular, if $P(F) = 1$ we say that "F occurs with probability 1", or "almost surely (a.s.)".

Given any family $\mathcal{U}$ of subsets of Ω there is a smallest σ-algebra $\mathcal{H}_{\mathcal{U}}$ containing $\mathcal{U}$, namely

$$\mathcal{H}_{\mathcal{U}} = \bigcap\{\mathcal{H}; \mathcal{H} \ \sigma\text{-algebra of } \Omega, \ \mathcal{U} \subset \mathcal{H}\} \ .$$

(See Exercise 2.3.)

We call $\mathcal{H}_{\mathcal{U}}$ *the σ-algebra generated by $\mathcal{U}$.*

For example, if $\mathcal{U}$ is the collection of all open subsets of a topological space Ω (e.g. $\Omega = \mathbf{R}^n$), then $\mathcal{B} = \mathcal{H}_{\mathcal{U}}$ is called the *Borel σ-algebra* on Ω and the elements $B \in \mathcal{B}$ are called *Borel sets*. $\mathcal{B}$ contains all open sets, all closed sets, all countable unions of closed sets, all countable intersections of such countable unions etc.

If $(\Omega, \mathcal{F}, P)$ is a given probability space, then a function $Y: \Omega \rightarrow \mathbf{R}^n$ is called $\mathcal{F}$-*measurable* if

$$Y^{-1}(U) := \{\omega \in \Omega; Y(\omega) \in U\} \in \mathcal{F}$$

for all open sets $U \in \mathbf{R}^n$ (or, equivalently, for all Borel sets $U \subset \mathbf{R}^n$).

If $X: \Omega \rightarrow \mathbf{R}^n$ is any function, then *the σ-algebra $\mathcal{H}_X$ generated by X is* the smallest σ-algebra on Ω containing all the sets

$$X^{-1}(U) \ ; \qquad U \subset \mathbf{R}^n \ \text{open} \ .$$

It is not hard to show that

$$\mathcal{H}_X = \{X^{-1}(B); \; B \in \mathcal{B}\} \, ,$$

where $\mathcal{B}$ is the Borel σ-algebra on $\mathbf{R}^n$. Clearly, X will then be $\mathcal{H}_X$-measurable and $\mathcal{H}_X$ is the smallest σ-algebra with this property.

The following result is useful. It is a special case of a result sometimes called the *Doob-Dynkin lemma*. See e.g. M. M. Rao (1984), Prop. 3, p. 7.

Lemma 2.1.2 *If $X, Y \colon \Omega \to \mathbf{R}^n$ are two given functions, then Y is $\mathcal{H}_X$-measurable if and only if there exists a Borel measurable function $g \colon \mathbf{R}^n \to \mathbf{R}^n$ such that*

$$Y = g(X) \, .$$

In the following we let $(\Omega, \mathcal{F}, P)$ denote a given complete probability space. A *random variable* X is an $\mathcal{F}$-measurable function $X \colon \Omega \to \mathbf{R}^n$. Every random variable induces a probability measure μ_X on $\mathbf{R}^n$, defined by

$$\mu_X(B) = P(X^{-1}(B)) \, .$$

μ_X is called the *distribution of X*.

If $\int_\Omega |X(\omega)| dP(\omega) < \infty$ then the number

$$E[X] := \int_\Omega X(\omega) dP(\omega) = \int_{\mathbf{R}^n} x \, d\mu_X(x)$$

is called *the expectation* of X (w.r.t. P).

More generally, if $f \colon \mathbf{R}^n \to \mathbf{R}$ is Borel measurable and $\int_\Omega |f(X(\omega))| dP(\omega) < \infty$ then we have

$$E[f(X)] := \int_\Omega f(X(\omega)) dP(\omega) = \int_{\mathbf{R}^n} f(x) \, d\mu_X(x) \, .$$

The L^p-spaces

If $X \colon \Omega \to \mathbf{R}^n$ is a random variable and $p \in [1, \infty)$ is a constant we define the L^p-norm of X, $\|X\|_p$, by

$$\|X\|_p = \|X\|_{L^p(P)} = \left(\int_\Omega |X(\omega)|^p dP(\omega) \right)^{\frac{1}{p}} .$$

If $p = \infty$ we set

$$\|X\|_\infty = \|X\|_{L^\infty(P)} = \inf\{N \in \mathbf{R}; |X(\omega)| \le N \text{ a. s.}\}.$$

The corresponding L^p-spaces are defined by

$$L^p(P) = L^p(\Omega) = \{X : \Omega \to \mathbf{R}^n; \|X\|_p < \infty\}.$$

With this norm the L^p-spaces are *Banach spaces*, i.e. complete (see Exercise 2.19), normed linear spaces. If $p = 2$ the space $L^2(P)$ is even a *Hilbert space*, i.e. a complete inner product space, with inner product

$$(X,Y)_{L^2(P)} := E[X \cdot Y]; \qquad X, Y \in L^2(P).$$

The mathematical model for *independence* is the following:

Definition 2.1.3 *Two subsets $A, B \in \mathcal{F}$ are called* independent *if*

$$P(A \cap B) = P(A) \cdot P(B) .$$

A collection $\mathcal{A} = \{\mathcal{H}_i; i \in I\}$ of families $\mathcal{H}_i$ of measurable sets is independent *if*

$$P(H_{i_1} \cap \cdots \cap H_{i_k}) = P(H_{i_1}) \cdots P(H_{i_k})$$

for all choices of $H_{i_1} \in \mathcal{H}_{i_1}, \cdots, H_{i_k} \in \mathcal{H}_{i_k}$ with different indices $i_1, \ldots, i_k$.
  A collection of random variables $\{X_i; i \in I\}$ is independent *if the collection of generated σ-algebras $\mathcal{H}_{X_i}$ is independent.*

If two random variables $X, Y \colon \Omega \to \mathbf{R}$ are independent then

$$E[XY] = E[X]E[Y] ,$$

provided that $E[|X|] < \infty$ and $E[|Y|] < \infty$. (See Exercise 2.5.)

Definition 2.1.4 *A* stochastic process *is a parametrized collection of random variables*

$$\{X_t\}_{t \in T}$$

defined on a probability space $(\Omega, \mathcal{F}, P)$ and assuming values in $\mathbf{R}^n$.

The parameter space T is usually (as in this book) the halfline $[0, \infty)$, but it may also be an interval $[a, b]$, the non-negative integers and even subsets of $\mathbf{R}^n$ for $n \geq 1$. Note that for each $t \in T$ fixed we have a random variable

$$\omega \to X_t(\omega) ; \qquad \omega \in \Omega .$$

On the other hand, fixing $\omega \in \Omega$ we can consider the function

$$t \to X_t(\omega) ; \qquad t \in T$$

which is called a *path* of X_t.

It may be useful for the intuition to think of t as "time" and each ω as an individual "particle" or "experiment". With this picture $X_t(\omega)$ would represent the position (or result) at time t of the particle (experiment) ω. Sometimes it is convenient to write $X(t, \omega)$ instead of $X_t(\omega)$. Thus we may also regard the process as a function of two variables

$$(t, \omega) \rightarrow X(t, \omega)$$

from $T \times \Omega$ into $\mathbf{R}^n$. This is often a natural point of view in stochastic analysis, because (as we shall see) there it is crucial to have $X(t, \omega)$ jointly measurable in (t, ω).

Finally we note that we may identify each ω with the function $t \rightarrow X_t(\omega)$ from T into $\mathbf{R}^n$. Thus we may regard Ω as a subset of the space $\widetilde{\Omega} = (\mathbf{R}^n)^T$ of all functions from T into $\mathbf{R}^n$. Then the σ-algebra $\mathcal{F}$ will contain the σ-algebra $\mathcal{B}$ generated by sets of the form

$$\{\omega; \omega(t_1) \in F_1, \cdots, \omega(t_k) \in F_k\}, \qquad F_i \subset \mathbf{R}^n \text{ Borel sets}$$

Therefore one may also adopt the point of view that a stochastic process is *a probability measure P on the measurable space $((\mathbf{R}^n)^T, \mathcal{B})$*.

The *(finite-dimensional) distributions* of the process $X = \{X_t\}_{t \in T}$ are the measures $\mu_{t_1, \ldots, t_k}$ defined on $\mathbf{R}^{nk}$, $k = 1, 2, \ldots$, by

$$\mu_{t_1, \ldots, t_k}(F_1 \times F_2 \times \cdots \times F_k) = P[X_{t_1} \in F_1, \cdots, X_{t_k} \in F_k]; \qquad t_i \in T .$$

Here $F_1, \ldots, F_k$ denote Borel sets in $\mathbf{R}^n$.

The family of all finite-dimensional distributions determines many (but not all) important properties of the process X.

Conversely, given a family $\{\nu_{t_1, \ldots, t_k}; k \in \mathbf{N}, t_i \in T\}$ of probability measures on $\mathbf{R}^{nk}$ it is important to be able to construct a stochastic process $Y = \{Y_t\}_{t \in T}$ having $\nu_{t_1, \ldots, t_k}$ as its finite-dimensional distributions. One of Kolmogorov's famous theorems states that this can be done provided $\{\nu_{t_1, \ldots, t_k}\}$ satisfies two natural consistency conditions: (See Lamperti (1977).)

Theorem 2.1.5 (Kolmogorov's extension theorem)
For all $t_1, \ldots, t_k \in T$, $k \in \mathbf{N}$ let $\nu_{t_1, \ldots, t_k}$ be probability measures on $\mathbf{R}^{nk}$ s.t.

$$\nu_{t_{\sigma(1)}, \cdots, t_{\sigma(k)}}(F_1 \times \cdots \times F_k) = \nu_{t_1, \cdots, t_k}(F_{\sigma^{-1}(1)} \times \cdots \times F_{\sigma^{-1}(k)}) \qquad \text{(K1)}$$

for all permutations σ on $\{1, 2, \ldots, k\}$ and

$$\nu_{t_1, \ldots, t_k}(F_1 \times \cdots \times F_k) = \nu_{t_1, \ldots, t_k, t_{k+1}, \ldots, t_{k+m}}(F_1 \times \cdots \times F_k \times \mathbf{R}^n \times \cdots \times \mathbf{R}^n) \qquad \text{(K2)}$$

for all $m \in \mathbf{N}$, where (of course) the set on the right hand side has a total of $k + m$ factors.

Then there exists a probability space $(\Omega, \mathcal{F}, P)$ and a stochastic process $\{X_t\}$ on Ω, $X_t : \Omega \rightarrow \mathbf{R}^n$, s.t.

$$\nu_{t_1, \ldots, t_k}(F_1 \times \cdots \times F_k) = P[X_{t_1} \in F_1, \cdots, X_{t_k} \in F_k] ,$$

for all $t_i \in T$, $k \in \mathbf{N}$ and all Borel sets F_i.

2.2 An Important Example: Brownian Motion

In 1828 the Scottish botanist Robert Brown observed that pollen grains suspended in liquid performed an irregular motion. The motion was later explained by the random collisions with the molecules of the liquid. To describe the motion mathematically it is natural to use the concept of a stochastic process $B_t(\omega)$, interpreted as the position at time t of the pollen grain ω. We will generalize slightly and consider an n-dimensional analog.

To construct $\{B_t\}_{t \geq 0}$ it suffices, by the Kolmogorov extension theorem, to specify a family $\{\nu_{t_1,\ldots,t_k}\}$ of probability measures satisfying (K1) and (K2). These measures will be chosen so that they agree with our observations of the pollen grain behaviour:

Fix $x \in \mathbf{R}^n$ and define

$$p(t, x, y) = (2\pi t)^{-n/2} \cdot \exp(-\frac{|x-y|^2}{2t}) \qquad \text{for } y \in \mathbf{R}^n, \ t > 0 \ .$$

If $0 \leq t_1 \leq t_2 \leq \cdots \leq t_k$ define a measure $\nu_{t_1,\ldots,t_k}$ on $\mathbf{R}^{nk}$ by

$$\nu_{t_1,\ldots,t_k}(F_1 \times \cdots \times F_k) = \tag{2.2.1}$$

$$= \int_{F_1 \times \cdots \times F_k} p(t_1, x, x_1) p(t_2 - t_1, x_1, x_2) \cdots p(t_k - t_{k-1}, x_{k-1}, x_k) dx_1 \cdots dx_k$$

where we use the notation $dy = dy_1 \cdots dy_k$ for Lebesgue measure and the convention that $p(0, x, y)dy = \delta_x(y)$, the unit point mass at x.

Extend this definition to all finite sequences of t_i's by using (K1). Since $\int_{\mathbf{R}^n} p(t, x, y)dy = 1$ for all $t \geq 0$, (K2) holds, so by Kolmogorov's theorem there exists a probability space $(\Omega, \mathcal{F}, P^x)$ and a stochastic process $\{B_t\}_{t \geq 0}$ on Ω such that the finite-dimensional distributions of B_t are given by (2.2.1), i.e.

$$P^x(B_{t_1} \in F_1, \cdots, B_{t_k} \in F_k) =$$

$$= \int_{F_1 \times \cdots \times F_k} p(t_1, x, x_1) \cdots p(t_k - t_{k-1}, x_{k-1}, x_k) dx_1 \ldots dx_k \ . \tag{2.2.2}$$

Definition 2.2.1 *Such a process is called (a version of) Brownian motion starting at x (observe that $P^x(B_0 = x) = 1$).*

The Brownian motion thus defined is not unique, i.e. there exist several quadruples $(B_t, \Omega, \mathcal{F}, P^x)$ such that (2.2.2) holds. However, for our purposes this is not important, we may simply choose any version to work with. As we shall soon see, the paths of a Brownian motion are (or, more correctly, can be chosen to be) continuous, a.s. Therefore we may identify (a.a.) $\omega \in \Omega$ with a continuous function $t \to B_t(\omega)$ from $[0, \infty)$ into $\mathbf{R}^n$. Thus we may adopt the point of view that Brownian motion is just the space $C([0, \infty), \mathbf{R}^n)$ equipped with certain probability measures P^x (given by (2.2.1) and (2.2.2) above).

This version is called the *canonical* Brownian motion. Besides having the advantage of being intuitive, this point of view is useful for the further analysis of measures on $C([0, \infty), \mathbf{R}^n)$, since this space is Polish (i.e. a complete separable metric space). See Stroock and Varadhan (1979).

We state some basic properties of Brownian motion:

(i) B_t is a *Gaussian process*, i.e. for all $0 \leq t_1 \leq \cdots \leq t_k$ the random variable $Z = (B_{t_1}, \ldots, B_{t_k}) \in \mathbf{R}^{nk}$ has a *(multi)normal distribution*. This means that there exists a vector $M \in \mathbf{R}^{nk}$ and a non-negative definite matrix $C = [c_{jm}] \in \mathbf{R}^{nk \times nk}$ (the set of all $nk \times nk$-matrices with real entries) such that

$$E^x\left[\exp\left(i\sum_{j=1}^{nk} u_j Z_j\right)\right] = \exp\left(-\tfrac{1}{2}\sum_{j,m} u_j c_{jm} u_m + i\sum_j u_j M_j\right)$$
(2.2.3)

for all $u = (u_1, \ldots, u_{nk}) \in \mathbf{R}^{nk}$, where $i = \sqrt{-1}$ is the imaginary unit and E^x denotes expectation with respect to P^x. Moreover, if (2.2.3) holds then

$$M = E^x[Z] \quad \text{is the mean value of } Z \qquad (2.2.4)$$

and

$$c_{jm} = E^x[(Z_j - M_j)(Z_m - M_m)] \quad \text{is the covariance matrix of } Z. \quad (2.2.5)$$

(See Appendix A).
To see that (2.2.3) holds for $Z = (B_{t_1}, \ldots, B_{t_k})$ we calculate its left hand side explicitly by using (2.2.2) (see Appendix A) and obtain (2.2.3) with

$$M = E^x[Z] = (x, x, \cdots, x) \in \mathbf{R}^{nk} \qquad (2.2.6)$$

and

$$C = \begin{pmatrix} t_1 I_n & t_1 I_n & \cdots & t_1 I_n \\ t_1 I_n & t_2 I_n & \cdots & t_2 I_n \\ \vdots & \vdots & & \vdots \\ t_1 I_n & t_2 I_n & \cdots & t_k I_n \end{pmatrix}. \qquad (2.2.7)$$

Hence

$$E^x[B_t] = x \qquad \text{for all } t \geq 0 \qquad (2.2.8)$$

and

$$E^x[(B_t - x)^2] = nt, \, E^x[(B_t - x)(B_s - x)] = n \, \min(s, t). \qquad (2.2.9)$$

Moreover,

$$E^x[(B_t - B_s)^2] = n(t - s) \text{ if } t \geq s, \qquad (2.2.10)$$

since

$$E^x[(B_t - B_s)^2] = E^x[(B_t - x)^2 - 2(B_t - x)(B_s - x) + (B_s - x)^2]$$
$$= n(t - 2s + s) = n(t - s), \text{when } t \geq s.$$

(ii) B_t has *independent increments*, i.e.

$$B_{t_1}, B_{t_2} - B_{t_1}, \cdots, B_{t_k} - B_{t_{k-1}} \text{ are independent}$$

$$\text{for all } 0 \leq t_1 < t_2 \cdots < t_k . \qquad (2.2.11)$$

To prove this we use the fact that normal random variables are independent iff they are uncorrelated. (See Appendix A). So it is enough to prove that

$$E^x[(B_{t_i} - B_{t_{i-1}})(B_{t_j} - B_{t_{j-1}})] = 0 \qquad \text{when } t_i < t_j , \qquad (2.2.12)$$

which follows from the form of C:

$$E^x[B_{t_i}B_{t_j} - B_{t_{i-1}}B_{t_j} - B_{t_i}B_{t_{j-1}} + B_{t_{i-1}}B_{t_{j-1}}]$$
$$= n(t_i - t_{i-1} - t_i + t_{i-1}) = 0 .$$

From this we deduce that $B_s - B_t$ is independent of $\mathcal{F}_t$ if $s > t$.

(iii)   Finally we ask: Is $t \to B_t(\omega)$ continuous for almost all ω? Stated like this the question does not make sense, because the set $H = \{\omega; t \to B_t(\omega) \text{ is continuous}\}$ is not measurable with respect to the Borel σ-algebra $\mathcal{B}$ on $(\mathbf{R}^n)^{[0,\infty)}$ mentioned above (H involves an uncountable number of t's). However, if modified slightly the question can be given a positive answer. To explain this we need the following important concept:

Definition 2.2.2 *Suppose that $\{X_t\}$ and $\{Y_t\}$ are stochastic processes on $(\Omega, \mathcal{F}, P)$. Then we say that $\{X_t\}$ is a version of (or a modification of) $\{Y_t\}$ if*

$$P(\{\omega; X_t(\omega) = Y_t(\omega)\}) = 1 \quad \text{for all } t .$$

Note that if X_t is a version of Y_t, then X_t and Y_t have the same finite-dimensional distributions. Thus from the point of view that a stochastic process is a probability law on $(\mathbf{R}^n)^{[0,\infty)}$ two such processes are the same, but nevertheless their path properties may be different. (See Exercise 2.9.)

The continuity question of Brownian motion can be answered by using another famous theorem of Kolmogorov:

Theorem 2.2.3 (Kolmogorov's continuity theorem) *Suppose that the process $X = \{X_t\}_{t \geq 0}$ satisfies the following condition: For all $T > 0$ there exist positive constants α, β, D such that*

$$E[|X_t - X_s|^\alpha] \leq D \cdot |t - s|^{1+\beta} ; \qquad 0 \leq s, t \leq T . \qquad (2.2.13)$$

Then there exists a continuous version of X.

For a proof see for example Stroock and Varadhan (1979, p. 51).
For Brownian motion B_t it is not hard to prove that (See Exercise 2.8)

$$E^x[|B_t - B_s|^4] = n(n+2)|t-s|^2 . \qquad (2.2.14)$$

So Brownian motion satisfies Kolmogorov's condition (2.2.13) with $\alpha = 4$, $D = n(n+2)$ and $\beta = 1$, and therefore it has a continuous version. From now on we will assume that B_t is such a continuous version.

Finally we note that

> If $B_t = (B_t^{(1)}, \cdots, B_t^{(n)})$ is n-dimensional Brownian motion, then
>
> the 1-dimensional processes $\{B_t^{(j)}\}_{t \geq 0}$, $1 \leq j \leq n$ are independent,
>
> 1-dimensional Brownian motions .   (2.2.15)

Exercises

2.1. Suppose that $X: \Omega \to \mathbf{R}$ is a function which assumes only countably many values $a_1, a_2, \ldots \in \mathbf{R}$.

a) Show that X is a random variable if and only if

$$X^{-1}(a_k) \in \mathcal{F} \qquad \text{for all } k = 1, 2, \ldots \qquad (2.2.16)$$

b) Suppose (2.2.16) holds. Show that

$$E[|X|] = \sum_{k=1}^{\infty} |a_k| P[X = a_k] . \qquad (2.2.17)$$

c) If (2.2.16) holds and $E[|X|] < \infty$, show that

$$E[X] = \sum_{k=1}^{\infty} a_k P[X = a_k] .$$

d) If (2.2.16) holds and $f: \mathbf{R} \to \mathbf{R}$ is measurable and bounded, show that

$$E[f(X)] = \sum_{k=1}^{\infty} f(a_k) P[X = a_k] .$$

2.2. Let $X: \Omega \to \mathbf{R}$ be a random variable. The *distribution function* F of X is defined by

$$F(x) = P[X \leq x] .$$

a) Prove that F has the following properties:
   (i) $0 \leq F \leq 1$, $\lim_{x \to -\infty} F(x) = 0$, $\lim_{x \to \infty} F(x) = 1$.
   (ii) F is increasing (= non-decreasing).
   (iii) F is right-continuous, i.e. $F(x) = \lim_{\substack{h \to 0 \\ h > 0}} F(x + h)$.

b)   Let $g: \mathbf{R} \to \mathbf{R}$ be measurable such that $E[|g(X)|] < \infty$. Prove that

$$E[g(X)] = \int_{-\infty}^{\infty} g(x)dF(x) \, ,$$

where the integral on the right is interpreted in the Lebesgue-Stieltjes sense.

c)   Let $p(x) \geq 0$ be a measurable function on $\mathbf{R}$. We say that X has the density p if

$$F(x) = \int_{-\infty}^{x} p(y)dy \qquad \text{for all } x \, .$$

Thus from $(2.2.1)$–$(2.2.2)$ we know that 1-dimensional Brownian motion B_t at time t with $B_0 = 0$ has the density

$$p(x) = \frac{1}{\sqrt{2\pi t}} \exp(-\frac{x^2}{2t}); \quad x \in \mathbf{R} \, .$$

Find the density of B_t^2.

2.3.   Let $\{\mathcal{H}_i\}_{i \in I}$ be a family of σ-algebras on Ω. Prove that

$$\mathcal{H} = \bigcap \{\mathcal{H}_i; i \in I\}$$

is again a σ-algebra.

2.4.*   a)   Let $X: \Omega \to \mathbf{R}^n$ be a random variable such that

$$E[|X|^p] < \infty \qquad \text{for some } p, \ 0 < p < \infty \, .$$

Prove *Chebychev's inequality*:

$$P[|X| \geq \lambda] \leq \frac{1}{\lambda^p} E[|X|^p] \qquad \text{for all } \lambda \geq 0 \, .$$

Hint: $\int_{\Omega} |X|^p dP \geq \int_A |X|^p dP$, where $A = \{\omega: |X| \geq \lambda\}$.

b)   Suppose there exists $k > 0$ such that

$$M = E[\exp(k|X|)] < \infty \, .$$

Prove that $P[|X| \geq \lambda] \leq M e^{-k\lambda}$ for all $\lambda \geq 0$.

2.5.   Let $X, Y: \Omega \to \mathbf{R}$ be two independent random variables and assume for simplicity that X and Y are bounded. Prove that

$$E[XY] = E[X]E[Y] \, .$$

$\Bigg($ Hint: Assume $|X| \le M$, $|Y| \le N$. Approximate X and Y by simple

functions $\varphi(\omega) = \sum\limits_{i=1}^{m-1} a_i \mathcal{X}_{F_i}(\omega)$, $\psi(\omega) = \sum\limits_{j=1}^{n-1} b_j \mathcal{X}_{G_j}(\omega)$, respectively,

where $F_i = X^{-1}([a_i, a_{i+1}))$, $G_j = Y^{-1}([b_j, b_{j+1}))$, $-M = a_0 < a_1 < \ldots < a_m = M$, $-N = b_0 < b_1 < \ldots < b_n = N$. Then

$$E[X] \approx E[\varphi] = \sum_i a_i P(F_i), \quad E[Y] \approx E[\psi] = \sum_j b_j P(G_j)$$

and
$$E[XY] \approx E[\varphi\psi] = \sum_{i,j} a_i b_j P(F_i \cap G_j) \ldots \Bigg) \,.$$

2.6.* Let $(\Omega, \mathcal{F}, P)$ be a probability space and let $A_1, A_2, \ldots$ be sets in $\mathcal{F}$ such that

$$\sum_{k=1}^{\infty} P(A_k) < \infty \,.$$

Prove the *Borel-Cantelli* lemma:

$$P(\bigcap_{m=1}^{\infty} \bigcup_{k=m}^{\infty} A_k) = 0 \,,$$

i.e. the probability that ω belongs to infinitely many $A'_k s$ is zero.

2.7.* a) Suppose $G_1, G_2, \ldots, G_n$ are disjoint subsets of Ω such that

$$\Omega = \bigcup_{i=1}^{n} G_i \,.$$

Prove that the family $\mathcal{G}$ consisting of $\emptyset$ and all unions of some (or all) of $G_1, \ldots, G_n$ constitutes a σ-algebra on Ω.

b) Prove that any *finite* σ-algebra $\mathcal{F}$ on Ω is of the type described in a).

c) Let $\mathcal{F}$ be a *finite* σ-algebra on Ω and let $X : \Omega \to \mathbf{R}$ be $\mathcal{F}$-measurable. Prove that X assumes only finitely many possible values. More precisely, there exists a disjoint family of subsets $F_1, \ldots, F_m \in \mathcal{F}$ and real numbers $c_1, \ldots, c_m$ such that

$$X(\omega) = \sum_{i=1}^{m} c_i \mathcal{X}_{F_i}(\omega) \,.$$

2.8. Let B_t be Brownian motion on $\mathbf{R}$, $B_0 = 0$. Put $E = E^0$.

a) Use (2.2.3) to prove that

$$E[e^{iuB_t}] = \exp(-\tfrac{1}{2}u^2 t) \quad \text{for all } u \in \mathbf{R} \,.$$

b)   Use the power series expansion of the exponential function on both sides, compare the terms with the same power of u and deduce that

$$E[B_t^4] = 3t^2$$

and more generally that

$$E\left[B_t^{2k}\right] = \frac{(2k)!}{2^k \cdot k!} t^k ; \qquad k \in \mathbf{N} .$$

c)   If you feel uneasy about the lack of rigour in the method in b), you can proceed as follows: Prove that (2.2.2) implies that

$$E[f(B_t)] = \frac{1}{\sqrt{2\pi t}} \int_{\mathbf{R}} f(x) e^{-\frac{x^2}{2t}} dx$$

for all functions f such that the integral on the right converges. Then apply this to $f(x) = x^{2k}$ and use integration by parts and induction on k.

d)   Prove (2.2.14), for example by using b) and induction on n.

2.9.* To illustrate that the (finite-dimensional) distributions alone do not give all the information regarding the continuity properties of a process, consider the following example:
Let $(\Omega, \mathcal{F}, P) = ([0, \infty), \mathcal{B}, \mu)$ where $\mathcal{B}$ denotes the Borel σ-algebra on $[0, \infty)$ and μ is a probability measure on $[0, \infty)$ with no mass on single points. Define

$$X_t(\omega) = \begin{cases} 1 & \text{if } t = \omega \\ 0 & \text{otherwise} \end{cases}$$

and

$$Y_t(\omega) = 0 \quad \text{for all } (t, \omega) \in [0, \infty) \times [0, \infty) .$$

Prove that $\{X_t\}$ and $\{Y_t\}$ have the same distributions and that X_t is a version of Y_t. And yet we have that $t \to Y_t(\omega)$ is continuous for all ω, while $t \to X_t(\omega)$ is discontinuous for all ω.

2.10. A stochastic process X_t is called *stationary* if $\{X_t\}$ has the same distribution as $\{X_{t+h}\}$ for any $h > 0$. Prove that Brownian motion B_t has stationary increments, i.e. that the process $\{B_{t+h} - B_t\}_{h \geq 0}$ has the same distribution for all t.

2.11. Prove (2.2.15).

2.12. Let B_t be Brownian motion and fix $t_0 \geq 0$. Prove that

$$\tilde{B}_t := B_{t_0+t} - B_{t_0} ; \qquad t \geq 0$$

is a Brownian motion.

2.13.* Let B_t be 2-dimensional Brownian motion and put

$$D_\rho = \{x \in \mathbf{R}^2; |x| < \rho\} \qquad \text{for } \rho > 0 .$$

Compute

$$P^0[B_t \in D_\rho] .$$

2.14.* Let B_t be n-dimensional Brownian motion and let $K \subset \mathbf{R}^n$ have zero n-dimensional Lebesgue measure. Prove that the expected total length of time that B_t spends in K is zero. (This implies that the *Green measure* associated with B_t is absolutely continuous with respect to Lebesgue measure. See Chapter 9).

2.15.* Let B_t be n-dimensional Brownian motion starting at 0 and let $U \in \mathbf{R}^{n \times n}$ be a (constant) orthogonal matrix, i.e. $UU^T = I$. Prove that

$$\widetilde{B}_t := UB_t$$

is also a Brownian motion.

2.16. (**Brownian scaling**). Let B_t be a 1-dimensional Brownian motion and let $c > 0$ be a constant. Prove that

$$\widehat{B}_t := \frac{1}{c} B_{c^2 t}$$

is also a Brownian motion.

2.17.* If $X_t(\cdot): \Omega \to \mathbf{R}$ is a continuous stochastic process, then for $p > 0$ the *p'th variation process* of X_t, $\langle X, X \rangle_t^{(p)}$ is defined by

$$\langle X, X \rangle_t^{(p)}(\omega) = \lim_{\Delta t_k \to 0} \sum_{t_k \leq t} \left| X_{t_{k+1}}(\omega) - X_{t_k}(\omega) \right|^p \quad \text{(limit in probability)}$$

where $0 = t_1 < t_2 < \ldots < t_n = t$ and $\Delta t_k = t_{k+1} - t_k$. In particular, if $p = 1$ this process is called the *total variation process* and if $p = 2$ this is called the *quadratic variation process*. (See Exercise 4.7.) For Brownian motion $B_t \in \mathbf{R}$ we now show that the quadratic variation process is simply

$$\langle B, B \rangle_t(\omega) = \langle B, B \rangle_t^{(2)}(\omega) = t \quad \text{a.s.}$$

Proceed as follows:

a)   Define

$$\Delta B_k = B_{t_{k+1}} - B_{t_k}$$

and put

$$Y(t, \omega) = \sum_{t_k \leq t} (\Delta B_k(\omega))^2 ,$$

Show that

$$E[(\sum_{t_k \leq t}(\Delta B_k)^2 - t)^2] = 2\sum_{t_k \leq t}(\Delta t_k)^2$$

and deduce that $Y(t, \cdot) \to t$ in $L^2(P)$ as $\Delta t_k \to 0$.

b) Use a) to prove that a.a. paths of Brownian motion do not have a bounded variation on $[0, t]$, i.e. the total variation of Brownian motion is infinite, a.s.

2.18. a) Let $\Omega = \{1, 2, 3, 4, 5\}$ and let $\mathcal{U}$ be the collection

$$\mathcal{U} = \{\{1, 2, 3\}, \{3, 4, 5\}\}$$

of subsets of Ω. Find the smallest σ-algebra containing $\mathcal{U}$ (i.e. the σ-algebra $\mathcal{H}_{\mathcal{U}}$ generated by $\mathcal{U}$).

b) Define $X : \Omega \to \mathbf{R}$ by

$$X(1) = X(2) = 0, \quad X(3) = 10, \quad X(4) = X(5) = 1.$$

Is X measurable with respect to $\mathcal{H}_{\mathcal{U}}$?

c) Define $Y : \Omega \to \mathbf{R}$ by

$$Y(1) = 0, \quad Y(2) = Y(3) = Y(4) = Y(5) = 1.$$

Find the σ-algebra $\mathcal{H}_Y$ generated by Y.

2.19. Let $(\Omega, \mathcal{F}, \mu)$ be a probability space and let $p \in [1, \infty]$. A sequence $\{f_n\}_{n=1}^{\infty}$ of functions $f_n \in L^p(\mu)$ is called a *Cauchy sequence* if

$$\|f_n - f_m\|_p \to 0 \qquad \text{as } n, m \to \infty.$$

The sequence is called *convergent* if there exists $f \in L^p(\mu)$ such that $f_n \to f$ in $L^p(\mu)$.

Prove that every convergent sequence is a Cauchy sequence.

A fundamental theorem in measure theory states that the converse is also true: *Every Cauchy sequence in $L^p(\mu)$ is convergent.* A normed linear space with this property is called *complete*. Thus the $L^p(\mu)$ spaces are complete.

2.20. Let B_t be 1-dimensional Brownian motion, $\sigma \in \mathbf{R}$ be constant and $0 \leq s < t$. Use (2.2.2) to prove that

$$E[\exp(\sigma(B_s - B_t))] = \exp\left(\tfrac{1}{2}\sigma^2(t - s)\right). \qquad (2.2.18)$$

Chapter 3
Itô Integrals

3.1 Construction of the Itô Integral

We now turn to the question of finding a reasonable mathematical interpretation of the "noise" term in the equation of Problem 1 in the Introduction:

$$\frac{dN}{dt} = (r(t) + \text{"noise"})N(t)$$

or more generally in equations of the form

$$\frac{dX}{dt} = b(t, X_t) + \sigma(t, X_t) \cdot \text{"noise"} , \tag{3.1.1}$$

where b and σ are some given functions. Let us first concentrate on the case when the noise is 1-dimensional. It is reasonable to look for some stochastic process W_t to represent the noise term, so that

$$\frac{dX}{dt} = b(t, X_t) + \sigma(t, X_t) \cdot W_t . \tag{3.1.2}$$

Based on many situations, for example in engineering, one is led to assume that W_t has, at least approximately, these properties:

(i) $t_1 \neq t_2 \Rightarrow W_{t_1}$ and W_{t_2} are independent.
(ii) $\{W_t\}$ is stationary, i.e. the (joint) distribution of $\{W_{t_1+t}, \dots, W_{t_k+t}\}$ does not depend on t.
(iii) $E[W_t] = 0$ for all t.

However, it turns out there does *not* exist any "reasonable" stochastic process satisfying (i) and (ii): Such a W_t cannot have continuous paths. (See Exercise 3.11.) If we require $E[W_t^2] = 1$ then the function $(t, \omega) \rightarrow W_t(\omega)$ cannot even be measurable, with respect to the σ-algebra $\mathcal{B} \times \mathcal{F}$, where $\mathcal{B}$ is the Borel σ-algebra on $[0, \infty]$. (See Kallianpur (1980, p. 10).)

Nevertheless it is possible to represent W_t as a generalized stochastic process called the *white noise process*.

That the process is *generalized* means that it can be constructed as a probability measure on the space $\mathcal{S}'$ of tempered distributions on $[0, \infty)$, and not as a probability measure on the much smaller space $\mathbf{R}^{[0,\infty)}$, like an ordinary process can. See e.g. Hida (1980), Adler (1981), Rozanov (1982), Hida, Kuo, Potthoff and Streit (1993), Kuo (1996) or Holden, Øksendal, Ubøe and Zhang (1996).

We will avoid this kind of construction and rather try to rewrite equation (3.1.2) in a form that suggests a replacement of W_t by a proper stochastic process: Let $0 = t_0 < t_1 < \cdots < t_m = t$ and consider a discrete version of (3.1.2):

$$X_{k+1} - X_k = b(t_k, X_k)\Delta t_k + \sigma(t_k, X_k)W_k\Delta t_k \ , \qquad (3.1.3)$$

where

$$X_j = X(t_j), \quad W_k = W_{t_k}, \quad \Delta t_k = t_{k+1} - t_k \ .$$

We abandon the W_k-notation and replace $W_k\Delta t_k$ by $\Delta V_k = V_{t_{k+1}} - V_{t_k}$, where $\{V_t\}_{t \geq 0}$ is some suitable stochastic process. The assumptions (i), (ii) and (iii) on W_t suggest that V_t should have *stationary independent increments with mean 0*. It turns out that the only such process with continuous paths is the Brownian motion B_t. (See Knight (1981)). Thus we put $V_t = B_t$ and obtain from (3.1.3):

$$X_k = X_0 + \sum_{j=0}^{k-1} b(t_j, X_j)\Delta t_j + \sum_{j=0}^{k-1} \sigma(t_j, X_j)\Delta B_j \ . \qquad (3.1.4)$$

Is it possible to prove that the limit of the right hand side of (3.1.4) exists, in some sense, when $\Delta t_j \to 0$? If so, then by applying the usual integration notation we should obtain

$$X_t = X_0 + \int_0^t b(s, X_s)ds + \text{``}\int_0^t \sigma(s, X_s)dB_s\text{''} \qquad (3.1.5)$$

and we would adopt as a convention that (3.1.2) really means that $X_t = X_t(\omega)$ is a stochastic process satisfying (3.1.5).

Thus, in the remainder of this chapter we will prove the existence, in a certain sense, of

$$\text{``}\int_0^t f(s, \omega)dB_s(\omega)\text{''}$$

where $B_t(\omega)$ is 1-dimensional Brownian motion starting at the origin, for a wide class of functions $f: [0, \infty] \times \Omega \to \mathbf{R}$. Then, in Chapter 5, we will return to the solution of (3.1.5).

Suppose $0 \leq S < T$ and $f(t, \omega)$ is given. We want to define

$$\int_{S}^{T} f(t,\omega)dB_t(\omega) \ . \tag{3.1.6}$$

It is reasonable to start with a definition for a simple class of functions f and then extend by some approximation procedure. Thus, let us first assume that f has the form

$$\phi(t,\omega) = \sum_{j \ge 0} e_j(\omega) \cdot \mathcal{X}_{[j \cdot 2^{-n},(j+1)2^{-n})}(t) \ , \tag{3.1.7}$$

where $\mathcal{X}$ denotes the characteristic (indicator) function and n is a natural number. For such functions it is reasonable to define

$$\int_{S}^{T} \phi(t,\omega)dB_t(\omega) = \sum_{j \ge 0} e_j(\omega)[B_{t_{j+1}} - B_{t_j}](\omega) \ , \tag{3.1.8}$$

where

$$t_k = t_k^{(n)} = \left\{ \begin{array}{lll} k \cdot 2^{-n} & \text{if} & S \le k \cdot 2^{-n} \le T \\ S & \text{if} & k \cdot 2^{-n} < S \\ T & \text{if} & k \cdot 2^{-n} > T \end{array} \right\}$$

However, without any further assumptions on the functions $e_j(\omega)$ this leads to difficulties, as the next example shows.

Here – and in the following – E means the same as E^0, the expectation w.r.t. the law P^0 for Brownian motion starting at 0. And P means the same as P^0.

Example 3.1.1 Choose

$$\phi_1(t,\omega) = \sum_{j \ge 0} B_{j \cdot 2^{-n}}(\omega) \cdot \mathcal{X}_{[j \cdot 2^{-n},(j+1)2^{-n})}(t)$$

$$\phi_2(t,\omega) = \sum_{j \ge 0} B_{(j+1)2^{-n}}(\omega) \cdot \mathcal{X}_{[j \cdot 2^{-n},(j+1)2^{-n})}(t) \ .$$

Then

$$E\left[\int_{0}^{T} \phi_1(t,\omega)dB_t(\omega)\right] = \sum_{j \ge 0} E[B_{t_j}(B_{t_{j+1}} - B_{t_j})] = 0 \ ,$$

since $\{B_t\}$ has independent increments. But

$$E\left[\int_{0}^{T} \phi_2(t,\omega)dB_t(\omega)\right] = \sum_{j \ge 0} E[B_{t_{j+1}} \cdot (B_{t_{j+1}} - B_{t_j})]$$

$$= \sum_{j \ge 0} E[(B_{t_{j+1}} - B_{t_j})^2] = T \ , \quad \text{by (2.2.10)} \ .$$

So, in spite of the fact that both ϕ_1 and ϕ_2 appear to be very reasonable approximations to

$$f(t,\omega) = B_t(\omega) \,,$$

their integrals according to (3.1.8) are not close to each other at all, no matter how large n is chosen.

This only reflects the fact that the variations of the paths of B_t are too big to enable us to define the integral (3.1.6) in the Riemann-Stieltjes sense. In fact, one can show that the paths $t \rightarrow B_t$ of Brownian motion are nowhere differentiable, almost surely (a.s.). (See Breiman (1968)). In particular, the total variation of the path is infinite, a.s.

In general it is natural to approximate a given function $f(t,\omega)$ by

$$\sum_j f(t_j^*,\omega) \cdot \mathcal{X}_{[t_j,t_{j+1})}(t)$$

where the points t_j^* belong to the intervals $[t_j, t_{j+1}]$, and then define $\int_S^T f(t,\omega)dB_t(\omega)$ as the limit (in a sense that we will explain) of

$$\sum_j f(t_j^*,\omega)[B_{t_{j+1}} - B_{t_j}](\omega) \text{ as } n \rightarrow \infty.$$ However, the example above shows that – unlike the Riemann-Stieltjes integral – it does make a difference here what points t_j^* we choose. The following two choices have turned out to be the most useful ones:

1) $t_j^* = t_j$ (the left end point), which leads to the *Itô integral*, from now on denoted by

$$\int_S^T f(t,\omega)dB_t(\omega) \,,$$

   and

2) $t_j^* = (t_j + t_{j+1})/2$ (the mid point), which leads to the *Stratonovich integral*, denoted by

$$\int_S^T f(t,\omega) \circ dB_t(\omega) \,.$$

(See Protter (2004, Th. V. 5.30)).

In the end of this chapter we will explain why these choices are the best and discuss the relations and distinctions between the corresponding integrals.

In any case one must restrict oneself to a special class of functions $f(t,\omega)$ in (3.1.6), also if they have the particular form (3.1.7), in order to obtain a reasonable definition of the integral. We will here present Itô's choice $t_j^* = t_j$. The approximation procedure indicated above will work out successfully provided that f has the property that each of the functions $\omega \rightarrow f(t_j,\omega)$ *only depends on the behaviour of $B_s(\omega)$ up to time t_j*. This leads to the following important concepts:

Definition 3.1.2 *Let $B_t(\omega)$ be n-dimensional Brownian motion. Then we define $\mathcal{F}_t = \mathcal{F}_t^{(n)}$ to be the σ-algebra generated by the random variables $\{B_i(s)\}_{1\leq i\leq n, 0\leq s\leq t}$. In other words, $\mathcal{F}_t$ is the smallest σ-algebra containing all sets of the form*

$$\{\omega; B_{t_1}(\omega) \in F_1, \cdots, B_{t_k}(\omega) \in F_k\} ,$$

where $t_j \leq t$ and $F_j \subset \mathbf{R}^n$ are Borel sets, $j \leq k = 1, 2, \ldots$ (We assume that all sets of measure zero are included in $\mathcal{F}_t$).

One often thinks of $\mathcal{F}_t$ as "the history of B_s up to time t". A function $h(\omega)$ will be $\mathcal{F}_t$-measurable if and only if h can be written as the pointwise a.e. limit of sums of functions of the form

$$g_1(B_{t_1})g_2(B_{t_2}) \cdots g_k(B_{t_k}) ,$$

where $g_1, \ldots, g_k$ are bounded continuous functions and $t_j \leq t$ for $j \leq k$, $k = 1, 2, \ldots$. (See Exercise 3.14.) Intuitively, that h is $\mathcal{F}_t$-measurable means that the value of $h(\omega)$ can be decided from the values of $B_s(\omega)$ for $s \leq t$. For example, $h_1(\omega) = B_{t/2}(\omega)$ is $\mathcal{F}_t$-measurable, while $h_2(\omega) = B_{2t}(\omega)$ is not.

Note that $\mathcal{F}_s \subset \mathcal{F}_t$ for $s < t$ (i.e. $\{\mathcal{F}_t\}$ is *increasing*) and that $\mathcal{F}_t \subset \mathcal{F}$ for all t.

Definition 3.1.3 *Let $\{\mathcal{N}_t\}_{t\geq 0}$ be an increasing family of σ-algebras of subsets of Ω. A process $g(t,\omega): [0,\infty) \times \Omega \to \mathbf{R}^n$ is called $\mathcal{N}_t$-adapted if for each $t \geq 0$ the function*

$$\omega \to g(t,\omega)$$

is $\mathcal{N}_t$-measurable.

Thus the process $h_1(t,\omega) = B_{t/2}(\omega)$ is $\mathcal{F}_t$-adapted, while $h_2(t,\omega) = B_{2t}(\omega)$ is not.

We now describe our class of functions for which the Itô integral will be defined:

Definition 3.1.4 *Let $\mathcal{V} = \mathcal{V}(S,T)$ be the class of functions*

$$f(t,\omega): [0,\infty) \times \Omega \to \mathbf{R}$$

such that

(i)   *$(t,\omega) \to f(t,\omega)$ is $\mathcal{B} \times \mathcal{F}$-measurable, where $\mathcal{B}$ denotes the Borel σ-algebra on $[0,\infty)$.*

(ii) *$f(t,\omega)$ is $\mathcal{F}_t$-adapted.*

(iii) *$E\left[\int_S^T f(t,\omega)^2 dt\right] < \infty.$*

The Itô Integral

For functions $f \in \mathcal{V}$ we will now show how to define the *Itô integral*

$$\mathcal{I}[f](\omega) = \int_S^T f(t,\omega)dB_t(\omega) \, ,$$

where B_t is 1-dimensional Brownian motion.

The idea is natural: First we define $\mathcal{I}[\phi]$ for a simple class of functions ϕ. Then we show that each $f \in \mathcal{V}$ can be approximated (in an appropriate sense) by such ϕ's and we use this to define $\int f dB$ as the limit of $\int \phi dB$ as $\phi \to f$.

We now give the details of this construction: A function $\phi \in \mathcal{V}$ is called *elementary* if it has the form

$$\phi(t,\omega) = \sum_j e_j(\omega) \cdot \mathcal{X}_{[t_j,t_{j+1})}(t) \, . \tag{3.1.9}$$

Note that since $\phi \in \mathcal{V}$ each function e_j must be $\mathcal{F}_{t_j}$-measurable. Thus in Example 3.1.1 above the function ϕ_1 is elementary while ϕ_2 is not.

For elementary functions $\phi(t,\omega)$ we define the integral according to (3.1.8), i.e.

$$\int_S^T \phi(t,\omega)dB_t(\omega) = \sum_{j \geq 0} e_j(\omega)[B_{t_{j+1}} - B_{t_j}](\omega) \, . \tag{3.1.10}$$

Now we make the following important observation:

Lemma 3.1.5 (The Itô isometry) *If $\phi(t,\omega)$ is bounded and elementary then*

$$E\left[\left(\int_S^T \phi(t,\omega)dB_t(\omega)\right)^2\right] = E\left[\int_S^T \phi(t,\omega)^2 dt\right] \, . \tag{3.1.11}$$

Proof. Put $\Delta B_j = B_{t_{j+1}} - B_{t_j}$. Then

$$E[e_i e_j \Delta B_i \Delta B_j] = \begin{cases} 0 & \text{if } i \neq j \\ E[e_j^2] \cdot (t_{j+1} - t_j) & \text{if } i = j \end{cases}$$

using that $e_i e_j \Delta B_i$ and ΔB_j are independent if $i < j$. Thus

$$E\left[\left(\int_S^T \phi dB\right)^2\right] = \sum_{i,j} E[e_i e_j \Delta B_i \Delta B_j] = \sum_j E[e_j^2] \cdot (t_{j+1} - t_j)$$

$$= E\left[\int_S^T \phi^2 dt\right] \, . \qquad \square$$

The idea is now to use the isometry (3.1.11) to extend the definition from elementary functions to functions in $\mathcal{V}$. We do this in several steps:

Step 1. *Let $g \in \mathcal{V}$ be bounded and $g(\cdot, \omega)$ continuous for each ω. Then there exist elementary functions $\phi_n \in \mathcal{V}$ such that*

$$E\left[\int_S^T (g - \phi_n)^2 dt \right] \to 0 \qquad \text{as } n \to \infty .$$

Proof. Define $\phi_n(t, \omega) = \sum_j g(t_j, \omega) \cdot \mathcal{X}_{[t_j, t_{j+1})}(t)$. Then ϕ_n is elementary since $g \in \mathcal{V}$, and

$$\int_S^T (g - \phi_n)^2 dt \to 0 \qquad \text{as } n \to \infty, \text{ for each } \omega ,$$

since $g(\cdot, \omega)$ is continuous for each ω. Hence $E[\int_S^T (g - \phi_n)^2 dt] \to 0$ as $n \to \infty$, by bounded convergence.   □

Step 2. *Let $h \in \mathcal{V}$ be bounded. Then there exist bounded functions $g_n \in \mathcal{V}$ such that $g_n(\cdot, \omega)$ is continuous for all ω and n, and*

$$E\left[\int_S^T (h - g_n)^2 dt \right] \to 0 .$$

Proof. Suppose $|h(t, \omega)| \leq M$ for all (t, ω). For each n let ψ_n be a non-negative, continuous function on $\mathbf{R}$ such that

(i) $\psi_n(x) = 0$ for $x \leq -\frac{1}{n}$ and $x \geq 0$

and

(ii) $\int_{-\infty}^{\infty} \psi_n(x) dx = 1$

Define

$$g_n(t, \omega) = \int_0^t \psi_n(s - t) h(s, \omega) ds .$$

Then $g_n(\cdot, \omega)$ is continuous for each ω and $|g_n(t, \omega)| \leq M$. Since $h \in \mathcal{V}$ we can show that $g_n(t, \cdot)$ is $\mathcal{F}_t$-measurable for all t. (This is a subtle point; see e.g. Karatzas and Shreve (1991), p. 133 for details.) Moreover,

$$\int_S^T (g_n(s, \omega) - h(s, \omega))^2 ds \to 0 \qquad \text{as } n \to \infty, \text{ for each } \omega ,$$

since $\{\psi_n\}_n$ constitutes an approximate identity. (See e.g. Hoffman (1962, p. 22).) So by bounded convergence

$$E\left[\int_S^T (h(t,\omega) - g_n(t,\omega))^2 dt\right] \to 0 \qquad \text{as } n \to \infty,$$

as asserted.   □

Step 3. *Let* $f \in \mathcal{V}$. *Then there exists a sequence* $\{h_n\} \subset \mathcal{V}$ *such that* h_n *is bounded for each* n *and*

$$E\left[\int_S^T (f - h_n)^2 dt\right] \to 0 \text{ as } n \to \infty.$$

Proof. Put

$$h_n(t,\omega) = \begin{cases} -n & \text{if} & f(t,\omega) < -n \\ f(t,\omega) & \text{if} & -n \le f(t,\omega) \le n \\ n & \text{if} & f(t,\omega) > n. \end{cases}$$

Then the conclusion follows by dominated convergence.

That completes the approximation procedure.   □

We are now ready to complete the definition of the Itô integral

$$\int_S^T f(t,\omega)dB_t(\omega) \quad \text{for } f \in \mathcal{V}.$$

If $f \in \mathcal{V}$ we choose, by Steps 1-3, elementary functions $\phi_n \in \mathcal{V}$ such that

$$E\left[\int_S^T |f - \phi_n|^2 dt\right] \to 0.$$

Then define

$$\mathcal{I}[f](\omega) := \int_S^T f(t,\omega)dB_t(\omega) := \lim_{n\to\infty} \int_S^T \phi_n(t,\omega)dB_t(\omega).$$

The limit exists as an element of $L^2(P)$, since $\left\{\int_S^T \phi_n(t,\omega)dB_t(\omega)\right\}$ forms a Cauchy sequence in $L^2(P)$, by (3.1.11).

We summarize this as follows:

Definition 3.1.6 (The Itô integral) *Let $f \in \mathcal{V}(S,T)$. Then the Itô integral of f (from S to T) is defined by*

$$\int_S^T f(t,\omega) dB_t(\omega) = \lim_{n\to\infty} \int_S^T \phi_n(t,\omega) dB_t(\omega) \qquad \text{(limit in } L^2(P)) \quad (3.1.12)$$

where $\{\phi_n\}$ is a sequence of elementary functions such that

$$E\left[\int_S^T (f(t,\omega) - \phi_n(t,\omega))^2 dt\right] \to 0 \qquad \text{as } n \to \infty . \quad (3.1.13)$$

Note that such a sequence $\{\phi_n\}$ satisfying (3.1.13) exists by Steps 1–3 above. Moreover, by (3.1.11) the limit in (3.1.12) exists and does not depend on the actual choice of $\{\phi_n\}$, as long as (3.1.13) holds. Furthermore, from (3.1.11) and (3.1.12) we get the following important

Corollary 3.1.7 (The Itô isometry)

$$E\left[\left(\int_S^T f(t,\omega) dB_t\right)^2\right] = E\left[\int_S^T f^2(t,\omega) dt\right] \quad \text{for all } f \in \mathcal{V}(S,T) . \quad (3.1.14)$$

Corollary 3.1.8 *If $f(t,\omega) \in \mathcal{V}(S,T)$ and $f_n(t,\omega) \in \mathcal{V}(S,T)$ for $n = 1,2,\dots$ and $E\left[\int_S^T (f_n(t,\omega) - f(t,\omega))^2 dt\right] \to 0$ as $n \to \infty$, then*

$$\int_S^T f_n(t,\omega) dB_t(\omega) \to \int_S^T f(t,\omega) dB_t(\omega) \qquad \text{in } L^2(P) \text{ as } n \to \infty .$$

We illustrate this integral with an example:

Example 3.1.9 Assume $B_0 = 0$. Then

$$\int_0^t B_s dB_s = \tfrac{1}{2}B_t^2 - \tfrac{1}{2}t .$$

Proof. Put $\phi_n(s,\omega) = \sum B_j(\omega) \cdot \mathcal{X}_{[t_j,t_{j+1})}(s)$, where $B_j = B_{t_j}$. Then

$$E\left[\int_0^t (\phi_n - B_s)^2 ds\right] = E\left[\sum_j \int_{t_j}^{t_{j+1}} (B_j - B_s)^2 ds\right]$$

$$= \sum_j \int_{t_j}^{t_{j+1}} (s - t_j) ds = \sum_j \tfrac{1}{2}(t_{j+1} - t_j)^2 \to 0 \qquad \text{as } \Delta t_j \to 0 .$$

So by Corollary 3.1.8

$$\int_0^t B_s dB_s = \lim_{\Delta t_j \to 0} \int_0^t \phi_n dB_s = \lim_{\Delta t_j \to 0} \sum_j B_j \Delta B_j \ .$$

(See also Exercise 3.13.) Now

$$\Delta(B_j^2) = B_{j+1}^2 - B_j^2 = (B_{j+1} - B_j)^2 + 2B_j(B_{j+1} - B_j)$$
$$= (\Delta B_j)^2 + 2B_j \Delta B_j \ ,$$

and therefore, since $B_0 = 0$,

$$B_t^2 = \sum_j \Delta(B_j^2) = \sum_j (\Delta B_j)^2 + 2\sum_j B_j \Delta B_j$$

or

$$\sum_j B_j \Delta B_j = \tfrac{1}{2} B_t^2 - \tfrac{1}{2} \sum_j (\Delta B_j)^2 \ .$$

Since $\sum_j (\Delta B_j)^2 \to t$ in $L^2(P)$ as $\Delta t_j \to 0$ (Exercise 2.17), the result follows.

□

The extra term $-\frac{1}{2}t$ shows that the Itô stochastic integral does not behave like ordinary integrals. In the next chapter we will establish the *Itô formula*, which explains the result in this example and which makes it easy to calculate many stochastic integrals.

3.2 Some properties of the Itô integral

First we observe the following:

Theorem 3.2.1 *Let* $f, g \in \mathcal{V}(0, T)$ *and let* $0 \leq S < U < T$. *Then*

(i) $\int_S^T f dB_t = \int_S^U f dB_t + \int_U^T f dB_t$ *for a.a.* ω

(ii) $\int_S^T (cf + g)dB_t = c \cdot \int_S^T f dB_t + \int_S^T g dB_t$ *(c constant) for a.a.* ω

(iii) $E\left[\int_S^T f dB_t\right] = 0$

(iv) $\int_S^T f dB_t$ *is* $\mathcal{F}_T$-measurable.

Proof. This clearly holds for all elementary functions, so by taking limits we obtain this for all $f, g \in \mathcal{V}(0, T)$.

An important property of the Itô integral is that it is a *martingale*:

Definition 3.2.2 *A* filtration *(on $(\Omega, \mathcal{F})$) is a family $\mathcal{M} = \{\mathcal{M}_t\}_{t \geq 0}$ of σ-algebras $\mathcal{M}_t \subset \mathcal{F}$ such that*

$$0 \leq s < t \Rightarrow \mathcal{M}_s \subset \mathcal{M}_t$$

(i.e. $\{\mathcal{M}_t\}$ is increasing*). An n-dimensional stochastic process $\{M_t\}_{t \geq 0}$ on $(\Omega, \mathcal{F}, P)$ is called a* martingale *with respect to a filtration $\{\mathcal{M}_t\}_{t \geq 0}$ (and with respect to P) if*

(i) M_t is $\mathcal{M}_t$-measurable for all t,
(ii) $E[|M_t|] < \infty$ for all t
       and
(iii) $E[M_s | \mathcal{M}_t] = M_t$ for all $s \geq t$.

Here the expectation in (ii) and the conditional expectation in (iii) are taken with respect to $P = P^0$. (See Appendix B for a survey of conditional expectation).

Example 3.2.3 Brownian motion B_t in $\mathbf{R}^n$ is a martingale w.r.t. the σ-algebras $\mathcal{F}_t$ generated by $\{B_s; s \leq t\}$, because

$$E[|B_t|]^2 \leq E[|B_t|^2] = |B_0|^2 + nt \qquad \text{and if } s \geq t \text{ then}$$
$$E[B_s | \mathcal{F}_t] = E[B_s - B_t + B_t | \mathcal{F}_t]$$
$$= E[B_s - B_t | \mathcal{F}_t] + E[B_t | \mathcal{F}_t] = 0 + B_t = B_t \ .$$

Here we have used that $E[(B_s - B_t) | \mathcal{F}_t] = E[B_s - B_t] = 0$ since $B_s - B_t$ is independent of $\mathcal{F}_t$ (see (2.2.11) and Theorem B.2.d)) and we have used that $E[B_t | \mathcal{F}_t] = B_t$ since B_t is $\mathcal{F}_t$-measurable (see Theorem B.2.c)).

For continuous martingales we have the following important inequality due to Doob: (See e.g. Stroock and Varadhan (1979), Theorem 1.2.3 or Revuz and Yor (1991), Theorem II.1.7)

Theorem 3.2.4 (Doob's martingale inequality) *If M_t is a martingale such that $t \to M_t(\omega)$ is continuous a.s., then for all $p \geq 1, T \geq 0$ and all $\lambda > 0$*

$$P[\sup_{0 \leq t \leq T} |M_t| \geq \lambda] \leq \frac{1}{\lambda^p} \cdot E[|M_T|^p] \ .$$

We now use this inequality to prove that the Itô integral

$$\int\limits_0^t f(s, \omega) dB_s$$

can be chosen to depend continuously on t :

Theorem 3.2.5 *Let $f \in \mathcal{V}(0,T)$. Then there exists a t-continuous version of*

$$\int_0^t f(s,\omega)dB_s(\omega) ; \qquad 0 \le t \le T ,$$

i.e. there exists a t-continuous stochastic process J_t on $(\Omega, \mathcal{F}, P)$ such that

$$P[J_t = \int_0^t fdB] = 1 \qquad \text{for all } t, 0 \le t \le T . \qquad (3.2.1)$$

Proof. Let $\phi_n = \phi_n(t,\omega) = \sum_j e_j^{(n)}(\omega)\mathcal{X}_{[t_j^{(n)},t_{j+1}^{(n)})}(t)$ be elementary functions such that

$$E\left[\int_0^T (f - \phi_n)^2 dt\right] \to 0 \qquad \text{when } n \to \infty .$$

Put

$$I_n(t,\omega) = \int_0^t \phi_n(s,\omega)dB_s(\omega)$$

and

$$I_t = I(t,\omega) = \int_0^t f(s,\omega)dB_s(\omega) ; \qquad 0 \le t \le T .$$

Then $I_n(\cdot,\omega)$ is continuous, for all n. Moreover, $I_n(t,\omega)$ is a martingale with respect to $\mathcal{F}_t$, for all n :

$$E[I_n(s,\omega)|\mathcal{F}_t] = E\left[\left(\int_0^t \phi_n dB + \int_t^s \phi_n dB\right)\Big|\mathcal{F}_t\right]$$

$$= \int_0^t \phi_n dB + E\left[\sum_{t \le t_j^{(n)} \le t_{j+1}^{(n)} \le s} e_j^{(n)}\Delta B_j|\mathcal{F}_t\right]$$

$$= \int_0^t \phi_n dB + \sum_j E\left[E[e_j^{(n)}\Delta B_j|\mathcal{F}_{t_j^{(n)}}]|\mathcal{F}_t\right]$$

$$= \int_0^t \phi_n dB + \sum_j E\left[e_j^{(n)} E[\Delta B_j|\mathcal{F}_{t_j^{(n)}}]|\mathcal{F}_t\right]$$

$$= \int_0^t \phi_n dB = I_n(t,\omega) \qquad (3.2.2)$$

when $t < s$, using Theorem B.3. and Theorem B.2.d).

Hence $I_n - I_m$ is also an $\mathcal{F}_t$-martingale, so by the martingale inequality (Theorem 3.2.4) it follows that

$$P\left[\sup_{0\leq t\leq T}|I_n(t,\omega) - I_m(t,\omega)| > \epsilon\right] \leq \frac{1}{\epsilon^2}\cdot E\left[|I_n(T,\omega) - I_m(T,\omega)|^2\right]$$

$$= \frac{1}{\epsilon^2}E\left[\int_0^T(\phi_n - \phi_m)^2 ds\right] \to 0 \qquad \text{as } m,n \to \infty .$$

Hence we may choose a subsequence $n_k \uparrow \infty$ s.t.

$$P\left[\sup_{0\leq t\leq T}|I_{n_{k+1}}(t,\omega) - I_{n_k}(t,\omega)| > 2^{-k}\right] < 2^{-k} .$$

By the Borel-Cantelli lemma

$$P\left[\sup_{0\leq t\leq T}|I_{n_{k+1}}(t,\omega) - I_{n_k}(t,\omega)| > 2^{-k} \quad \text{for infinitely many } k\right] = 0 .$$

So for a.a. ω there exists $k_1(\omega)$ such that

$$\sup_{0\leq t\leq T}|I_{n_{k+1}}(t,\omega) - I_{n_k}(t,\omega)| \leq 2^{-k} \qquad \text{for } k \geq k_1(\omega) .$$

Therefore $I_{n_k}(t,\omega)$ is uniformly convergent for $t \in [0,T]$, for a.a. ω and so the limit, denoted by $J_t(\omega)$, is t-continuous for $t \in [0,T]$, a.s. Since $I_{n_k}(t,\cdot) \to I(t,\cdot)$ in $L^2[P]$ for all t, we must have

$$I_t = J_t \text{ a.s. }, \qquad \text{for all } t \in [0,T] .$$

That completes the proof. $\square$

From now on we shall always assume that $\int_0^t f(s,\omega)dB_s(\omega)$ means a t-continuous version of the integral.

Corollary 3.2.6 *Let $f(t,\omega) \in \mathcal{V}(0,T)$ for all T. Then*

$$M_t(\omega) = \int_0^t f(s,\omega)dB_s$$

is a martingale w.r.t. $\mathcal{F}_t$ and

$$P\left[\sup_{0\leq t\leq T}|M_t| \geq \lambda\right] \leq \frac{1}{\lambda^2}\cdot E\left[\int_0^T f(s,\omega)^2 ds\right] ; \qquad \lambda, T > 0 . \qquad (3.2.3)$$

Proof. This follows from (3.2.2), the a.s. t-continuity of M_t and the martingale inequality (Theorem 3.2.4), combined with the Itô isometry (3.1.14). $\square$

3.3 Extensions of the Itô integral

The Itô integral $\int f dB$ can be defined for a larger class of integrands f than $\mathcal{V}$. First, the measurability condition (ii) of Definition 3.1.4 can be relaxed to the following:

(ii)'   There exists an increasing family of σ-algebras $\mathcal{H}_t; t \geq 0$ such that
    a) B_t is a martingale with respect to $\mathcal{H}_t$ and
    b) f_t is $\mathcal{H}_t$-adapted.

Note that a) implies that $\mathcal{F}_t \subset \mathcal{H}_t$. The essence of this extension is that we can allow f_t to depend on more than $\mathcal{F}_t$ as long as B_t remains a martingale with respect to the "history" of f_s; $s \leq t$. If (ii)' holds, then $E[B_s - B_t | \mathcal{H}_t] = 0$ for all $s > t$ and if we inspect our proofs above, we see that this is sufficient to carry out the construction of the Itô integral as before.

The most important example of a situation where (ii)' applies (and (ii) doesn't) is the following:

Suppose $B_t(\omega) = B_k(t, \omega)$ is the k'th coordinate of n-dimensional Brownian motion $(B_1, \ldots, B_n)$. Let $\mathcal{F}_t^{(n)}$ be the σ-algebra generated by $B_1(s_1, \cdot), \cdots, B_n(s_n, \cdot); s_k \leq t$. Then $B_k(t, \omega)$ is a martingale with respect to $\mathcal{F}_t^{(n)}$ because $B_k(s, \cdot) - B_k(t, \cdot)$ is independent of $\mathcal{F}_t^{(n)}$ when $s > t$. Hence we can choose $\mathcal{H}_t = \mathcal{F}_t^{(n)}$ in (ii)' above. Thus we have now defined $\int_0^t f(s, \omega) dB_k(s, \omega)$ for $\mathcal{F}_t^{(n)}$-adapted integrands $f(t, \omega)$. That includes integrals like

$$\int B_2 dB_1 \qquad \text{or} \qquad \int \sin(B_1^2 + B_2^2)\, dB_2$$

involving several components of n-dimensional Brownian motion. (Here we have used the notation $dB_1 = dB_1(t, \omega)$ etc.)

This allows us to define the **multi-dimensional Itô integral** as follows:

Definition 3.3.1 Let $B = (B_1, B_2, \ldots, B_n)$ be n-dimensional Brownian motion. Then $\mathcal{V}_{\mathcal{H}}^{m \times n}(S, T)$ denotes the set of $m \times n$ matrices $v = [v_{ij}(t, \omega)]$ where each entry $v_{ij}(t, \omega)$ satisfies (i) and (iii) of Definition 3.1.4 and (ii)' above, with respect to some filtration $\mathcal{H} = \{\mathcal{H}_t\}_{t \geq 0}$.

If $v \in \mathcal{V}_{\mathcal{H}}^{m \times n}(S, T)$ we define, using matrix notation

$$\int_S^T v dB = \int_S^T \begin{pmatrix} v_{11} & \cdots & v_{1n} \\ \vdots & & \vdots \\ v_{m1} & \cdots & v_{mn} \end{pmatrix} \begin{pmatrix} dB_1 \\ \vdots \\ dB_n \end{pmatrix}$$

to be the $m \times 1$ matrix (column vector) whose i'th component is the following sum of (extended) 1-dimensional Itô integrals:

$$\sum_{j=1}^{n} \int_S^T v_{ij}(s, \omega) dB_j(s, \omega).$$

If $\mathcal{H} = \mathcal{F}^{(n)} = \{\mathcal{F}_t^{(n)}\}_{t \geq 0}$ *we write* $\mathcal{V}^{m \times n}(S, T)$ *and if* $m = 1$ *we write* $\mathcal{V}_{\mathcal{H}}^n(S, T)$ *(respectively* $\mathcal{V}^n(S, T)$*) instead of* $\mathcal{V}_{\mathcal{H}}^{1 \times n}(S, T)$ *(respectively* $\mathcal{V}^{1 \times n}(S, T)$*). We also put*

$$\mathcal{V}^{m \times n} = \mathcal{V}^{m \times n}(0, \infty) = \bigcap_{T > 0} \mathcal{V}^{m \times n}(0, T) \, .$$

The next extension of the Itô integral consists of weakening condition (iii) of Definition 3.1.4 to

(iii)' $\qquad P\left[\int\limits_S^T f(s, \omega)^2 ds < \infty \right] = 1 \, .$

Definition 3.3.2 $\mathcal{W}_{\mathcal{H}}(S, T)$ *denotes the class of processes* $f(t, \omega) \in \mathbf{R}$ *satisfying (i) of Definition 3.1.4 and (ii)', (iii)' above. Similarly to the notation for* $\mathcal{V}$ *we put* $\mathcal{W}_{\mathcal{H}} = \bigcap\limits_{T > 0} \mathcal{W}_{\mathcal{H}}(0, T)$ *and in the matrix case we write* $\mathcal{W}_{\mathcal{H}}^{m \times n}(S, T)$ *etc. If* $\mathcal{H} = \mathcal{F}^{(n)}$ *we write* $\mathcal{W}(S, T)$ *instead of* $\mathcal{W}_{\mathcal{F}^{(n)}}(S, T)$ *etc. If the dimension is clear from the context we sometimes drop the superscript and write* $\mathcal{F}$ *for* $\mathcal{F}^{(n)}$ *and so on.*

Let B_t denote 1-dimensional Brownian motion. If $f \in \mathcal{W}_{\mathcal{H}}$ one can show that for all t there exist step functions $f_n \in \mathcal{W}_{\mathcal{H}}$ such that $\int\limits_0^t |f_n - f|^2 ds \to 0$ in probability, i.e. in measure with respect to P. For such a sequence one has that $\int\limits_0^t f_n(s, \omega) dB_s$ converges in probability to some random variable and the limit only depends on f, not on the sequence $\{f_n\}$. Thus we may define

$$\int\limits_0^t f(s, \omega) dB_s(\omega) = \lim_{n \to \infty} \int\limits_0^t f_n(s, \omega) dB_s(\omega) \text{ (limit in probability) for } f \in \mathcal{W}_{\mathcal{H}} \, .$$

(3.3.1)

As before there exists a t-continuous version of this integral. See Friedman (1975, Chap. 4) or McKean (1969, Chap. 2) for details. Note, however, that this integral is not in general a martingale. See for example Dudley's Theorem (Theorem 12.1.5). It is, however, a *local* martingale. See Karatzas and Shreve (1991), p. 146. See also Exercise 7.12.

A comparison of Itô and Stratonovich integrals

Let us now return to our original question in this chapter: We have argued that the mathematical interpretation of the white noise equation

$$\frac{dX}{dt} = b(t, X_t) + \sigma(t, X_t) \cdot W_t \tag{3.3.2}$$

is that X_t is a solution of the integral equation

$$X_t = X_0 + \int_0^t b(s, X_s)ds + \text{``} \int_0^t \sigma(s, X_s)dB_s\text{''} , \qquad (3.3.3)$$

for some suitable interpretation of the last integral in (3.3.3). However, as indicated earlier, the Itô interpretation of an integral of the form

$$\text{``} \int_0^t f(s, \omega)dB_s(\omega)\text{''} \qquad (*)$$

is just one of several reasonable choices. For example, the Stratonovich integral is another possibility, leading (in general) to a different result. So the question still remains: *Which interpretation of (∗) makes (3.3.3) the "right" mathematical model for the equation (3.3.2)?* Here is an argument that indicates that the *Stratonovich* interpretation in some situations may be the most appropriate: Choose t-continuously differentiable processes $B_t^{(n)}$ such that for a.a. ω

$$B^{(n)}(t, \omega) \to B(t, \omega) \qquad \text{as } n \to \infty$$

uniformly (in t) in bounded intervals. For each ω let $X_t^{(n)}(\omega)$ be the solution of the corresponding (deterministic) differential equation

$$\frac{dX_t}{dt} = b(t, X_t) + \sigma(t, X_t)\frac{dB_t^{(n)}}{dt} . \qquad (3.3.4)$$

Then $X_t^{(n)}(\omega)$ converges to some function $X_t(\omega)$ in the same sense: For a.a. ω we have that $X_t^{(n)}(\omega) \to X_t(\omega)$ as $n \to \infty$, uniformly (in t) in bounded intervals.

It turns out (see Wong and Zakai (1969) and Sussman (1978)) that this solution X_t coincides with the solution of (3.3.3) obtained by using *Stratonovich* integrals, i.e.

$$X_t = X_0 + \int_0^t b(s, X_s)ds + \int_0^t \sigma(s, X_s) \circ dB_s . \qquad (3.3.5)$$

This implies that X_t is the solution of the following *modified Itô equation*:

$$X_t = X_0 + \int_0^t b(s, X_s)ds + \tfrac{1}{2}\int_0^t \sigma'(s, X_s)\sigma(s, X_s)ds + \int_0^t \sigma(s, X_s)dB_s , \quad (3.3.6)$$

where σ' denotes the derivative of $\sigma(t, x)$ with respect to x. (See Stratonovich (1966)).

Therefore, from this point of view it seems reasonable to use (3.3.6) (i.e. the Stratonovich interpretation) – and not the Itô interpretation

$$X_t = X_0 + \int\limits_0^t b(s, X_s)ds + \int\limits_0^t \sigma(s, X_s)dB_s \qquad (3.3.7)$$

as the model for the original white noise equation (3.3.2).

On the other hand, the specific feature of the Itô model of "not looking into the future" (as explained after Example 3.1.1) seems to be a reason for choosing the Itô interpretation in many cases, for example in biology (see the discussion in Turelli (1977)). The difference between the two interpretations is illustrated in Example 5.1.1. Note that (3.3.6) and (3.3.7) coincide if $\sigma(t, x)$ does not depend on x. For example, this is the situation in the linear case handled in the filtering problem in Chapter 6.

In any case, because of the explicit connection (3.3.6) between the two models (and a similar connection in higher dimensions – see (6.1.3)), it will for many purposes suffice to do the general mathematical treatment for one of the two types of integrals. In general one can say that the Stratonovich integral has the advantage of leading to ordinary chain rule formulas under a transformation (change of variable), i.e. there are no second order terms in the Stratonovich analogue of the Itô transformation formula (see Theorems 4.1.2 and 4.2.1). This property makes the Stratonovich integral natural to use for example in connection with stochastic differential equations on manifolds (see Elworthy (1982) or Ikeda and Watanabe (1989)).

However, Stratonovich integrals are not martingales, as we have seen that Itô integrals are. This gives the Itô integral an important computational advantage, even though it does not behave so nicely under transformations (as Example 3.1.9 shows). For our purposes the Itô integral will be most convenient, so we will base our discussion on that from now on.

Exercises

Unless otherwise stated B_t denotes Brownian motion in $\mathbf{R}$, $B_0 = 0$.

3.1.* Prove directly from the definition of Itô integrals (Definition 3.1.6) that

$$\int\limits_0^t sdB_s = tB_t - \int\limits_0^t B_s ds \ .$$

(Hint: Note that

$$\sum_j \Delta(s_j B_j) = \sum_j s_j \Delta B_j + \sum_j B_{j+1}\Delta s_j \ .)$$

3.2.   Prove directly from the definition of Itô integrals that

$$\int\limits_0^t B_s^2 dB_s = \tfrac{1}{3} B_t^3 - \int\limits_0^t B_s ds \ .$$

3.3.*   If $X_t : \Omega \to \mathbf{R}^n$ is a stochastic process, let $\mathcal{H}_t = \mathcal{H}_t^{(X)}$ denote the σ-algebra generated by $\{X_s(\cdot);\ s \leq t\}$ (i.e. $\{\mathcal{H}_t^{(X)}\}_{t \geq 0}$ is the *filtration of the process* $\{X_t\}_{t \geq 0}$).

a)   Show that if X_t is a martingale w.r.t. *some* filtration $\{\mathcal{N}_t\}_{t \geq 0}$, then X_t is also a martingale w.r.t. its own filtration $\{\mathcal{H}_t^{(X)}\}_{t \geq 0}$.

b)   Show that if X_t is a martingale w.r.t $\mathcal{H}_t^{(X)}$, then

$$E[X_t] = E[X_0] \qquad \text{for all } t \geq 0 \ . \tag{$*$}$$

c)   Give an example of a stochastic process X_t satisfying $(*)$ and which is *not* a martingale w.r.t. its own filtration.

3.4.*   Check whether the following processes X_t are martingales w.r.t. $\{\mathcal{F}_t\}$:

(i) $X_t = B_t + 4t$
(ii) $X_t = B_t^2$
(iii) $X_t = t^2 B_t - 2 \int\limits_0^t s B_s ds$
(iv) $X_t = B_1(t) B_2(t)$, where $(B_1(t), B_2(t))$ is 2-dimensional Brownian motion.

3.5.*   Prove directly (without using Example 3.1.9) that

$$M_t = B_t^2 - t$$

is an $\mathcal{F}_t$-martingale.

3.6.*   Prove that $N_t = B_t^3 - 3t B_t$ is a martingale.

3.7.   A famous result of Itô (1951) gives the following formula for n times *iterated Itô integrals*:

$$n! \int\limits_{0 \leq u_1 \leq \cdots \leq u_n \leq t} \cdots (\int (\int dB_{u_1}) dB_{u_2}) \cdots dB_{u_n} = t^{\frac{n}{2}} h_n\left(\frac{B_t}{\sqrt{t}}\right) \tag{3.3.8}$$

where h_n is the *Hermite polynomial* of degree n, defined by

$$h_n(x) = (-1)^n e^{\frac{x^2}{2}} \frac{d^n}{dx^n}\left(e^{-\frac{x^2}{2}}\right) ; \qquad n = 0, 1, 2, \ldots$$

(Thus $h_0(x) = 1$, $h_1(x) = x$, $h_2(x) = x^2 - 1$, $h_3(x) = x^3 - 3x$.)

    a)   Verify that in each of these n Itô integrals the integrand satisfies
        the requirements in Definition 3.1.4.

    b)   Verify formula (3.3.8) for $n = 1, 2, 3$ by combining Example 3.1.9
        and Exercise 3.2.

    c)   Use b) to give a new proof of the statement in Exercise 3.6.

3.8.*   a)   Let Y be a real valued random variable on $(\Omega, \mathcal{F}, P)$ such that

$$E[|Y|] < \infty .$$

Define
$$M_t = E[Y|\mathcal{F}_t] ; \qquad t \geq 0 .$$

Show that M_t is an $\mathcal{F}_t$-martingale.

    b)   Conversely, let M_t; $t \geq 0$ be a real valued $\mathcal{F}_t$-martingale such
        that

$$\sup_{t \geq 0} E[|M_t|^p] < \infty \qquad \text{for some } p > 1 .$$

Show that there exists $Y \in L^1(P)$ such that

$$M_t = E[Y|\mathcal{F}_t] .$$

(Hint: Use Corollary C.7.)

3.9.*   Suppose $f \in \mathcal{V}(0, T)$ and that $t \to f(t, \omega)$ is continuous for a.a. ω.
Then we have shown that

$$\int_0^T f(t, \omega)dB_t(\omega) = \lim_{\Delta t_j \to 0} \sum_j f(t_j, \omega)\Delta B_j \qquad \text{in } L^2(P) .$$

Similarly we define the *Stratonovich integral* of f by

$$\int_0^T f(t, \omega) \circ dB_t(\omega) = \lim_{\Delta t_j \to 0} \sum_j f(t_j^*, \omega)\Delta B_j, \quad \text{where } t_j^* = \tfrac{1}{2}(t_j + t_{j+1}),$$

whenever the limit exists in $L^2(P)$. In general these integrals are
different. For example, compute

$$\int_0^T B_t \circ dB_t$$

and compare with Example 3.1.9.

3.10.   If the function f in Exercise 3.9 varies "smoothly" with t then in
fact the Itô and Stratonovich integrals of f coincide. More precisely,
assume that there exists $K < \infty$ and $\epsilon > 0$ such that

$$E[|f(s, \cdot) - f(t, \cdot)|^2] \leq K|s - t|^{1+\epsilon} ; \qquad 0 \leq s, \ t \leq T .$$

Prove that then we have

$$\int_0^T f(t,\omega)dB_t = \lim_{\Delta t_j \to 0} \sum_j f(t'_j,\omega)\Delta B_j \qquad \text{(limit in } L^1(P))$$

for any choice of $t'_j \in [t_j, t_{j+1}]$. In particular,

$$\int_0^T f(t,\omega)dB_t = \int_0^T f(t,\omega) \circ dB_t \;.$$

(Hint: Consider $E\big[|\sum_j f(t_j,\omega)\Delta B_j - \sum_j f(t'_j,\omega)\Delta B_j|\big]$.)

3.11. Let W_t be a stochastic process satisfying (i), (ii) and (iii) (below (3.1.2)). Prove that W_t cannot have continuous paths. (Hint: Consider $E[(W_t^{(N)} - W_s^{(N)})^2]$, where

$$W_t^{(N)} = (-N) \vee (N \wedge W_t), \; N = 1,2,3,\ldots) \;.$$

3.12.* As in Exercise 3.9 we let $\circ dB_t$ denote Stratonovich differentials.

  (i)   Use (3.3.6) to transform the following Stratonovich differential equations into Itô differential equations:
        (a) $dX_t = \gamma X_t dt + \alpha X_t \circ dB_t$
        (b) $dX_t = \sin X_t \cos X_t dt + (t^2 + \cos X_t) \circ dB_t$
  (ii) Transform the following Itô differential equations into Stratonovich differential equations:
        (a) $dX_t = rX_t dt + \alpha X_t dB_t$
        (b) $dX_t = 2e^{-X_t}dt + X_t^2 dB_t$

3.13. A stochastic process $X_t(\cdot)\colon \Omega \to \mathbf{R}$ is *continuous in mean square* if $E[X_t^2] < \infty$ for all t and

$$\lim_{s \to t} E[(X_s - X_t)^2] = 0 \qquad \text{for all } t \geq 0 \;.$$

a) Prove that Brownian motion B_t is continuous in mean square.
b) Let $f\colon \mathbf{R} \to \mathbf{R}$ be a Lipschitz continuous function, i.e. there exists $C < \infty$ such that

$$|f(x) - f(y)| \leq C|x - y| \qquad \text{for all } x, y \in \mathbf{R} \;.$$

Prove that
$$Y_t := f(B_t)$$

is continuous in mean square.

c) Let X_t be a stochastic process which is continuous in mean square and assume that $X_t \in \mathcal{V}(S,T)$, $T < \infty$. Show that

$$\int_S^T X_t dB_t = \lim_{n \to \infty} \int_S^T \phi_n(t, \omega) dB_t(\omega) \qquad (\text{limit in } L^2(P))$$

where

$$\phi_n(t, \omega) = \sum_j X_{t_j^{(n)}}(\omega) \mathcal{X}_{[t_j^{(n)}, t_{j+1}^{(n)})}(t) , \qquad T < \infty .$$

(Hint: Consider

$$E\left[\int_S^T (X_t - \phi_n(t))^2 dt\right] = E\left[\sum_j \int_{t_j^{(n)}}^{t_{j+1}^{(n)}} (X_t - X_{t_j^{(n)}})^2 dt\right] .$$

3.14. Show that a function $h(\omega)$ is $\mathcal{F}_t$-measurable if and only if h is a pointwise limit (for a.a. ω) of sums of functions of the form

$$g_1(B_{t_1}) \cdot g_2(B_{t_2}) \cdots g_k(B_{t_k})$$

where $g_1, \ldots, g_k$ are bounded continuous functions and $t_j \le t$ for $j \le k$, $k = 1, 2, \ldots$
Hint: Complete the following steps:

a) We may assume that h is bounded.

b) For $n = 1, 2, \ldots$ and $j = 1, 2, \ldots$ put $t_j = t_j^{(n)} = j \cdot 2^{-n}$. For fixed n let $\mathcal{H}_n$ be the σ-algebra generated by $\{B_{t_j}(\cdot)\}_{t_j \le t}$. Then by Corollary C.9

$$h = E[h|\mathcal{F}_t] = \lim_{n \to \infty} E[h|\mathcal{H}_n] \qquad (\text{pointwise a.e. limit})$$

c) Define $h_n := E[h|\mathcal{H}_n]$. Then by the Doob-Dynkin lemma (Lemma 2.1.2) we have

$$h_n(\omega) = G_n(B_{t_1}(\omega), \ldots, B_{t_k}(\omega))$$

for some Borel function $G_n : \mathbf{R}^k \to \mathbf{R}$, where $k = \max\{j; j \cdot 2^{-n} \le t\}$. Now use that any Borel function $G : \mathbf{R}^k \to \mathbf{R}$ can be approximated pointwise a.e. by a continuous function $F : \mathbf{R}^k \to \mathbf{R}$ and complete the proof by applying the Stone-Weierstrass theorem.

3.15.* Suppose $f, g \in \mathcal{V}(S,T)$ and that there exist constants C, D such that

$$C + \int_S^T f(t, \omega) dB_t(\omega) = D + \int_S^T g(t, \omega) dB_t(\omega) \qquad \text{for a.a. } \omega \in \Omega .$$

Show that

$$C = D$$

and

$$f(t, \omega) = g(t, \omega) \qquad \text{for a.a. } (t, \omega) \in [S, T] \times \Omega .$$

3.16. Let $X \colon \Omega \to \mathbf{R}$ be a random variable such that $E[X^2] < \infty$ and let $\mathcal{H} \subset \mathcal{F}$ be a σ-algebra. Show that

$$E\big[(E[X|\mathcal{H}])^2\big] \leq E[X^2] .$$

(See Lemma 6.1.1. See also the Jensen inequality for conditional expectation (Appendix B).)

3.17. Let $(\Omega, \mathcal{F}, P)$ be a probability space and let $X \colon \Omega \to \mathbf{R}$ be a random variable with $E[|X|] < \infty$. If $\mathcal{G} \subset \mathcal{F}$ is a *finite* σ-algebra, then by Exercise 2.7 there exists a partition $\Omega = \bigcup\limits_{i=1}^{n} G_i$ such that $\mathcal{G}$ consists of $\emptyset$ and unions of some (or all) of $G_1, \ldots, G_n$.

a) Explain why $E[X|\mathcal{G}](\omega)$ is constant on each G_i. (See Exercise 2.7 c).)

b) Assume that $P[G_i] > 0$. Show that

$$E[X|\mathcal{G}](\omega) = \frac{\int_{G_i} X dP}{P(G_i)} \qquad \text{for } \omega \in G_i .$$

c) Suppose X assumes only finitely many values $a_1, \ldots, a_m$. Then from elementary probability theory we know that (see Exercise 2.1)

$$E[X|G_i] = \sum_{k=1}^{m} a_k P[X = a_k | G_i] .$$

Compare with b) and verify that

$$E[X|G_i] = E[X|\mathcal{G}](\omega) \qquad \text{for } \omega \in G_i .$$

Thus we may regard the conditional expectation as defined in Appendix B as a (substantial) generalization of the conditional expectation in elementary probability theory.

3.18. Let B_t be 1-dimensional Brownian motion and let $\sigma \in \mathbf{R}$ be constant. Prove directly from the definition that

$$M_t := \exp(\sigma B_t - \tfrac{1}{2}\sigma^2 t); \qquad t \geq 0$$

is an $\mathcal{F}_t$-martingale.

(Hint: If $s > t$ then $E[\exp(\sigma B_t - \tfrac{1}{2}\sigma^2 s)|\mathcal{F}_t] = E[\exp(\sigma(B_s - B_t)) \times \exp(\sigma B_t - \tfrac{1}{2}\sigma^2 s)|\mathcal{F}_t]$. Now use Theorem B.2 e), Theorem B.2 d) and Exercise 2.20.)

Chapter 4
The Itô Formula and the Martingale Representation Theorem

4.1 The 1-dimensional Itô formula

Example 3.1.9 illustrates that the basic definition of Itô integrals is not very useful when we try to evaluate a given integral. This is similar to the situation for ordinary Riemann integrals, where we do not use the basic definition but rather the fundamental theorem of calculus plus the chain rule in the explicit calculations.

In this context, however, we have no differentiation theory, only integration theory. Nevertheless it turns out that it is possible to establish an Itô integral version of the chain rule, called the Itô formula. The Itô formula is, as we will show by examples, very useful for evaluating Itô integrals.

From the example

$$\int_0^t B_s dB_s = \tfrac{1}{2}B_t^2 - \tfrac{1}{2}t \qquad \text{or} \qquad \tfrac{1}{2}B_t^2 = \tfrac{1}{2}t + \int_0^t B_s dB_s \, , \qquad (4.1.1)$$

we see that the image of the Itô integral $B_t = \int_0^t dB_s$ by the map $g(x) = \tfrac{1}{2}x^2$ is not again an Itô integral of the form

$$\int_0^t f(s,\omega)dB_s(\omega)$$

but a combination of a dB_s-and a ds-integral:

$$\tfrac{1}{2}B_t^2 = \int_0^t \tfrac{1}{2}ds + \int_0^t B_s dB_s \, . \qquad (4.1.2)$$

It turns out that if we introduce *Itô processes* (also called *stochastic integrals*) as sums of a dB_s-and a ds-integral then this family of integrals is stable under smooth maps. Thus we define

Definition 4.1.1 (1-dimensional Itô processes)
Let B_t be 1-dimensional Brownian motion on $(\Omega, \mathcal{F}, P)$. A (1-dimensional) Itô process (or stochastic integral) is a stochastic process X_t on $(\Omega, \mathcal{F}, P)$ of the form

$$X_t = X_0 + \int_0^t u(s, \omega)ds + \int_0^t v(s, \omega)dB_s , \qquad (4.1.3)$$

where $v \in \mathcal{W}_{\mathcal{H}}$, so that

$$P\left[\int_0^t v(s,\omega)^2 ds < \infty \text{ for all } t \geq 0\right] = 1 \qquad (4.1.4)$$

(see Definition 3.3.2). We also assume that u is $\mathcal{H}_t$-adapted (where $\mathcal{H}_t$ is as in (ii)', Section 3.3) and

$$P\left[\int_0^t |u(s,\omega)|ds < \infty \text{ for all } t \geq 0\right] = 1 . \qquad (4.1.5)$$

If X_t is an Itô process of the form (4.1.3) the equation (4.1.3) is sometimes written in the shorter differential form

$$dX_t = udt + vdB_t . \qquad (4.1.6)$$

For example, (4.1.1) (or (4.1.2)) may be represented by

$$d\left(\tfrac{1}{2}B_t^2\right) = \tfrac{1}{2}dt + B_t dB_t .$$

We are now ready to state the first main result in this chapter:

Theorem 4.1.2 (The 1-dimensional Itô formula)
Let X_t be an Itô process given by

$$dX_t = udt + vdB_t .$$

Let $g(t, x) \in C^2([0, \infty) \times \mathbf{R})$ (i.e. g is twice continuously differentiable on $[0, \infty) \times \mathbf{R}$). Then

$$Y_t = g(t, X_t)$$

is again an Itô process, and

$$dY_t = \frac{\partial g}{\partial t}(t, X_t)dt + \frac{\partial g}{\partial x}(t, X_t)dX_t + \frac{1}{2}\frac{\partial^2 g}{\partial x^2}(t, X_t) \cdot (dX_t)^2 , \qquad (4.1.7)$$

where $(dX_t)^2 = (dX_t) \cdot (dX_t)$ is computed according to the rules

$$dt \cdot dt = dt \cdot dB_t = dB_t \cdot dt = 0 , \quad dB_t \cdot dB_t = dt . \qquad (4.1.8)$$

Before we prove Itô's formula let us look at some examples.

Example 4.1.3 Let us return to the integral

$$I = \int_0^t B_s dB_s \quad \text{from Chapter 3 .}$$

Choose $X_t = B_t$ and $g(t,x) = \frac{1}{2}x^2$. Then

$$Y_t = g(t, B_t) = \tfrac{1}{2}B_t^2 .$$

Then by Itô's formula,

$$dY_t = \frac{\partial g}{\partial t}dt + \frac{\partial g}{\partial x}dB_t + \frac{1}{2}\frac{\partial^2 g}{\partial x^2}(dB_t)^2 = B_t dB_t + \tfrac{1}{2}(dB_t)^2 = B_t dB_t + \tfrac{1}{2}dt .$$

Hence

$$d(\tfrac{1}{2}B_t^2) = B_t dB_t + \tfrac{1}{2}dt .$$

In other words,

$$\tfrac{1}{2}B_t^2 = \int_0^t B_s dB_s + \tfrac{1}{2}t, \quad \text{as in Chapter 3 .}$$

Example 4.1.4 What is

$$\int_0^t s dB_s \ ?$$

From classical calculus it seems reasonable that a term of the form tB_t should appear, so we put

$$g(t,x) = tx$$

and

$$Y_t = g(t, B_t) = tB_t .$$

Then by Itô's formula,

$$dY_t = B_t dt + t dB_t + 0$$

i.e.

$$d(tB_t) = B_t dt + t dB_t$$

or

$$tB_t = \int_0^t B_s ds + \int_0^t s dB_s$$

or

$$\int\limits_0^t s dB_s = tB_t - \int\limits_0^t B_s ds ,$$

which is reasonable from an integration-by-parts point of view.

More generally, the same method gives

Theorem 4.1.5 (Integration by parts) *Suppose $f(s,\omega)$ is continuous and of bounded variation with respect to $s \in [0,t]$, for a.a. ω. (See Exercise 2.17.) Then*

$$\int\limits_0^t f(s) dB_s = f(t)B_t - \int\limits_0^t B_s df_s .$$

Note that it is crucial for the result to hold that f is of bounded variation. (See Exercise 4.3 for the general case.)

Sketch of proof of the Itô formula. First observe that if we substitute

$$dX_t = udt + vdB_t$$

in (4.1.7) and use (4.1.8) we get the equivalent expression

$$g(t, X_t) = g(0, X_0) + \int\limits_0^t \left(\frac{\partial g}{\partial s}(s, X_s) + u_s \frac{\partial g}{\partial x}(s, X_s) + \tfrac{1}{2} v_s^2 \cdot \frac{\partial^2 g}{\partial x^2}(s, X_s) \right) ds$$

$$+ \int\limits_0^t v_s \cdot \frac{\partial g}{\partial x}(s, X_s) dB_s \quad \text{where } u_s = u(s, \omega), \ v_s = v(s, \omega) . \quad (4.1.9)$$

Note that (4.1.9) is an Itô process in the sense of Definition 4.1.1.

We may assume that $g, \frac{\partial g}{\partial t}, \frac{\partial g}{\partial x}$ and $\frac{\partial^2 g}{\partial x^2}$ are bounded, for if (4.1.9) is proved in this case we obtain the general case by approximating by C^2 functions g_n such that $g_n, \frac{\partial g_n}{\partial t}, \frac{\partial g_n}{\partial x}$ and $\frac{\partial^2 g_n}{\partial x^2}$ are bounded for each n and converge uniformly on compact subsets of $[0, \infty) \times \mathbf{R}$ to $g, \frac{\partial g}{\partial t}, \frac{\partial g}{\partial x}, \frac{\partial^2 g}{\partial x^2}$, respectively. (See Exercise 4.9.) Moreover, from (3.3.1) we see that we may assume that $u(t, \omega)$ and $v(t, \omega)$ are elementary functions. Using Taylor's theorem we get

$$g(t, X_t) = g(0, X_0) + \sum_j \Delta g(t_j, X_j) = g(0, X_0) + \sum_j \frac{\partial g}{\partial t} \Delta t_j + \sum_j \frac{\partial g}{\partial x} \Delta X_j$$

$$+ \tfrac{1}{2} \sum_j \frac{\partial^2 g}{\partial t^2} (\Delta t_j)^2 + \sum_j \frac{\partial^2 g}{\partial t \partial x} (\Delta t_j)(\Delta X_j) + \tfrac{1}{2} \sum_j \frac{\partial^2 g}{\partial x^2} (\Delta X_j)^2 + \sum_j R_j ,$$

where $\frac{\partial g}{\partial t}, \frac{\partial g}{\partial x}$ etc. are evaluated at the points (t_j, X_{t_j}),

$$\Delta t_j = t_{j+1} - t_j, \ \Delta X_j = X_{t_{j+1}} - X_{t_j}, \ \Delta g(t_j, X_j) = g(t_{j+1}, X_{t_{j+1}}) - g(t_j, X_j)$$

and $R_j = o(|\Delta t_j|^2 + |\Delta X_j|^2)$ for all j.
  If $\Delta t_j \to 0$ then

$$\sum_j \frac{\partial g}{\partial t} \Delta t_j = \sum_j \frac{\partial g}{\partial t}(t_j, X_j) \Delta t_j \to \int_0^t \frac{\partial g}{\partial t}(s, X_s) ds \qquad (4.1.10)$$

$$\sum_j \frac{\partial g}{\partial x} \Delta X_j = \sum_j \frac{\partial g}{\partial x}(t_j, X_j) \Delta X_j \to \int_0^t \frac{\partial g}{\partial x}(s, X_s) dX_s . \qquad (4.1.11)$$

Moreover, since u and v are elementary we get

$$\sum_j \frac{\partial^2 g}{\partial x^2}(\Delta X_j)^2 = \sum_j \frac{\partial^2 g}{\partial x^2} u_j^2 (\Delta t_j)^2 + 2 \sum_j \frac{\partial^2 g}{\partial x^2} u_j v_j (\Delta t_j)(\Delta B_j)$$

$$+ \sum_j \frac{\partial^2 g}{\partial x^2} v_j^2 \cdot (\Delta B_j)^2, \quad \text{where } u_j = u(t_j, \omega), \ v_j = v(t_j, \omega) . \qquad (4.1.12)$$

The first two terms here tend to 0 as $\Delta t_j \to 0$. For example,

$$E\left[\left(\sum_j \frac{\partial^2 g}{\partial x^2} u_j v_j (\Delta t_j)(\Delta B_j) \right)^2 \right] =$$

$$= \sum_j E\left[\left(\frac{\partial^2 g}{\partial x^2} u_j v_j \right)^2 \right] (\Delta t_j)^3 \to 0 \quad \text{as } \Delta t_j \to 0 .$$

We claim that the last term tends to

$$\int_0^t \frac{\partial^2 g}{\partial x^2} v^2 ds \qquad \text{in } L^2(P), \text{ as } \Delta t_j \to 0 .$$

To prove this, put $a(t) = \frac{\partial^2 g}{\partial x^2}(t, X_t) v^2(t, \omega)$, $a_j = a(t_j)$ and consider

$$E\left[\left(\sum_j a_j (\Delta B_j)^2 - \sum_j a_j \Delta t_j \right)^2 \right] = \sum_{i,j} E[a_i a_j ((\Delta B_i)^2 - \Delta t_i)((\Delta B_j)^2 - \Delta t_j)] .$$

If $i < j$ then $a_i a_j ((\Delta B_i)^2 - \Delta t_i)$ and $(\Delta B_j)^2 - \Delta t_j$ are independent so the terms vanish in this case, and similarly if $i > j$. So we are left with

$$\sum_j E[a_j^2 ((\Delta B_j)^2 - \Delta t_j)^2] = \sum_j E[a_j^2] \cdot E[(\Delta B_j)^4 - 2(\Delta B_j)^2 \Delta t_j + (\Delta t_j)^2]$$

$$= \sum_j E[a_j^2] \cdot (3(\Delta t_j)^2 - 2(\Delta t_j)^2 + (\Delta t_j)^2) = 2 \sum_j E[a_j^2] \cdot (\Delta t_j)^2$$

$$\to 0 \qquad \text{as } \Delta t_j \to 0 .$$

In other words, we have established that

$$\sum_j a_j (\Delta B_j)^2 \to \int_0^t a(s)ds \qquad \text{in } L^2(P) \text{ as } \Delta t_j \to 0$$

and this is often expressed shortly by the striking formula

$$(dB_t)^2 = dt . \tag{4.1.13}$$

The argument above also proves that $\sum R_j \to 0$ as $\Delta t_j \to 0$. That completes the proof of the Itô formula.   □

Remark. Note that it is enough that $g(t,x)$ is C^2 on $[0,\infty) \times U$, if $U \subset \mathbf{R}$ is an open set such that $X_t(\omega) \in U$ for all $t \geq 0, \omega \in \Omega$. Moreover, it is sufficient that $g(t,x)$ is C^1 w.r.t. t and C^2 w.r.t. x.

4.2 The Multi-dimensional Itô Formula

We now turn to the situation in higher dimensions: Let $B(t,\omega) = (B_1(t,\omega), \ldots, B_m(t,\omega))$ denote m-dimensional Brownian motion. If each of the processes $u_i(t,\omega)$ and $v_{ij}(t,\omega)$ satisfies the conditions given in Definition 4.1.1 ($1 \leq i \leq n$, $1 \leq j \leq m$) then we can form the following n Itô processes

$$\begin{cases} dX_1 = u_1 dt + v_{11} dB_1 + \cdots + v_{1m} dB_m \\ \ \vdots \quad \ \vdots \qquad\qquad\qquad\qquad \vdots \\ dX_n = u_n dt + v_{n1} dB_1 + \cdots + v_{nm} dB_m \end{cases} \tag{4.2.1}$$

Or, in matrix notation simply

$$dX(t) = udt + vdB(t) , \tag{4.2.2}$$

where

$$X(t) = \begin{pmatrix} X_1(t) \\ \vdots \\ X_n(t) \end{pmatrix}, \ u = \begin{pmatrix} u_1 \\ \vdots \\ u_n \end{pmatrix}, \ v = \begin{pmatrix} v_{11} \cdots v_{1m} \\ \vdots \qquad \vdots \\ v_{n1} \cdots v_{nm} \end{pmatrix}, \ dB(t) = \begin{pmatrix} dB_1(t) \\ \vdots \\ dB_m(t) \end{pmatrix} \tag{4.2.3}$$

Such a process $X(t)$ is called an n-**dimensional Itô process** (or just an Itô process).

We now ask: What is the result of applying a smooth function to X? The answer is given by

Theorem 4.2.1 (The general Itô formula)
Let

$$dX(t) = udt + vdB(t)$$

be an n-dimensional Itô process as above. Let $g(t,x) = (g_1(t,x), \ldots, g_p(t,x))$ be a C^2 map from $[0,\infty) \times \mathbf{R}^n$ into $\mathbf{R}^p$. Then the process

$$Y(t,\omega) = g(t, X(t))$$

is again an Itô process, whose component number k, Y_k, is given by

$$dY_k = \frac{\partial g_k}{\partial t}(t,X)dt + \sum_i \frac{\partial g_k}{\partial x_i}(t,X)dX_i + \tfrac{1}{2}\sum_{i,j} \frac{\partial^2 g_k}{\partial x_i \partial x_j}(t,X)dX_i dX_j$$

where $dB_i dB_j = \delta_{ij}dt, dB_i dt = dtdB_i = 0$.

The proof is similar to the 1-dimensional version (Theorem 4.1.2) and is omitted.

Example 4.2.2 Let $B = (B_1, \ldots, B_n)$ be Brownian motion in $\mathbf{R}^n$, $n \geq 2$, and consider

$$R(t,\omega) = |B(t,\omega)| = (B_1^2(t,\omega) + \cdots + B_n^2(t,\omega))^{\frac{1}{2}} ,$$

i.e. the distance to the origin of $B(t,\omega)$. The function $g(t,x) = |x|$ is not C^2 at the origin, but since B_t never hits the origin, a.s. when $n \geq 2$ (see Exercise 9.7) Itô's formula still works and we get

$$dR = \sum_{i=1}^{n} \frac{B_i dB_i}{R} + \frac{n-1}{2R}dt .$$

The process R is called the *n-dimensional Bessel process* because its generator (Chapter 7) is the Bessel differential operator $Af(x) = \tfrac{1}{2}f''(x) + \frac{n-1}{2x}f'(x)$. See Example 8.4.1.

4.3 The Martingale Representation Theorem

Let $B(t) = (B_1(t), \ldots, B_n(t))$ be n-dimensional Brownian motion. In Chapter 3 (Corollary 3.2.6) we proved that if $v \in \mathcal{V}^n$ then the Itô integral

$$X_t = X_0 + \int_0^t v(s,\omega)dB(s) ; \qquad t \geq 0$$

is always a martingale w.r.t. filtration $\mathcal{F}_t^{(n)}$ (and w.r.t. the probability measure P). In this section we will prove that the converse is also true: Any $\mathcal{F}_t^{(n)}$-martingale (w.r.t. P) can be represented as an Itô integral (Theorem 4.3.4). This result, called the *martingale representation theorem*, is important for many applications, for example in mathematical finance. See Chapter 12. For simplicity we prove the result only when $n = 1$, but the reader can easily verify that essentially the same proof works for arbitrary n.

We first establish some auxiliary results.

Lemma 4.3.1 *Fix $T > 0$. The set of random variables*

$$\{\phi(B_{t_1}, \ldots, B_{t_n}); \ t_i \in [0, T], \ \phi \in C_0^\infty(\mathbf{R}^n), \ n = 1, 2, \ldots\}$$

is dense in $L^2(\mathcal{F}_T, P)$.

Proof. Let $\{t_i\}_{i=1}^\infty$ be a dense subset of $[0, T]$ and for each $n = 1, 2, \ldots$ let $\mathcal{H}_n$ be the σ-algebra generated by $B_{t_1}(\cdot), \ldots, B_{t_n}(\cdot)$. Then clearly

$$\mathcal{H}_n \subset \mathcal{H}_{n+1}$$

and $\mathcal{F}_T$ is the smallest σ-algebra containing all the $\mathcal{H}_n$'s. Choose $g \in L^2(\mathcal{F}_T, P)$. Then by the martingale convergence theorem Corollary C.9 (Appendix C) we have that

$$g = E[g|\mathcal{F}_T] = \lim_{n\to\infty} E[g|\mathcal{H}_n] \ .$$

The limit is pointwise a.e. (P) and in $L^2(\mathcal{F}_T, P)$. By the Doob-Dynkin Lemma (Lemma 2.1.2) we can write, for each n,

$$E[g|\mathcal{H}_n] = g_n(B_{t_1}, \ldots, B_{t_n})$$

for some Borel measurable function $g_n : \mathbf{R}^n \to \mathbf{R}$. Each such $g_n(B_{t_1}, \ldots, B_{t_n})$ can be approximated in $L^2(\mathcal{F}_T, P)$ by functions $\phi_n(B_{t_1}, \ldots, B_{t_n})$ where $\phi_n \in C_0^\infty(\mathbf{R}^n)$ and the result follows.   □

For an alternative proof of the next result see Exercise 4.17.

Lemma 4.3.2 *The linear span of random variables of the type*

$$\exp\left\{\int_0^T h(t)dB_t(\omega) - \frac{1}{2}\int_0^T h^2(t)dt\right\}; \quad h \in L^2[0, T] \ (deterministic) \qquad (4.3.1)$$

is dense in $L^2(\mathcal{F}_T, P)$.

Proof. Suppose $g \in L^2(\mathcal{F}_T, P)$ is orthogonal (in $L^2(\mathcal{F}_T, P)$) to all functions of the form (4.3.1). Then in particular

$$G(\lambda) := \int_\Omega \exp\{\lambda_1 B_{t_1}(\omega) + \cdots + \lambda_n B_{t_n}(\omega)\}g(\omega)dP(\omega) = 0 \qquad (4.3.2)$$

for all $\lambda = (\lambda_1, \ldots, \lambda_n) \in \mathbf{R}^n$ and all $t_1, \ldots, t_n \in [0, T]$. The function $G(\lambda)$ is real analytic in $\lambda \in \mathbf{R}^n$ and hence G has an analytic extension to the complex space $\mathbf{C}^n$ given by

$$G(z) = \int_\Omega \exp\{z_1 B_{t_1}(\omega) + \cdots + z_n B_{t_n}(\omega)\} g(\omega) dP(\omega) \qquad (4.3.3)$$

for all $z = (z_1, \ldots, z_n) \in \mathbf{C}^n$. (See the estimates in Exercise 2.8 b).) Since $G = 0$ on $\mathbf{R}^n$ and G is analytic, $G = 0$ on $\mathbf{C}^n$. In particular, $G(iy_1, iy_2, \ldots, iy_n) = 0$ for all $y = (y_1, \ldots, y_n) \in \mathbf{R}^n$. But then we get, for $\phi \in C_0^\infty(\mathbf{R}^n)$,

$$\int_\Omega \phi(B_{t_1}, \ldots, B_{t_n}) g(\omega) dP(\omega)$$

$$= \int_\Omega (2\pi)^{-n/2} \left(\int_{\mathbf{R}^n} \widehat{\phi}(y) e^{i(y_1 B_{t_1} + \cdots + y_n B_{t_n})} dy \right) g(\omega) dP(\omega)$$

$$= (2\pi)^{-n/2} \int_{\mathbf{R}^n} \widehat{\phi}(y) \left(\int_\Omega e^{i(y_1 B_{t_1} + \cdots + y_n B_{t_n})} g(\omega) dP(\omega) \right) dy$$

$$= (2\pi)^{-n/2} \int_{\mathbf{R}^n} \widehat{\phi}(y) G(iy) dy = 0 , \qquad (4.3.4)$$

where

$$\widehat{\phi}(y) = (2\pi)^{-n/2} \int_{\mathbf{R}^n} \phi(x) e^{-i\,x\cdot y} dx$$

is the Fourier transform of ϕ and we have used the inverse Fourier transform theorem

$$\phi(x) = (2\pi)^{-n/2} \int_{\mathbf{R}^n} \widehat{\phi}(y) e^{i\,x\cdot y} dy$$

(see e.g. Folland (1984)).

By (4.3.4) and Lemma 4.3.1 g is orthogonal to a dense subset of $L^2(\mathcal{F}_T, P)$ and we conclude that $g = 0$. Therefore the linear span of the functions in (4.3.1) must be dense in $L^2(\mathcal{F}_T, P)$ as claimed. $\square$

Suppose $B(t) = (B_1(t), \ldots, B_n(t))$ is n-dimensional. If $v(s, \omega) \in \mathcal{V}^n(0, T)$ then the random variable

$$V(\omega) := \int_0^T v(t, \omega) dB(t) \qquad (4.3.5)$$

is $\mathcal{F}_T^{(n)}$-measurable and by the Itô isometry

$$E[V^2] = \int_0^T E[v^2(t,\cdot)]dt < \infty , \qquad \text{so } V \in L^2(\mathcal{F}_T^{(n)}, P) .$$

The next result states that any $F \in L^2(\mathcal{F}_T^{(n)}, P)$ can be represented this way:

Theorem 4.3.3 (The Itô representation theorem)

Let $F \in L^2(\mathcal{F}_T^{(n)}, P)$. Then there exists a unique stochastic process $f(t,\omega) \in V^n(0,T)$ such that

$$F(\omega) = E[F] + \int_0^T f(t,\omega)dB(t) . \qquad (4.3.6)$$

Proof. Again we consider only the case $n = 1$. (The proof in the general case is similar.) First assume that F has the form (4.3.1), i.e.

$$F(\omega) = \exp\left\{ \int_0^T h(t)dB_t(\omega) - \tfrac{1}{2}\int_0^T h^2(t)dt \right\}$$

for some $h(t) \in L^2[0,T]$.

Define

$$Y_t(\omega) = \exp\left\{ \int_0^t h(s)dB_s(\omega) - \tfrac{1}{2}\int_0^t h^2(s)ds \right\} ; \qquad 0 \le t \le T .$$

Then by Itô's formula

$$dY_t = Y_t(h(t)dB_t - \tfrac{1}{2}h^2(t)dt) + \tfrac{1}{2}Y_t(h(t)dB_t)^2 = Y_th(t)dB_t$$

so that

$$Y_t = 1 + \int_0^t Y_sh(s)dB_s ; \qquad t \in [0,T] .$$

Therefore

$$F = Y_T = 1 + \int_0^T Y_sh(s)dB_s$$

and hence $E[F] = 1$. So (4.3.6) holds in this case. If $F \in L^2(\mathcal{F}_T, P)$ is arbitrary, we can by Lemma 4.3.2 approximate F in $L^2(\mathcal{F}_T, P)$ by linear combinations F_n of functions of the form (4.3.1). By linearity (4.3.6) also holds for linear combinations of functions of the form (4.3.1). Then for each n we have

$$F_n(\omega) = E[F_n] + \int_0^T f_n(s, \omega) dB_s(\omega), \qquad \text{where } f_n \in \mathcal{V}(0, T) \, .$$

By the Itô isometry

$$E[(F_n - F_m)^2] = E\left[(E[F_n - F_m] + \int_0^T (f_n - f_m) dB)^2\right]$$

$$= (E[F_n - F_m])^2 + \int_0^T E[(f_n - f_m)^2] dt \to 0 \qquad \text{as } n, m \to \infty$$

so $\{f_n\}$ is a Cauchy sequence in $L^2([0, T] \times \Omega)$ and hence converges to some $f \in L^2([0, T] \times \Omega)$. Since $f_n \in \mathcal{V}(0, T)$ we have $f \in \mathcal{V}(0, T)$. (A subsequence of $\{f_n(t, \omega)\}$ converges to $f(t, \omega)$ for a.a. $(t, \omega) \in [0, T] \times \Omega$. Therefore $f(t, \cdot)$ is $\mathcal{F}_t$-measurable for a.a. t. So by modifying $f(t, \omega)$ on a t-set of measure 0 we can obtain that $f(t, \omega)$ is $\mathcal{F}_t$-adapted.) Again using the Itô isometry we see that

$$F = \lim_{n \to \infty} F_n = \lim_{n \to \infty} \left(E[F_n] + \int_0^T f_n dB\right) = E[F] + \int_0^T f dB \, ,$$

the limit being taken in $L^2(\mathcal{F}_T, P)$. Hence the representation (4.3.6) holds for all $F \in L^2(\mathcal{F}_T, P)$.

The uniqueness follows from the Itô isometry: Suppose

$$F(\omega) = E[F] + \int_0^T f_1(t, \omega) dB_t(\omega) = E[F] + \int_0^T f_2(t, \omega) dB_t(\omega)$$

with $f_1, f_2 \in \mathcal{V}(0, T)$. Then

$$0 = E[(\int_0^T (f_1(t, \omega) - f_2(t, \omega)) dB_t(\omega))^2] = \int_0^T E[(f_1(t, \omega) - f_2(t, \omega))^2] dt$$

and therefore $f_1(t, \omega) = f_2(t, \omega)$ for a.a. $(t, \omega) \in [0, T] \times \Omega$. □

Remark. The process $f(t, \omega)$ can be expressed in terms of the Fréchet derivative and also in terms of the Malliavin derivative of $F(\omega)$. See Clark (1970/71), Davis (1980) and Ocone (1984).

Theorem 4.3.4 (The martingale representation theorem)
Let $B(t) = (B_1(t), \ldots, B_n(t))$ be n-dimensional. Suppose M_t is an $\mathcal{F}_t^{(n)}$-martingale (w.r.t. P) and that $M_t \in L^2(P)$ for all $t \geq 0$. Then there exists a unique stochastic process $g(s, \omega)$ such that $g \in \mathcal{V}^{(n)}(0, t)$ for all $t \geq 0$ and

$$M_t(\omega) = E[M_0] + \int_0^t g(s,\omega)dB(s) \qquad a.s., \; for \; all \; t \geq 0 \;.$$

Proof (n = 1). By Theorem 4.3.3 applied to $T = t$, $F = M_t$, we have that for all t there exists a unique $f^{(t)}(s,\omega) \in L^2(\mathcal{F}_t, P)$ such that

$$M_t(\omega) = E[M_t] + \int_0^t f^{(t)}(s,\omega)dB_s(\omega) = E[M_0] + \int_0^t f^{(t)}(s,\omega)dB_s(\omega) \;.$$

Now assume $0 \leq t_1 < t_2$. Then

$$M_{t_1} = E[M_{t_2}|\mathcal{F}_{t_1}] = E[M_0] + E\left[\int_0^{t_2} f^{(t_2)}(s,\omega)dB_s(\omega)|\mathcal{F}_{t_1} \right]$$

$$= E[M_0] + \int_0^{t_1} f^{(t_2)}(s,\omega)dB_s(\omega) \;. \tag{4.3.7}$$

But we also have

$$M_{t_1} = E[M_0] + \int_0^{t_1} f^{(t_1)}(s,\omega)dB_s(\omega) \;. \tag{4.3.8}$$

Hence, comparing (4.3.7) and (4.3.8) we get that

$$0 = E\left[\left(\int_0^{t_1} (f^{(t_2)} - f^{(t_1)})dB \right)^2 \right] = \int_0^{t_1} E[(f^{(t_2)} - f^{(t_1)})^2]ds$$

and therefore

$$f^{(t_1)}(s,\omega) = f^{(t_2)}(s,\omega) \qquad for \; a.a. \; (s,\omega) \in [0,t_1] \times \Omega \;.$$

So we can define $f(s,\omega)$ for a.a. $s \in [0,\infty) \times \Omega$ by setting

$$f(s,\omega) = f^{(N)}(s,\omega) \qquad if \; s \in [0,N]$$

and then we get

$$M_t = E[M_0] + \int_0^t f^{(t)}(s,\omega)dB_s(\omega) = E[M_0] + \int_0^t f(s,\omega)dB_s(\omega) \quad for \; all \; t \geq 0 \;.$$

$\square$

Exercises

4.1.* Use Itô's formula to write the following stochastic processes Y_t on the standard form

$$dY_t = u(t, \omega)dt + v(t, \omega)dB_t$$

for suitable choices of $u \in \mathbf{R}^n$, $v \in \mathbf{R}^{n \times m}$ and dimensions n, m:

a) $Y_t = B_t^2$, where B_t is 1-dimensional
b) $Y_t = 2 + t + e^{B_t}$ (B_t is 1-dimensional)
c) $Y_t = B_1^2(t) + B_2^2(t)$ where (B_1, B_2) is 2-dimensional
d) $Y_t = (t_0 + t, B_t)$ (B_t is 1-dimensional)
e) $Y_t = (B_1(t) + B_2(t) + B_3(t), B_2^2(t) - B_1(t)B_3(t))$, where (B_1, B_2, B_3) is 3-dimensional.

4.2.* Use Itô's formula to prove that

$$\int_0^t B_s^2 dB_s = \tfrac{1}{3}B_t^3 - \int_0^t B_s ds .$$

4.3.* Let X_t, Y_t be Itô processes in $\mathbf{R}$. Prove that

$$d(X_t Y_t) = X_t dY_t + Y_t dX_t + dX_t \cdot dY_t .$$

Deduce the following general *integration by parts formula*

$$\int_0^t X_s dY_s = X_t Y_t - X_0 Y_0 - \int_0^t Y_s dX_s - \int_0^t dX_s \cdot dY_s .$$

4.4. (Exponential martingales)
Suppose $\theta(t, \omega) = (\theta_1(t, \omega), \ldots, \theta_n(t, \omega)) \in \mathbf{R}^n$ with $\theta_k(t, \omega) \in \mathcal{V}[0, T]$ for $k = 1, \ldots, n$, where $T \leq \infty$. Define

$$Z_t = \exp\left\{ \int_0^t \theta(s, \omega)dB(s) - \tfrac{1}{2} \int_0^t \theta^2(s, \omega)ds \right\}; \qquad 0 \leq t \leq T$$

where $B(s) \in \mathbf{R}^n$ and $\theta^2 = \theta \cdot \theta$ (dot product).

a) Use Itô's formula to prove that

$$dZ_t = Z_t \theta(t, \omega)dB(t) .$$

b) Deduce that Z_t is a martingale for $t \leq T$, provided that

$$Z_t \theta_k(t, \omega) \in \mathcal{V}[0, T] \qquad \text{for } 1 \leq k \leq n .$$

Remark. A sufficient condition that Z_t be a martingale is the *Kazamaki condition*

$$E\left[\exp\left(\tfrac{1}{2}\int_0^t \theta(s,\omega)dB(s)\right)\right] < \infty \qquad \text{for all } t \leq T . \qquad (4.3.9)$$

This condition is implied by the following (stronger) *Novikov condition*

$$E\left[\exp\left(\tfrac{1}{2}\int_0^T \theta^2(s,\omega)ds\right)\right] < \infty . \qquad (4.3.10)$$

See e.g. Ikeda & Watanabe (1989), Section III.5, and the references therein. See also Section 8.6.

4.5.*   Let $B_t \in \mathbf{R}$, $B_0 = 0$. Define

$$\beta_k(t) = E[B_t^k] ; \qquad k = 0, 1, 2, \ldots; \ t \geq 0 .$$

Use Itô's formula to prove that

$$\beta_k(t) = \tfrac{1}{2}k(k-1)\int_0^t \beta_{k-2}(s)ds ; \qquad k \geq 2 .$$

a)   Deduce that

$$E[B_t^4] = 3t^2 \qquad (\text{see } (2.2.14))$$

and find

$$E[B_t^6] .$$

b)   Show that

$$E[B(t)^{2k+1}] = 0$$

and

$$E[B(t)^{2k}] = \frac{(2k)!t^k}{2^k k!} ; \qquad k = 1, 2, \ldots$$

(Compare with Exercise 2.8.)

4.6.   a)   For c, α constants, $B_t \in \mathbf{R}$ define

$$X_t = e^{ct + \alpha B_t} .$$

Prove that

$$dX_t = (c + \tfrac{1}{2}\alpha^2)X_t dt + \alpha X_t dB_t .$$

b)   For $c, \alpha_1, \ldots, \alpha_n$ constants, $B_t = (B_1(t), \ldots, B_n(t)) \in \mathbf{R}^n$ define

$$X_t = \exp\left(ct + \sum_{j=1}^{n} \alpha_j B_j(t)\right).$$

Prove that

$$dX_t = \left(c + \tfrac{1}{2}\sum_{j=1}^{n} \alpha_j^2\right)X_t dt + X_t\left(\sum_{j=1}^{n} \alpha_j dB_j\right).$$

4.7. Let X_t be an Itô integral

$$dX_t = v(t,\omega)dB_t(\omega) \quad \text{where } v \in \mathcal{V}^n(0,T),\ B_t \in \mathbf{R}^n,\ 0 \le t \le T.$$

a) Give an example to show that X_t^2 is not in general a martingale.

b) Prove that if v is bounded then

$$M_t := X_t^2 - \int_0^t |v_s|^2 ds \quad \text{is a martingale}.$$

The process $\langle X, X\rangle_t := \int_0^t |v_s|^2 ds$ is called the *quadratic varia-tion process* of the martingale X_t. For general processes X_t it is defined by

$$\langle X, X\rangle_t = \lim_{\Delta t_k \to 0} \sum_{t_k \le t} |X_{t_{k+1}} - X_{t_k}|^2 \quad \text{(limit in probability)}$$

$$(4.3.11)$$

where $0 = t_1 < t_2 \cdots < t_n = t$ and $\Delta t_k = t_{k+1} - t_k$. The limit can be shown to exist for continuous square integrable martingales X_t. See e.g. Karatzas and Shreve (1991).

4.8. a) Let B_t denote n-dimensional Brownian motion and let $f : \mathbf{R}^n \to \mathbf{R}$ be C^2. Use Itô's formula to prove that

$$f(B_t) = f(B_0) + \int_0^t \nabla f(B_s)dB_s + \tfrac{1}{2}\int_0^t \Delta f(B_s)ds,$$

where $\Delta = \sum_{i=1}^{n} \frac{\partial^2}{\partial x_i^2}$ is the Laplace operator.

b) Assume that $g : \mathbf{R} \to \mathbf{R}$ is C^1 everywhere and C^2 out-side finitely many points $z_1, \ldots, z_N$ with $|g''(x)| \le M$ for $x \notin \{z_1, \ldots, z_N\}$. Let B_t be 1-dimensional Brownian motion. Prove that the 1-dimensional version of a) still holds, i.e.

$$g(B_t) = g(B_0) + \int_0^t g'(B_s)dB_s + \frac{1}{2}\int_0^t g''(B_s)ds.$$

(Hint: Choose $f_k \in C^2(\mathbf{R})$ s.t. $f_k \to g$ uniformly, $f_k' \to g'$ uniformly and $|f_k''| \leq M, f_k'' \to g''$ outside $z_1, \ldots, z_N$. Apply a) to f_k and let $k \to \infty$).

4.9. Prove that we may assume that g and its first two derivatives are bounded in the proof of the Itô formula (Theorem 4.1.2) by proceeding as follows: For fixed $t \geq 0$ and $n = 1, 2, \ldots$ choose g_n as in the statement such that $g_n(s, x) = g(s, x)$ for all $s \leq t$ and all $|x| \leq n$. Suppose we have proved that (4.1.9) holds for each g_n. Define the stochastic time

$$\tau_n = \tau_n(\omega) = \inf\{s > 0; |X_s(\omega)| \geq n\}$$

(τ_n is called a *stopping* time (See Chapter 7)) and prove that

$$\left(\int\limits_0^t v \frac{\partial g_n}{\partial x}(s, X_s) \mathcal{X}_{s \leq \tau_n} dB_s := \right)$$

$$\int\limits_0^{t \wedge \tau_n} v \frac{\partial g_n}{\partial x}(s, X_s) dB_s = \int\limits_0^{t \wedge \tau_n} v \frac{\partial g}{\partial x}(s, X_s) dB_s$$

for each n. This gives that

$$g(t \wedge \tau_n, X_{t \wedge \tau_n}) = g(0, X_0)$$
$$+ \int\limits_0^{t \wedge \tau_n} \left(\frac{\partial g}{\partial s} + u \frac{\partial g}{\partial x} + \tfrac{1}{2} v^2 \frac{\partial^2 g}{\partial x^2} \right) ds + \int\limits_0^{t \wedge \tau_n} v \frac{\partial g}{\partial x} dB_s$$

and since

$$P[\tau_n > t] \to 1 \qquad \text{as } n \to \infty$$

we can conclude that (4.1.9) holds (a.s.) for g.

4.10. (Tanaka's formula and local time).
What happens if we try to apply the Itô formula to $g(B_t)$ when B_t is 1-dimensional and $g(x) = |x|$? In this case g is not C^2 at $x = 0$, so we modify $g(x)$ near $x = 0$ to $g_\epsilon(x)$ as follows:

$$g_\epsilon(x) = \begin{cases} |x| & \text{if } |x| \geq \epsilon \\ \tfrac{1}{2}(\epsilon + \frac{x^2}{\epsilon}) & \text{if } |x| < \epsilon \end{cases}$$

where $\epsilon > 0$.

a) Apply Exercise 4.8 b) to show that

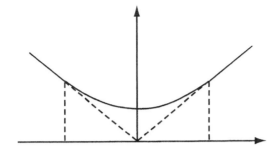

$$g_\epsilon(B_t) = g_\epsilon(B_0) + \int\limits_0^t g_\epsilon'(B_s)dB_s + \frac{1}{2\epsilon} \cdot |\{s \in [0,t]; B_s \in (-\epsilon, \epsilon)\}|$$

where $|F|$ denotes the Lebesgue measure of the set F.

b)   Prove that

$$\int\limits_0^t g_\epsilon'(B_s) \cdot \mathcal{X}_{B_s \in (-\epsilon, \epsilon)} dB_s = \int\limits_0^t \frac{B_s}{\epsilon} \cdot \mathcal{X}_{B_s \in (-\epsilon, \epsilon)} dB_s \to 0$$

in $L^2(P)$ as $\epsilon \to 0$.

(Hint: Apply the Itô isometry to

$$E\left[\left(\int\limits_0^t \frac{B_s}{\epsilon} \cdot \mathcal{X}_{B_s \in (-\epsilon, \epsilon)} dB_s\right)^2\right].$$

c)   By letting $\epsilon \to 0$ prove that

$$|B_t| = |B_0| + \int\limits_0^t \operatorname{sign}(B_s)dB_s + L_t(\omega), \qquad (4.3.12)$$

where

$$L_t = \lim_{\epsilon \to 0} \frac{1}{2\epsilon} \cdot |\{s \in [0,t]; B_s \in (-\epsilon, \epsilon)\}| \quad (\text{limit in } L^2(P))$$

and

$$\operatorname{sign}(x) = \begin{cases} -1 & \text{for} \quad x \le 0 \\ 1 & \text{for} \quad x > 0 \end{cases}.$$

L_t is called the *local time* for Brownian motion at 0 and (4.3.12) is the *Tanaka formula* (for Brownian motion). (See e.g. Rogers and Williams (1987)).

4.11.*   Use Itô's formula (for example in the form of Exercise 4.3) to prove that the following stochastic processes are $\{\mathcal{F}_t\}$-martingales:

a) $X_t = e^{\frac{1}{2}t} \cos B_t$ $(B_t \in \mathbf{R})$
b) $X_t = e^{\frac{1}{2}t} \sin B_t$ $(B_t \in \mathbf{R})$
c) $X_t = (B_t + t)\exp(-B_t - \frac{1}{2}t)$ $(B_t \in \mathbf{R})$.

4.12.   Let $dX_t = u(t,\omega)dt + v(t,\omega)dB_t$ be an Itô process in $\mathbf{R}^n$ such that

$$E\left[\int_0^t |u(r,\omega)|dr\right] + E\left[\int_0^t |vv^T(r,\omega)|dr\right] < \infty \qquad \text{for all } t \geq 0 \,.$$

Suppose X_t is an $\{\mathcal{F}_t^{(n)}\}$-martingale. Prove that

$$u(s,\omega) = 0 \qquad \text{for a.a. } (s,\omega) \in [0,\infty) \times \Omega \,. \qquad (4.3.13)$$

Remarks:

1)   This result may be regarded as a special case of the Martingale Representation Theorem.

2)   The conclusion (4.3.13) does not hold if the filtration $\mathcal{F}_t^{(n)}$ is replaced by the σ-algebras $\mathcal{M}_t$ generated by $X_s(\cdot)$; $s \leq t$, i.e. if we only assume that X_t is a martingale w.r.t. its own filtration. See e.g. the Brownian motion characterization in Chapter 8.

Hint for the solution:
   If X_t is an $\mathcal{F}_t^{(n)}$-martingale, then deduce that

$$E\left[\int_t^s u(r,\omega)dr\Big|\mathcal{F}_t^{(n)}\right] = 0 \qquad \text{for all } s \geq t \,.$$

Differentiate w.r.t. s to deduce that

$$E[u(s,\omega)|\mathcal{F}_t^{(n)}] = 0 \qquad \text{a.s., for a.a. } s > t \,.$$

Then let $t \uparrow s$ and apply Corollary C.9.

4.13.   Let $dX_t = u(t,\omega)dt + dB_t$ $(u \in \mathbf{R}, \ B_t \in \mathbf{R})$ be an Itô process and assume for simplicity that u is bounded. Then from Exercise 4.12 we know that unless $u = 0$ the process X_t is not an $\mathcal{F}_t$-martingale. However, it turns out that we can construct an $\mathcal{F}_t$-martingale from X_t by multiplying by a suitable exponential martingale. More precisely, define

$$Y_t = X_t M_t$$

where

$$M_t = \exp\left(-\int_0^t u(r,\omega)dB_r - \frac{1}{2}\int_0^t u^2(r,\omega)dr\right).$$

Use Itô's formula to prove that

$$Y_t \quad \text{is an } \mathcal{F}_t\text{-martingale} .$$

Remarks:

a) Compare with Exercise 4.11 c).

b) This result is a special case of the important *Girsanov Theorem*. It can be interpreted as follows: $\{X_t\}_{t\le T}$ is a martingale w.r.t the measure Q defined on $\mathcal{F}_T$ by

$$dQ = M_T dP \qquad (T < \infty) .$$

See Section 8.6.

4.14.* In each of the cases below find the process $f(t,\omega) \in \mathcal{V}[0,T]$ such that (4.3.6) holds, i.e.

$$F(\omega) = E[F] + \int\limits_0^T f(t,\omega)dB_t(\omega) .$$

a) $F(\omega) = B_T(\omega)$ b) $F(\omega) = \int\limits_0^T B_t(\omega)dt$

c) $F(\omega) = B_T^2(\omega)$ d) $F(\omega) = B_T^3(\omega)$

e) $F(\omega) = e^{B_T(\omega)}$ f) $F(\omega) = \sin B_T(\omega)$

4.15. Let $x > 0$ be a constant and define

$$X_t = (x^{1/3} + \tfrac{1}{3}B_t)^3 ; \qquad t \ge 0 .$$

Show that

$$dX_t = \tfrac{1}{3}X_t^{1/3}dt + X_t^{2/3}dB_t ; \qquad X_0 = x .$$

4.16. By Exercise 3.8 we know that if Y is an $\mathcal{F}_T$-measurable random variable such that $E[|Y|^2] < \infty$ then the process

$$M_t := E[Y|\mathcal{F}_t] ; \qquad 0 \le t \le T$$

is a martingale with respect to $\{\mathcal{F}_t\}_{0\le t\le T}$.

a) Show that $E[M_t^2] < \infty$ for all $t \in [0,T]$. (Hint: Use Exercise 3.16.)

b) According to the martingale representation theorem (Theorem 4.3.4) there exists a unique process $g(t,\omega) \in \mathcal{V}(0,T)$ such that

$$M_t = E[M_0] + \int_0^t g(s, \omega)dB(s) \; ; \qquad t \in [0, T].$$

Find g in the following cases:
(i) $Y(\omega) = B^2(T)$
(ii) $Y(\omega) = B^3(T)$
(iii) $Y(\omega) = \exp(\sigma B(T)); \; \sigma \in \mathbf{R}$ is constant.
       (Hint: Use that $\exp(\sigma B(t) - \frac{1}{2}\sigma^2 t)$ is a martingale.)

4.17. Here is an alternative proof of Theorem 4.3.3 which, in particular, does not use the complex analysis argument of Lemma 4.3.2. The idea of the proof is taken from Davis (1980), where it is extended to give a proof of the Clark representation formula. (See the Remark before Theorem 4.3.4.):

In view of Lemma 4.3.1 it is enough to prove the following:

Let $Y = \phi(B_{t_1}, \dots, B_{t_n})$ where $0 \le t_1 < t_2 < \cdots < t_n \le T$ and $\phi \in C_0^\infty(\mathbf{R}^n)$. We want to prove that there exists $f(t, \omega) \in \mathcal{V}(0, T)$ such that

$$Y = E[Y] + \int_0^T f(t)dB(t). \qquad (4.3.14)$$

a)   Use the Itô formula to prove that if $w = w(t, x_1, \dots, x_k)$: $[t_{k-1}, t_k] \times \mathbf{R}^k \to \mathbf{R}$ is once continuously differentiable with respect to t and twice with respect to x_k, then

$$w(t, B(t_1), \dots, B(t_{k-1}), B(t))$$
$$= w(t_{k-1}, B(t_1), \dots, B(t_{k-1}), B(t_{k-1}))$$
$$+ \int_{t_{k-1}}^t \frac{\partial w}{\partial x_k}(s, B(t_1), \dots, B(t_{k-1}), B(s))dB(s)$$
$$+ \int_{t_{k-1}}^t \left(\frac{\partial w}{\partial s} + \frac{1}{2}\frac{\partial^2 w}{\partial x_k^2}\right)(s, B(t_1), \dots, B(t_{k-1}), B(s))ds; \; t \in [t_{k-1}, t_k].$$

b)   For $k = 1, \dots, n$ define functions v_k : $[t_{k-1}, t_k] \times \mathbf{R}^k \to \mathbf{R}$ inductively as follows:

$$\begin{cases} \frac{\partial v_n}{\partial t} + \frac{1}{2}\frac{\partial^2 v_n}{\partial x_n^2} = 0 & ; \quad t_{n-1} < t < t_n \\ v_n(t_n, x_1, \dots, x_n) = \phi(x_1, \dots, x_n) & ; \end{cases} \qquad (4.3.15)$$

and, for $k = n - 1, n - 2, \dots, 1$,

$$\begin{cases} \frac{\partial v_k}{\partial t} + \frac{1}{2}\frac{\partial^2 v_k}{\partial x_k^2} = 0 & ; \ t_{k-1} < t < t_k \\ v_k(t_k, x_1, \dots, x_k) = v_{k+1}(t_k, x_1, \dots, x_k, x_k) \end{cases}$$

$$(4.3.16)$$

Verify that the solution of (4.3.16) is, for $t \in [t_{k-1}, t_k]$

$$v_k(t, x_1, \dots, x_k) \hspace{5cm} (4.3.17)$$

$$= (2\pi(t_k - t))^{-1/2} \int_{\mathbf{R}} v_{k+1}(t_k, x_1, \dots, x_k, y) \exp\left(-\frac{(x_k-y)^2}{2(t_k-t)}\right) dy$$

$$\left(= E\left[v_{k+1}(t_k, x_1, \dots, x_k, B_{t_k-t}^{(x_k)})\right]\right) \quad \text{(compare with Theorem 8.1.1))}.$$

In particular, $w = v_k$ satisfies the smoothenss conditions of a).

c)   Show that the representation (4.3.14) holds with

$$f(t, \omega) = \frac{\partial v_k}{\partial x_k}(t, B(t_1), \dots, B(t_{k-1}), B(t)) \qquad \text{for } t \in [t_{k-1}, t_k].$$

[Hint: By (4.3.15) and a) we have

$$\phi(B(t_1), \dots, B(t_n)) = v_n(t_n, B(t_1), \dots, B(t_n))$$
$$= v_n(t_{n-1}, B(t_1), \dots, B(t_{n-1}), B(t_{n-1}))$$
$$+ \int_{t_{n-1}}^{t_n} \frac{\partial v_n}{\partial x_n}(s, B(t_1), \dots, B(t_{n-1}), B(s)) dB(s)$$
$$= v_{n-1}(t_{n-1}, B(t_1), \dots, B(t_{n-1}))$$
$$+ \int_{t_{n-1}}^{t_n} \frac{\partial v_n}{\partial x_n}(s, B(t_1), \dots, B(t_{n-1}), B(s)) dB(s) .$$

Now repeat the procedure with $v_{n-1}(t_{n-1}, B(t_1), \dots, B(t_{n-1}))$ and proceed by induction.]

Chapter 5
Stochastic Differential Equations

5.1 Examples and Some Solution Methods

We now return to the possible solutions $X_t(\omega)$ of the stochastic differential equation

$$\frac{dX_t}{dt} = b(t, X_t) + \sigma(t, X_t)W_t, \quad b(t, x) \in \mathbf{R}, \sigma(t, x) \in \mathbf{R} \qquad (5.1.1)$$

where W_t is 1-dimensional "white noise". As discussed in Chapter 3 the Itô interpretation of (5.1.1) is that X_t satisfies the stochastic integral equation

$$X_t = X_0 + \int_0^t b(s, X_s)ds + \int_0^t \sigma(s, X_s)dB_s$$

or in differential form

$$dX_t = b(t, X_t)dt + \sigma(t, X_t)dB_t \ . \qquad (5.1.2)$$

Therefore, to get from (5.1.1) to (5.1.2) we formally just replace the white noise W_t by $\frac{dB_t}{dt}$ in (5.1.1) and multiply by dt. It is natural to ask:

(A)   Can one obtain existence and uniqueness theorems for such equations? What are the properties of the solutions?

(B)   How can one solve a given such equation?

We will first consider question (B) by looking at some simple examples, and then in Section 5.2 we will discuss (A).

It is the Itô formula that is the key to the solution of many stochastic differential equations. The method is illustrated in the following examples.

Example 5.1.1 Let us return to the population growth model in Chapter 1:

$$\frac{dN_t}{dt} = a_t N_t \ , \qquad N_0 \text{ given}$$

where $a_t = r_t + \alpha W_t$, $W_t =$ white noise, $\alpha =$ constant.

Let us assume that $r_t = r =$ constant. By the Itô interpretation (5.1.2) this equation is equivalent to (here $\sigma(t,x) = \alpha x$)

$$dN_t = rN_t dt + \alpha N_t dB_t \tag{5.1.3}$$

or

$$\frac{dN_t}{N_t} = rdt + \alpha dB_t \ .$$

Hence

$$\int_0^t \frac{dN_s}{N_s} = rt + \alpha B_t \quad (B_0 = 0) \ . \tag{5.1.4}$$

To evaluate the integral on the left hand side we use the Itô formula for the function

$$g(t,x) = \ln x \ ; \qquad x > 0$$

and obtain

$$\begin{aligned} d(\ln N_t) &= \frac{1}{N_t} \cdot dN_t + \frac{1}{2}\left(-\frac{1}{N_t^2}\right)(dN_t)^2 \\ &= \frac{dN_t}{N_t} - \frac{1}{2N_t^2} \cdot \alpha^2 N_t^2 dt = \frac{dN_t}{N_t} - \frac{1}{2}\alpha^2 dt \ . \end{aligned}$$

Hence

$$\frac{dN_t}{N_t} = d(\ln N_t) + \frac{1}{2}\alpha^2 dt$$

so from (5.1.4) we conclude

$$\ln \frac{N_t}{N_0} = (r - \frac{1}{2}\alpha^2)t + \alpha B_t$$

or

$$N_t = N_0 \exp((r - \frac{1}{2}\alpha^2)t + \alpha B_t) \ . \tag{5.1.5}$$

For comparison, referring to the discussion at the end of Chapter 3, the *Stratonovich* interpretation of (5.1.3),

$$d\overline{N}_t = r\overline{N}_t dt + \alpha \overline{N}_t \circ dB_t \ ,$$

would have given the solution

$$\overline{N}_t = N_0 \exp(rt + \alpha B_t) \ . \tag{5.1.6}$$

The solutions $N_t, \overline{N}_t$ are both processes of the type

$$X_t = X_0 \exp(\mu t + \alpha B_t) \qquad (\mu, \alpha \text{ constants}) \ .$$

Such processes are called *geometric Brownian motions*. They are important also as models for stochastic prices in economics. See Chapters 10, 11, 12.

Remark. It seems reasonable that if B_t is independent of N_0 we should have

$$E[N_t] = E[N_0]e^{rt}, \qquad (*)$$

i.e. the same as when there is no noise in a_t. To see if this is indeed the case, we let

$$Y_t = e^{\alpha B_t}$$

and apply Itô's formula:

$$dY_t = \alpha e^{\alpha B_t} dB_t + \tfrac{1}{2}\alpha^2 e^{\alpha B_t} dt$$

or

$$Y_t = Y_0 + \alpha \int_0^t e^{\alpha B_s} dB_s + \tfrac{1}{2}\alpha^2 \int_0^t e^{\alpha B_s} ds .$$

Since $E[\int_0^t e^{\alpha B_s} dB_s] = 0$ (Theorem 3.2.1 (iii)), we get

$$E[Y_t] = E[Y_0] + \tfrac{1}{2}\alpha^2 \int_0^t E[Y_s]ds$$

i.e.

$$\frac{d}{dt}E[Y_t] = \tfrac{1}{2}\alpha^2 E[Y_t], E[Y_0] = 1 .$$

So

$$E[Y_t] = e^{\frac{1}{2}\alpha^2 t} ,$$

and therefore – as anticipated – we obtain

$$E[N_t] = E[N_0]e^{rt} .$$

For the *Stratonovich* solution, however, the same calculation gives

$$E[\overline{N}_t] = E[N_0]e^{(r+\frac{1}{2}\alpha^2)t} .$$

Now that we have found the explicit solutions N_t and $\overline{N}_t$ in (5.1.5), (5.1.6) we can use our knowledge about the behaviour of B_t to gain information on these solutions. For example, for the Itô solution N_t we get the following:

(i)   If $r > \tfrac{1}{2}\alpha^2$ then $N_t \to \infty$ as $t \to \infty$, a.s.

(ii)   If $r < \tfrac{1}{2}\alpha^2$ then $N_t \to 0$ as $t \to \infty$, a.s.

(iii) If $r = \tfrac{1}{2}\alpha^2$ then N_t will fluctuate between arbitrary large and arbitrary small values as $t \to \infty$, a.s.

These conclusions are direct consequences of the formula (5.1.5) for N_t together with the following basic result about 1-dimensional Brownian motion B_t:

Theorem 5.1.2 (The law of iterated logarithm)

$$\limsup_{t \to \infty} \frac{B_t}{\sqrt{2t \log \log t}} = 1 \ a.s.$$

For a proof we refer to Lamperti (1977), §22.

For the *Stratonovich* solution $\overline{N}_t$ we get by the same argument that $\overline{N}_t \to 0$ a.s. if $r < 0$ and $\overline{N}_t \to \infty$ a.s. if $r > 0$.

Thus the two solutions have fundamentally different properties and it is an interesting question what solution gives the best description of the situation.

Example 5.1.3 Let us return to the equation in Problem 2 of Chapter 1:

$$LQ_t'' + RQ_t' + \frac{1}{C}Q_t = F_t = G_t + \alpha W_t \ . \tag{5.1.7}$$

We introduce the vector

$$X = X(t, \omega) = \begin{pmatrix} X_1 \\ X_2 \end{pmatrix} = \begin{pmatrix} Q_t \\ Q_t' \end{pmatrix} \qquad \text{and obtain}$$

$$\begin{cases} X_1' = X_2 \\ LX_2' = -RX_2 - \frac{1}{C}X_1 + G_t + \alpha W_t \end{cases} \tag{5.1.8}$$

or, in matrix notation,

$$dX = dX(t) = AX(t)dt + H(t)dt + K dB_t \tag{5.1.9}$$

where

$$dX = \begin{pmatrix} dX_1 \\ dX_2 \end{pmatrix}, \ A = \begin{pmatrix} 0 & 1 \\ -\frac{1}{CL} & -\frac{R}{L} \end{pmatrix}, \ H(t) = \begin{pmatrix} 0 \\ \frac{1}{L}G_t \end{pmatrix}, \ K = \begin{pmatrix} 0 \\ \frac{\alpha}{L} \end{pmatrix}, \tag{5.1.10}$$

and B_t is a 1-dimensional Brownian motion.

Thus we are led to a *2-dimensional stochastic differential equation*. We rewrite (5.1.9) as

$$\exp(-At)dX(t) - \exp(-At)AX(t)dt = \exp(-At)[H(t)dt + K dB_t] \ , \tag{5.1.11}$$

where for a general $n \times n$ matrix F we define $\exp(F)$ to be the $n \times n$ matrix given by $\exp(F) = \sum_{n=0}^{\infty} \frac{1}{n!}F^n$. Here it is tempting to relate the left hand side to

$$d(\exp(-At)X(t)) \ .$$

To do this we use a 2-dimensional version of the Itô formula (Theorem 4.2.1).

Applying this result to the two coordinate functions g_1, g_2 of

$$g\colon [0, \infty) \times \mathbf{R}^2 \to \mathbf{R}^2 \qquad \text{given by} \quad g(t, x_1, x_2) = \exp(-At) \begin{pmatrix} x_1 \\ x_2 \end{pmatrix},$$

we obtain that

$$d(\exp(-At)X(t)) = (-A)\exp(-At)X(t)dt + \exp(-At)dX(t).$$

Substituted in (5.1.11) this gives

$$\exp(-At)X(t) - X(0) = \int_0^t \exp(-As)H(s)ds + \int_0^t \exp(-As)KdB_s$$

or

$$
\begin{aligned}
X(t) = \exp(At)[X(0) &+ \exp(-At)KB_t \\
&+ \int_0^t \exp(-As)[H(s) + AKB_s]ds],
\end{aligned}
\tag{5.1.12}
$$

by integration by parts (Theorem 4.1.5).

Example 5.1.4 Choose $X_t = B_t$, 1-dimensional Brownian motion, and

$$g(t, x) = e^{ix} = (\cos x, \sin x) \in \mathbf{R}^2 \qquad \text{for } x \in \mathbf{R}.$$

Then

$$Y(t) = g(t, X_t) = e^{iB_t} = (\cos B_t, \sin B_t)$$

is by Itô's formula again an Itô process.
 Its coordinates Y_1, Y_2 satisfy

$$\begin{cases} dY_1(t) = -\sin(B_t)dB_t - \frac{1}{2}\cos(B_t)dt \\ dY_2(t) = \cos(B_t)dB_t - \frac{1}{2}\sin(B_t)dt. \end{cases}$$

Thus the process $Y = (Y_1, Y_2)$, which we could call *Brownian motion on the unit circle*, is the solution of the stochastic differential equations

$$\begin{cases} dY_1 = -\frac{1}{2}Y_1 dt - Y_2 dB_t \\ dY_2 = -\frac{1}{2}Y_2 dt + Y_1 dB_t. \end{cases} \tag{5.1.13}$$

Or, in matrix notation,

$$dY(t) = -\frac{1}{2}Y(t)dt + KY(t)dB_t, \qquad \text{where } K = \begin{pmatrix} 0 & -1 \\ 1 & 0 \end{pmatrix}.$$

Other examples and solution methods can be found in the exercises of this chapter.

For a comprehensive description of reduction methods for 1-dimensional stochastic differential equations see Gard (1988), Chapter 4.

5.2 An Existence and Uniqueness Result

We now turn to the existence and uniqueness question (A) above.

Theorem 5.2.1 (Existence and uniqueness theorem for stochastic differential equations).

Let $T > 0$ and $b(\cdot,\cdot)\colon [0,T] \times \mathbf{R}^n \to \mathbf{R}^n, \sigma(\cdot,\cdot)\colon [0,T] \times \mathbf{R}^n \to \mathbf{R}^{n\times m}$ be measurable functions satisfying

$$|b(t,x)| + |\sigma(t,x)| \leq C(1 + |x|) ; \qquad x \in \mathbf{R}^n, \ t \in [0,T] \qquad (5.2.1)$$

for some constant C, (where $|\sigma|^2 = \sum|\sigma_{ij}|^2$) and such that

$$|b(t,x) - b(t,y)| + |\sigma(t,x) - \sigma(t,y)| \leq D|x-y| ; \quad x,y \in \mathbf{R}^n, \ t \in [0,T] \quad (5.2.2)$$

for some constant D. Let Z be a random variable which is independent of the σ-algebra $\mathcal{F}_\infty^{(m)}$ generated by $B_s(\cdot)$, $s \geq 0$ and such that

$$E[|Z|^2] < \infty .$$

Then the stochastic differential equation

$$dX_t = b(t, X_t)dt + \sigma(t, X_t)dB_t , \qquad 0 \leq t \leq T, X_0 = Z \qquad (5.2.3)$$

has a unique t-continuous solution $X_t(\omega)$ with the property that

$$X_t(\omega) \text{ is adapted to the filtration } \mathcal{F}_t^Z \text{ generated by } Z \text{ and } B_s(\cdot); \ s \leq t \qquad (5.2.4)$$

and

$$E\left[\int_0^T |X_t|^2 dt\right] < \infty . \qquad (5.2.5)$$

Remarks. Conditions (5.2.1) and (5.2.2) are natural in view of the following two simple examples from deterministic differential equations (i.e. $\sigma = 0$):

a)   The equation

$$\frac{dX_t}{dt} = X_t^2, \quad X_0 = 1 \qquad (5.2.6)$$

corresponding to $b(x) = x^2$ (which does not satisfy (5.2.1)) has the (unique) solution

$$X_t = \frac{1}{1-t} \; ; \qquad 0 \le t < 1 .$$

Thus it is impossible to find a global solution (defined for all t) in this case.

More generally, condition (5.2.1) ensures that the solution $X_t(\omega)$ of (5.2.3) does not *explode*, i.e. that $|X_t(\omega)|$ does not tend to ∞ in a finite time.

b) The equation

$$\frac{dX_t}{dt} = 3X_t^{2/3} \; ; \qquad X_0 = 0 \tag{5.2.7}$$

has more than one solution. In fact, for any $a > 0$ the function

$$X_t = \begin{cases} 0 & \text{for} \quad t \le a \\ (t-a)^3 & \text{for} \quad t > a \end{cases}$$

solves (5.2.7). In this case $b(x) = 3x^{2/3}$ does not satisfy the Lipschitz condition (5.2.2) at $x = 0$.

Thus condition (5.2.2) guarantees that equation (5.2.3) has a *unique* solution. Here uniqueness means that if $X_1(t, \omega)$ and $X_2(t, \omega)$ are two t-continuous processes satisfying (5.2.3), (5.2.4) and (5.2.5) then

$$X_1(t, \omega) = X_2(t, \omega) \qquad \text{for all } t \le T, \text{ a.s.} \tag{5.2.8}$$

Proof of Theorem 5.2.1. The uniqueness follows from the Itô isometry (Corollary 3.1.7) and the Lipschitz property (5.2.2): Let $X_1(t, \omega) = X_t(\omega)$ and $X_2(t, \omega) = \widehat{X}_t(\omega)$ be solutions with initial values $Z, \widehat{Z}$ respectively, i.e. $X_1(0, \omega) = Z(\omega), X_2(0, \omega) = \widehat{Z}(\omega), \omega \in \Omega$. For our purposes here we only need the case $Z = \widehat{Z}$, but the following more general estimate will be useful for us later, in connection with Feller continuity (Chapter 8).

Put $a(s, \omega) = b(s, X_s) - b(s, \widehat{X}_s)$ and $\gamma(s, \omega) = \sigma(s, X_s) - \sigma(s, \widehat{X}_s)$. Then

$$E[|X_t - \widehat{X}_t|^2] = E\left[\left(Z - \widehat{Z} + \int_0^t a\, ds + \int_0^t \gamma dB_s \right)^2\right]$$

$$\le 3E[|Z - \widehat{Z}|^2] + 3E\left[\left(\int_0^t a\, ds\right)^2\right] + 3E\left[\left(\int_0^t \gamma dB_s\right)^2\right]$$

$$\le 3E[|Z - \widehat{Z}|^2] + 3tE\left[\int_0^t a^2 ds\right] + 3E\left[\int_0^t \gamma^2 ds\right]$$

$$\le 3E[|Z - \widehat{Z}|^2] + 3(1+t)D^2 \int_0^t E[|X_s - \widehat{X}_s|^2]ds .$$

So the function

$$v(t) = E[|X_t - \widehat{X}_t|^2] \; ; \qquad 0 \le t \le T$$

satisfies

$$v(t) \le F + A \int_0^t v(s)ds \; , \tag{5.2.9}$$

where $F = 3E[|Z - \widehat{Z}|^2]$ and $A = 3(1 + T)D^2$.

By the Gronwall inequality (Exercise 5.17) we conclude that

$$v(t) \le F \exp(At) \; . \tag{5.2.10}$$

Now assume that $Z = \widehat{Z}$. Then $F = 0$ and so $v(t) = 0$ for all $t \ge 0$. Hence

$$P[|X_t - \widehat{X}_t| = 0 \qquad \text{for all } t \in \mathbf{Q} \cap [0, T]] = 1 \; ,$$

where $\mathbf{Q}$ denotes the rational numbers.

By continuity of $t \to |X_t - \widehat{X}_t|$ it follows that

$$P[|X_1(t, \omega) - X_2(t, \omega)| = 0 \qquad \text{for all } t \in [0, T]] = 1 \; , \tag{5.2.11}$$

and the uniqueness is proved.

The proof of the existence is similar to the familiar existence proof for ordinary differential equations: Define $Y_t^{(0)} = X_0$ and $Y_t^{(k)} = Y_t^{(k)}(\omega)$ inductively as follows

$$Y_t^{(k+1)} = X_0 + \int_0^t b(s, Y_s^{(k)})ds + \int_0^t \sigma(s, Y_s^{(k)})dB_s \; . \tag{5.2.12}$$

Then, similar computation as for the uniqueness above gives

$$E[|Y_t^{(k+1)} - Y_t^{(k)}|^2] \le (1 + T)3D^2 \int_0^t E[|Y_s^{(k)} - Y_s^{(k-1)}|^2]ds \; ,$$

for $k \ge 1$, $t \le T$ and

$$E[|Y_t^{(1)} - Y_t^{(0)}|^2] \le 2C^2 t^2 E[(1 + |X_0|)^2]$$
$$+ 2C^2 t(1 + E[|X_0|^2]) \le A_1 t$$

where the constant A_1 only depends on C, T and $E[|X_0|^2]$. So by induction on k we obtain

$$E[|Y_t^{(k+1)} - Y_t^{(k)}|^2] \leq \frac{A_2^{k+1} t^{k+1}}{(k+1)!} \; ; \qquad k \geq 0, \; t \in [0, T] \qquad (5.2.13)$$

for some suitable constant A_2 depending only on C, D, T and $E[|X_0|^2]$.

Hence, if λ denotes Lebesgue measure on $[0, T]$ and $m > n \geq 0$ we get

$$\left\| Y_t^{(m)} - Y_t^{(n)} \right\|_{L^2(\lambda \times P)} = \left\| \sum_{k=n}^{m-1} Y_t^{(k+1)} - Y_t^{(k)} \right\|_{L^2(\lambda \times P)}$$

$$\leq \sum_{k=n}^{m-1} \left\| Y_t^{(k+1)} - Y_t^{(k)} \right\|_{L^2(\lambda \times P)} = \sum_{k=n}^{m-1} \left(E\left[\int_0^T |Y_t^{(k+1)} - Y_t^{(k)}|^2 dt \right] \right)^{1/2}$$

$$\leq \sum_{k=n}^{m-1} \left(\int_0^T \frac{A_2^{k+1} t^{k+1}}{(k+1)!} dt \right)^{1/2} = \sum_{k=n}^{m-1} \left(\frac{A_2^{k+1} T^{k+2}}{(k+2)!} \right)^{1/2} \to 0 \qquad (5.2.14)$$

as $m, n \to \infty$.

Therefore $\{Y_t^{(n)}\}_{n=0}^{\infty}$ is a Cauchy sequence in $L^2(\lambda \times P)$. Hence $\{Y_t^{(n)}\}_{n=0}^{\infty}$ is convergent in $L^2(\lambda \times P)$. Define

$$X_t := \lim_{n \to \infty} Y_t^{(n)} \qquad \text{(limit in } L^2(\lambda \times P)\text{)}.$$

Then X_t is $\mathcal{F}_t^Z$-measurable for all t, since this holds for each $Y_t^{(n)}$. We prove that X_t satisfies (5.2.3):

For all n and all $t \in [0, T]$ we have

$$Y_t^{(n+1)} = X_0 + \int_0^t b(s, Y_s^{(n)}) ds + \int_0^t \sigma(s, Y_s^{(n)}) dB_s \; .$$

Now let $n \to \infty$. Then by the Hölder inequality we get that

$$\int_0^t b(s, Y_s^{(n)}) ds \to \int_0^t b(s, X_s) ds \qquad \text{in } L^2(P)$$

and by the Itô isometry it follows that

$$\int_0^t \sigma(s, Y_s^{(n)}) dB_s \to \int_0^t \sigma(s, X_s) dB_s \qquad \text{in } L^2(P).$$

We conclude that for all $t \in [0, T]$ we have

$$X_t = X_0 + \int_0^t b(s, X_s)ds + \int_0^t \sigma(s, X_s)dB_s \quad \text{a.s.} \qquad (5.2.15)$$

i.e. X_t satisfies (5.2.3).

It remains to prove that X_t can be chosen to be continuous. By Theorem 3.2.5 there is a continuous version of the right hand side of (5.2.15). Denote this version by $\widetilde{X}_t$. Then $\widetilde{X}_t$ is continuous and

$$\widetilde{X}_t = X_0 + \int_0^t b(s, X_s)ds + \int_0^t \sigma(s, X_s)dB_s \qquad \text{for a.a. } \omega$$

$$= \widetilde{X}_0 + \int_0^t b(s, \widetilde{X}_s)ds + \int_0^t \sigma(s, \widetilde{X}_s)dB_s \qquad \text{for a.a. } \omega \ .$$

$$\square$$

5.3 Weak and Strong Solutions

The solution X_t found above is called a *strong* solution, because the version B_t of Brownian motion is given in advance and the solution X_t constructed from it is $\mathcal{F}_t^Z$-adapted. If we are only given the functions $b(t, x)$ and $\sigma(t, x)$ and ask for a pair of processes $((\widetilde{X}_t, \widetilde{B}_t), \mathcal{H}_t)$ on a probability space $(\Omega, \mathcal{H}, P)$ such that (5.2.3) holds, then the solution $\widetilde{X}_t$ (or more precisely $(\widetilde{X}_t, \widetilde{B}_t)$) is called a *weak* solution. Here $\mathcal{H}_t$ is an increasing family of σ-algebras such that $\widetilde{X}_t$ is $\mathcal{H}_t$-adapted and $\widetilde{B}_t$ is an $\mathcal{H}_t$-*Brownian motion*, i.e. $\widetilde{B}_t$ is a Brownian motion, and $\widetilde{B}_t$ is a martingale w.r.t. $\mathcal{H}_t$ (and so $E[\widetilde{B}_{t+h} - \widetilde{B}_t | \mathcal{H}_t] = 0$ for all $t, h \geq 0$). Recall from Chapter 3 that this allows us to define the Itô integral on the right hand side of (5.2.3) exactly as before, even though $\widetilde{X}_t$ need *not* be $\mathcal{F}_t^Z$-adapted.

A strong solution is of course also a weak solution, but the converse is not true in general. See Example 5.3.2 below.

The uniqueness (5.2.8) that we obtain above is called *strong* or *pathwise* uniqueness, while *weak* uniqueness simply means that any two solutions (weak or strong) are identical in law, i.e. have the same finite-dimensional distributions. See Stroock and Varadhan (1979) for results about existence and uniqueness of weak solutions. A general discussion about strong and weak solutions can be found in Krylov and Zvonkin (1981).

Lemma 5.3.1 *If b and σ satisfy the conditions of Theorem 5.2.1 then we have*

A solution (weak or strong) of (5.2.3) is weakly unique .

Sketch of proof. Let $((\widetilde{X}_t, \widetilde{B}_t), \widetilde{\mathcal{H}}_t)$ and $((\widehat{X}_t, \widehat{B}_t), \widehat{\mathcal{H}}_t)$ be two weak solutions. Let X_t and Y_t be the *strong* solutions constructed from $\widetilde{B}_t$ and $\widehat{B}_t$, respectively, as above. Then the same uniqueness argument as above applies to show that $X_t = \widetilde{X}_t$ and $Y_t = \widehat{X}_t$ for all t, a.s. Therefore it suffices to show that X_t and Y_t must be identical in law. We show this by proving by induction that if $X_t^{(k)}, Y_t^{(k)}$ are the processes in the Picard iteration defined by (5.2.12) with Brownian motions $\widetilde{B}_t$ and $\widehat{B}_t$, then

$$(X_t^{(k)}, \widetilde{B}_t) \quad \text{and} \quad (Y_t^{(k)}, \widehat{B}_t)$$

have the same law for all k.   □

This observation will be useful for us in Chapter 7 and later, where we will investigate further the properties of processes which are solutions of stochastic differential equations (Itô diffusions).

From a modelling point of view the weak solution concept is often natural, because it does not specify beforehand the explicit representation of the white noise. Moreover, the concept is convenient for mathematical reasons, because there are stochastic differential equations which have *no strong solutions* but still a (weakly) *unique weak solution*. Here is a simple example:

Example 5.3.2 (The Tanaka equation) Consider the 1-dimensional stochastic differential equation

$$dX_t = \text{sign}(X_t)dB_t ; \qquad X_0 = 0 . \tag{5.3.1}$$

where

$$\text{sign}(x) = \begin{cases} +1 & \text{if } x \geq 0 \\ -1 & \text{if } x < 0 . \end{cases}$$

Note that here $\sigma(t, x) = \sigma(x) = \text{sign}(x)$ does not satisfy the Lipschitz condition (5.2.2), so Theorem 5.2.1 does not apply. Indeed, *the equation (5.3.1) has no strong solution.* To see this, let $\widehat{B}_t$ be a Brownian motion generating the filtration $\widehat{\mathcal{F}}_t$ and define

$$Y_t = \int_0^t \text{sign}(\widehat{B}_s)d\widehat{B}_s .$$

By the Tanaka formula (4.3.12) (Exercise 4.10) we have

$$Y_t = |\widehat{B}_t| - |\widehat{B}_0| - \widehat{L}_t(\omega) ,$$

where $\widehat{L}_t(\omega)$ is the local time for $\widehat{B}_t(\omega)$ at 0. It follows that Y_t is measurable w.r.t. the σ-algebra $\mathcal{G}_t$ generated by $|\widehat{B}_s(\cdot)|$; $s \leq t$, which is clearly strictly contained in $\widehat{\mathcal{F}}_t$. Hence the σ-algebra $\mathcal{N}_t$ generated by $Y_s(\cdot)$; $s \leq t$ is also strictly contained in $\widehat{\mathcal{F}}_t$.

Now suppose X_t is a strong solution of (5.3.1). Then by Theorem 8.4.2 it follows that X_t is a Brownian motion w.r.t. the measure P. (In case the reader is worried about the possibility of a circular argument, we point out that the proof of Theorem 8.4.2 is independent of this example!) Let $\mathcal{M}_t$ be the σ-algebra generated by $X_s(\cdot)$; $s \le t$. Since $(\text{sign}(x))^2 = 1$ we can rewrite (5.3.1) as

$$dB_t = \text{sign}(X_t)dX_t \ .$$

By the above argument applied to $\widehat{B}_t = X_t$, $Y_t = B_t$ we conclude that $\mathcal{F}_t$ is strictly contained in $\mathcal{M}_t$.

But this contradicts that X_t is a strong solution. Hence strong solutions of (5.3.1) do not exist.

To find a weak solution of (5.3.1) we simply choose X_t to be *any* Brownian motion $\widehat{B}_t$. Then we define $\widetilde{B}_t$ by

$$\widetilde{B}_t = \int_0^t \text{sign}(\widehat{B}_s)d\widehat{B}_s = \int_0^t \text{sign}(X_s)dX_s$$

i.e.

$$d\widetilde{B}_t = \text{sign}(X_t)dX_t \ .$$

Then

$$dX_t = \text{sign}(X_t)d\widetilde{B}_t \ ,$$

so X_t *is a weak solution.*

Finally, *weak uniqueness* follows from Theorem 8.4.2, which – as noted above – implies that any weak solution X_t must be a Brownian motion w.r.t. P.

Exercises

5.1. Verify that the given processes solve the given corresponding stochastic differential equations: (B_t denotes 1-dimensional Brownian motion)

(i) $X_t = e^{B_t}$ solves $dX_t = \frac{1}{2}X_t dt + X_t dB_t$

(ii) $X_t = \frac{B_t}{1+t}$; $B_0 = 0$ solves

$$dX_t = -\frac{1}{1+t}X_t dt + \frac{1}{1+t}dB_t \ ; \qquad X_0 = 0$$

(iii) $X_t = \sin B_t$ with $B_0 = a \in (-\frac{\pi}{2}, \frac{\pi}{2})$ solves

$$dX_t = -\tfrac{1}{2}X_t dt + \sqrt{1 - X_t^2}dB_t \ \text{ for } \ t < \inf\left\{s > 0; B_s \notin \left[-\tfrac{\pi}{2}, \tfrac{\pi}{2}\right]\right\}$$

(iv) $(X_1(t), X_2(t)) = (t, e^t B_t)$ solves

$$\begin{bmatrix} dX_1 \\ dX_2 \end{bmatrix} = \begin{bmatrix} 1 \\ X_2 \end{bmatrix} dt + \begin{bmatrix} 0 \\ e^{X_1} \end{bmatrix} dB_t$$

(v) $(X_1(t), X_2(t)) = (\cosh(B_t), \sinh(B_t))$ solves

$$\begin{bmatrix} dX_1 \\ dX_2 \end{bmatrix} = \frac{1}{2} \begin{bmatrix} X_1 \\ X_2 \end{bmatrix} dt + \begin{bmatrix} X_2 \\ X_1 \end{bmatrix} dB_t \ .$$

5.2.   A natural candidate for what we could call *Brownian motion on the ellipse*

$$\left\{ (x, y); \frac{x^2}{a^2} + \frac{y^2}{b^2} = 1 \right\} \qquad \text{where } a > 0, b > 0$$

is the process $X_t = (X_1(t), X_2(t))$ defined by

$$X_1(t) = a \cos B_t \ , \quad X_2(t) = b \sin B_t$$

where B_t is 1-dimensional Brownian motion. Show that X_t is a solution of the stochastic differential equation

$$dX_t = -\tfrac{1}{2} X_t dt + M X_t dB_t$$

where $M = \begin{bmatrix} 0 & -\frac{a}{b} \\ \frac{b}{a} & 0 \end{bmatrix}$.

5.3.*   Let $(B_1, \ldots, B_n)$ be Brownian motion in $\mathbf{R}^n$, $\alpha_1, \ldots, \alpha_n$ constants. Solve the stochastic differential equation

$$dX_t = r X_t dt + X_t \Big(\sum_{k=1}^{n} \alpha_k dB_k(t) \Big); \qquad X_0 > 0 \ .$$

(This is a model for exponential growth with several independent white noise sources in the relative growth rate).

5.4.*   Solve the following stochastic differential equations:

(i) $\begin{bmatrix} dX_1 \\ dX_2 \end{bmatrix} = \begin{bmatrix} 1 \\ 0 \end{bmatrix} dt + \begin{bmatrix} 1 & 0 \\ 0 & X_1 \end{bmatrix} \begin{bmatrix} dB_1 \\ dB_2 \end{bmatrix}$

(ii) $dX_t = X_t dt + dB_t$
       (Hint: Multiply both sides with "the integrating factor" e^{-t} and compare with $d(e^{-t} X_t)$)

(iii) $dX_t = -X_t dt + e^{-t} dB_t$.

5.5.   a)   Solve the *Ornstein-Uhlenbeck equation* (or *Langevin equation*)

$$dX_t = \mu X_t dt + \sigma dB_t$$

where μ, σ are real constants, $B_t \in \mathbf{R}$.
The solution is called the *Ornstein-Uhlenbeck process*. (Hint: See
Exercise 5.4 (ii).)

b)   Find $E[X_t]$ and $\mathrm{Var}[X_t] := E[(X_t - E[X_t])^2]$.

5.6.* Solve the stochastic differential equation

$$dY_t = r\,dt + \alpha Y_t dB_t$$

where r, α are real constants, $B_t \in \mathbf{R}$.
(Hint: Multiply the equation by the 'integrating factor'

$$F_t = \exp\left(-\alpha B_t + \tfrac{1}{2}\alpha^2 t\right).\,)$$

5.7.* The *mean-reverting Ornstein-Uhlenbeck process* is the solution X_t of
the stochastic differential equation

$$dX_t = (m - X_t)dt + \sigma dB_t$$

where m, σ are real constants, $B_t \in \mathbf{R}$.

a)   Solve this equation by proceeding as in Exercise 5.5 a).
b)   Find $E[X_t]$ and $\mathrm{Var}[X_t] := E[(X_t - E[X_t])^2]$.

5.8.* Solve the (2-dimensional) stochastic differential equation

$$dX_1(t) = X_2(t)dt + \alpha dB_1(t)$$
$$dX_2(t) = -X_1(t)dt + \beta dB_2(t)$$

where $(B_1(t), B_2(t))$ is 2-dimensional Brownian motion and α, β are
constants.
This is a model for a vibrating string subject to a stochastic force.
See Example 5.1.3.

5.9. Show that there is a unique strong solution X_t of the 1-dimensional
stochastic differential equation

$$dX_t = \ln(1 + X_t^2)dt + \mathcal{X}_{\{X_t > 0\}} X_t dB_t\,, \qquad X_0 = a \in \mathbf{R}\,.$$

5.10. Let b, σ satisfy (5.2.1), (5.2.2) and let X_t be the unique strong solution of (5.2.3). Show that

$$E[|X_t|^2] \le K_1 \cdot \exp(K_2 t) \qquad \text{for } t \le T \qquad (5.3.2)$$

where $K_1 = 3E[|Z|^2] + 6C^2 T(T+1)$ and $K_2 = 6(1+T)C^2$.
(Hint: Use the argument in the proof of (5.2.10)).

Remark. With global estimates of the growth of b and σ in (5.2.1) it is possible to improve (5.3.2) to a global estimate of $E[|X_t|^2]$. See Exercise 7.5.

5.11.* **(The Brownian bridge).**

For fixed $a, b \in \mathbf{R}$ consider the following 1-dimensional equation

$$dY_t = \frac{b - Y_t}{1 - t} dt + dB_t ; \qquad 0 \le t < 1 , \ Y_0 = a . \qquad (5.3.3)$$

Verify that

$$Y_t = a(1 - t) + bt + (1 - t) \int\limits_0^t \frac{dB_s}{1 - s} ; \qquad 0 \le t < 1 \qquad (5.3.4)$$

solves the equation and prove that $\lim\limits_{t \to 1} Y_t = b$ a.s. The process Y_t is called *the Brownian bridge* (from a to b). For other characterizations of Y_t see Rogers and Williams (1987, pp. 86–89).

5.12.* To describe the motion of a pendulum with small, random perturbations in its environment we try an equation of the form

$$y''(t) + (1 + \epsilon W_t)y = 0 ; \qquad y(0), y'(0) \text{ given} ,$$

where $W_t = \frac{dB_t}{dt}$ is 1-dimensional white noise, $\epsilon > 0$ is constant.

a) Discuss this equation, for example by proceeding as in Example 5.1.3.

b) Show that $y(t)$ solves a *stochastic Volterra equation* of the form

$$y(t) = y(0) + y'(0) \cdot t + \int\limits_0^t a(t, r)y(r)dr + \int\limits_0^t \gamma(t, r)y(r)dB_r$$

where $a(t, r) = r - t$, $\gamma(t, r) = \epsilon(r - t)$.

5.13. As a model for the horizontal slow drift motions of a moored floating platform or ship responding to incoming irregular waves John Grue (1989) introduced the equation

$$x_t'' + a_0 x_t' + w^2 x_t = (T_0 - \alpha_0 x_t')\eta W_t , \qquad (5.3.5)$$

where W_t is 1-dimensional white noise, a_0, w, T_0, α_0 and η are constants.

(i) Put $X_t = \begin{bmatrix} x_t \\ x_t' \end{bmatrix}$ and rewrite the equation in the form

$$dX_t = AX_t dt + K X_t dB_t + M dB_t \ ,$$

where

$$A = \begin{bmatrix} 0 & 1 \\ -w^2 & -a_0 \end{bmatrix}, \quad K = \alpha_0 \eta \begin{bmatrix} 0 & 0 \\ 0 & -1 \end{bmatrix} \quad \text{and} \quad M = T_0 \eta \begin{bmatrix} 0 \\ 1 \end{bmatrix}.$$

(ii)   Show that X_t satisfies the integral equation

$$X_t = \int_0^t e^{A(t-s)} K X_s dB_s + \int_0^t e^{A(t-s)} M dB_s \qquad \text{if } X_0 = 0 \ .$$

(iii) Verify that

$$e^{At} = \frac{e^{-\lambda t}}{\xi} \{ (\xi \cos \xi t + \lambda \sin \xi t) I + A \sin \xi t \}$$

where $\lambda = \frac{a_0}{2}, \xi = (w^2 - \frac{a_0^2}{4})^{\frac{1}{2}}$ and use this to prove that

$$x_t = \eta \int_0^t (T_0 - \alpha_0 y_s) g_{t-s} dB_s \qquad (5.3.6)$$

and

$$y_t = \eta \int_0^t (T_0 - \alpha_0 y_s) h_{t-s} dB_s \ , \qquad \text{with } y_t := x_t' \ , \qquad (5.3.7)$$

where

$$g_t = \frac{1}{\xi} \mathrm{Im}(e^{\zeta t})$$

$$h_t = \frac{1}{\xi} \mathrm{Im}(\zeta e^{\bar{\zeta} t}) \ , \qquad \zeta = -\lambda + i\xi \quad (i = \sqrt{-1}) \ .$$

So we can solve for y_t first in (5.3.7) and then substitute in (5.3.6) to find x_t.

5.14.   If (B_1, B_2) denotes 2-dimensional Brownian motion we may introduce complex notation and put

$$\mathbf{B}(t) := B_1(t) + i B_2(t) \quad (i = \sqrt{-1}) \ .$$

$\mathbf{B}(t)$ is called *complex Brownian motion*.

(i) If $F(z) = u(z) + iv(z)$ is an *analytic* function i.e. F satisfies the Cauchy-Riemann equations

$$\frac{\partial u}{\partial x} = \frac{\partial v}{\partial y} \ , \quad \frac{\partial u}{\partial y} = -\frac{\partial v}{\partial x} \ ; \qquad z = x + iy$$

and we define

$$Z_t = F(\mathbf{B}(t))$$

prove that

$$dZ_t = F'(\mathbf{B}(t))d\mathbf{B}(t) \ , \qquad\qquad (5.3.8)$$

where F' is the (complex) derivative of F. (Note that the usual second order terms in the (real) Itô formula are not present in (5.3.8)!)

(ii) Solve the complex stochastic differential equation

$$dZ_t = \alpha Z_t d\mathbf{B}(t) \quad (\alpha \text{ constant}) \ .$$

For more information about complex stochastic calculus involving analytic functions see e.g. Ubøe (1987).

5.15. (Population growth in a stochastic, crowded environment)
The nonlinear stochastic differential equation

$$dX_t = rX_t(K - X_t)dt + \beta X_t dB_t \ ; \qquad X_0 = x > 0 \qquad (5.3.9)$$

is often used as a model for the growth of a population of size X_t in a stochastic, crowded environment. The constant $K > 0$ is called the *carrying capacity* of the environment, the constant $r \in \mathbf{R}$ is a measure of the quality of the environment and the constant $\beta \in \mathbf{R}$ is a measure of the size of the noise in the system.
Verify that

$$X_t = \frac{\exp\{(rK - \frac{1}{2}\beta^2)t + \beta B_t\}}{x^{-1} + r \int\limits_0^t \exp\{(rK - \frac{1}{2}\beta^2)s + \beta B_s\}ds} \ ; \qquad t \geq 0 \quad (5.3.10)$$

is the unique (strong) solution of (5.3.9). (This solution can be found by performing a substitution (change of variables) which reduces (5.3.9) to a linear equation. See Gard (1988), Chapter 4 for details.)

5.16.* The technique used in Exercise 5.6 can be applied to more general nonlinear stochastic differential equations of the form

$$dX_t = f(t, X_t)dt + c(t)X_t dB_t \ , \qquad X_0 = x \qquad (5.3.11)$$

where $f: \mathbf{R} \times \mathbf{R} \to \mathbf{R}$ and $c: \mathbf{R} \to \mathbf{R}$ are given continuous (deterministic) functions. Proceed as follows:

a) Define the 'integrating factor'

$$F_t = F_t(\omega) = \exp\left(-\int_0^t c(s)dB_s + \tfrac{1}{2}\int_0^t c^2(s)ds \right). \qquad (5.3.12)$$

Show that (5.3.11) can be written

$$d(F_t X_t) = F_t \cdot f(t, X_t)dt . \qquad (5.3.13)$$

b) Now define

$$Y_t(\omega) = F_t(\omega)X_t(\omega) \qquad (5.3.14)$$

so that

$$X_t = F_t^{-1}Y_t . \qquad (5.3.15)$$

Deduce that equation (5.3.13) gets the form

$$\frac{dY_t(\omega)}{dt} = F_t(\omega) \cdot f(t, F_t^{-1}(\omega)Y_t(\omega)) ; \qquad Y_0 = x . \qquad (5.3.16)$$

Note that this is just a *deterministic* differential equation in the function $t \to Y_t(\omega)$, for each $\omega \in \Omega$. We can therefore solve (5.3.16) with ω as a parameter to find $Y_t(\omega)$ and then obtain $X_t(\omega)$ from (5.3.15).

c) Apply this method to solve the stochastic differential equation

$$dX_t = \frac{1}{X_t}dt + \alpha X_t dB_t ; \qquad X_0 = x > 0 \qquad (5.3.17)$$

where α is constant.

d) Apply the method to study the solutions of the stochastic differential equation

$$dX_t = X_t^{\gamma}dt + \alpha X_t dB_t ; \qquad X_0 = x > 0 \qquad (5.3.18)$$

where α and γ are constants.
For what values of γ do we get explosion?

5.17. (The Gronwall inequality)
Let $v(t)$ be a nonnegative function such that

$$v(t) \leq C + A\int_0^t v(s)ds \qquad \text{for } 0 \leq t \leq T$$

for some constants C, A where $A \geq 0$. Prove that

$$v(t) \leq C \exp(At) \qquad \text{for } 0 \leq t \leq T . \qquad (5.3.19)$$

[Hint: We may assume $A \neq 0$. Define $w(t) = \int\limits_0^t v(s)ds$. Then $w'(t) \leq C + Aw(t)$. Deduce that

$$w(t) \leq \frac{C}{A}(\exp(At) - 1) \qquad (5.3.20)$$

by considering $f(t) := w(t) \exp(-At)$.
Use (5.3.20) to deduce (5.3.19.)]

5.18. The *geometric mean reversion process* X_t is defined as the solution of the stochastic differential equation

$$dX_t = \kappa(\alpha - \log X_t)X_t dt + \sigma X_t dB_t ; \qquad X_0 = x > 0 \qquad (5.3.21)$$

where κ, α, σ and x are positive constants.
This process was used by J. Tvedt (1995) to model the spot freight rate in shipping.

a)   Show that the solution of (5.3.21) is given by

$$X_t := \exp\left(e^{-\kappa t} \ln x + \left(\alpha - \tfrac{\sigma^2}{2\kappa}\right)(1 - e^{\kappa t}) + \sigma e^{-\kappa t} \int\limits_0^t e^{\kappa s} dB_s\right). \qquad (5.3.22)$$

[Hint: The substitution

$$Y_t = \log X_t$$

transforms (5.3.21) into a linear equation for Y_t]

b)   Show that

$$E[X_t] = \exp\left(e^{-\kappa t} \ln x + \left(\alpha - \frac{\sigma^2}{2\kappa}\right)(1 - e^{-\kappa t}) + \frac{\sigma^2(1 - e^{-2\kappa t})}{2\kappa}\right).$$

5.19. Let $Y_t^{(k)}$ be the process defined inductively by (5.2.12). Show that $\{Y_t^{(n)}\}_{n=0}^{\infty}$ is uniformly convergent for $t \in [0, T]$, for a.a. ω. Since each $Y_t^{(n)}$ is continuous, this gives a direct proof that X_t can be chosen to be continuous in Theorem 5.2.1.

[Hint: Note that

$$\sup_{0 \le t \le T} |Y_t^{(k+1)} - Y_t^{(k)}| \le \int_0^T |b(s, Y_s^{(k)}) - b(s, Y_s^{(k-1)}| ds$$

$$+ \sup_{0 \le t \le T} \left| \int_0^t (\sigma(s, Y^{(k)}(s)) - \sigma(s, Y_s^{(k-1)})) dB_s \right|$$

Hence

$$P\left[\sup_{0 \le t \le T} |Y_t^{(k+1)} - Y_t^{(k)}| > 2^{-k} \right]$$

$$\le P\left[\int_0^T |b(s, Y_s^{(k)}) - b(s, Y_s^{(k-1)})| ds > 2^{-k-1} \right]$$

$$+ P\left[\sup_{0 \le t \le T} \left| \int_0^t (\sigma(s, Y_s^{(k)}) - \sigma(s, Y_s^{(k-1)})) dB_s \right| > 2^{-k-1} \right].$$

Now use the Chebychev inequality, the Hölder inequality and the martingale inequality (Theorem 3.2.4), respectively, combined with (5.2.13), to prove that

$$P\left[\sup_{0 \le t \le T} |Y_t^{(k+1)} - Y_t^{(k)}| > 2^{-k} \right] \le \frac{(A_3 T)^{k+1}}{(k+1)!}$$

for some constant $A < \infty$. Therefore the result follows by the Borel-Cantelli lemma.]

Chapter 6
The Filtering Problem

6.1 Introduction

Problem 3 in the introduction is a special case of the following general *filtering problem*:

Suppose the state $X_t \in \mathbf{R}^n$ at time t of a system is given by a stochastic differential equation

$$\frac{dX_t}{dt} = b(t, X_t) + \sigma(t, X_t)W_t , \qquad t \geq 0 , \tag{6.1.1}$$

where $b: \mathbf{R}^{n+1} \to \mathbf{R}^n$, $\sigma: \mathbf{R}^{n+1} \to \mathbf{R}^{n \times p}$ satisfy conditions (5.2.1), (5.2.2) and W_t is p-dimensional white noise. As discussed earlier the Itô interpretation of this equation is

(system) $\qquad dX_t = b(t, X_t)dt + \sigma(t, X_t)dU_t ,$ \hfill (6.1.2)

where U_t is p-dimensional Brownian motion. We also assume that the distribution of X_0 is known and independent of U_t. Similarly to the 1-dimensional situation (3.3.6) there is an explicit several-dimensional formula which expresses the *Stratonovich* interpretation of (6.1.1):

$$dX_t = b(t, X_t)dt + \sigma(t, X_t) \circ dU_t$$

in terms of Itô integrals as follows:

$$dX_t = \widetilde{b}(t, X_t)dt + \sigma(t, X_t)dU_t , \qquad \text{where}$$

$$\widetilde{b}_i(t, x) = b_i(t, x) + \tfrac{1}{2} \sum_{j=1}^{p} \sum_{k=1}^{n} \frac{\partial \sigma_{ij}}{\partial x_k} \sigma_{kj} ; \qquad 1 \leq i \leq n . \tag{6.1.3}$$

(See Stratonovich (1966)). From now on we will use the Itô interpretation (6.1.2).

In the continuous version of the filtering problem we assume that the observations $H_t \in \mathbf{R}^m$ are performed continuously and are of the form

$$H_t = c(t, X_t) + \gamma(t, X_t) \cdot \widetilde{W}_t \ , \qquad (6.1.4)$$

where $c \colon \mathbf{R}^{n+1} \to \mathbf{R}^m$, $\gamma \colon \mathbf{R}^{n+1} \to \mathbf{R}^{m \times r}$ are functions satisfying (5.2.1) and $\widetilde{W}_t$ denotes r-dimensional white noise, independent of U_t and X_0.

To obtain a tractable mathematical interpretation of (6.1.4) we introduce

$$Z_t = \int\limits_0^t H_s ds \qquad (6.1.5)$$

and thereby we obtain the stochastic integral representation

(observations) $dZ_t = c(t, X_t)dt + \gamma(t, X_t)dV_t \ , \quad Z_0 = 0$ \qquad (6.1.6)

where V_t is r-dimensional Brownian motion, independent of U_t and X_0.

Note that if H_s is known for $0 \leq s \leq t$, then Z_s is also known for $0 \leq s \leq t$ and conversely. So no information is lost or gained by considering Z_t as our "observations" instead of H_t. But this allows us to obtain a well-defined mathematical model of the situation.

The filtering problem is the following:

Given the observations Z_s satisfying (6.1.6) for $0 \leq s \leq t$, what is the best estimate $\widehat{X}_t$ of the state X_t of the system (6.1.2) based on these observations?

As we have pointed out earlier, it is necessary to find a precise mathematical formulation of this problem: By saying that the estimate $\widehat{X}_t$ *is based on the observations* $\{Z_s; s \leq t\}$ we mean that

$$\widehat{X}_t(\cdot) \quad \text{is } \mathcal{G}_t\text{-measurable,}$$
$$\text{where } \mathcal{G}_t \text{ is the } \sigma\text{-algebra generated by } \{Z_s(\cdot), s \leq t\} \ . \quad (6.1.7)$$

By saying that $\widehat{X}_t$ is *the best such estimate* we mean that

$$\int\limits_\Omega |X_t - \widehat{X}_t|^2 dP = E[|X_t - \widehat{X}_t|^2] = \inf\{E[|X_t - Y|^2]; \ Y \in \mathcal{K}\} \ . \quad (6.1.8)$$

Here – and in the rest of this chapter – $(\Omega, \mathcal{F}, P)$ is the probability space corresponding to the $(p+r)$-dimensional Brownian motion (U_t, V_t) starting at 0, E denotes expectation w.r.t. P and

$$\mathcal{K} := \mathcal{K}_t := \mathcal{K}(Z, t) := \{Y \colon \Omega \to \mathbf{R}^n; \ Y \in L^2(P) \text{ and } Y \text{ is } \mathcal{G}_t\text{-measurable}\} \ , \quad (6.1.9)$$

where $L^2(P) = L^2(\Omega, P)$.

Having found the mathematical formulation of our problem, we now start to study the properties of the solution $\widehat{X}_t$.

We first establish the following useful connection between conditional expectation and projection:

Lemma 6.1.1 *Let* $\mathcal{H} \subset \mathcal{F}$ *be a* σ-*algebra and let* $X \in L^2(P)$ *be* $\mathcal{F}$-*measurable. Put* $\mathcal{N} = \{Y \in L^2(P); Y$ *is* $\mathcal{H}$-*measurable}* *and let* $\mathcal{P}_\mathcal{N}$ *denote the (orthogonal) projection from the Hilbert space* $L^2(P)$ *into the subspace* $\mathcal{N}$. *Then*

$$\mathcal{P}_\mathcal{N}(X) = E[X|\mathcal{H}] \ .$$

Proof. Recall (see Appendix B) that $E[X|\mathcal{H}]$ is by definition the P-unique function from Ω to $\mathbf{R}$ such that

(i) $E[X|\mathcal{H}]$ is $\mathcal{H}$-measurable
(ii) $\int\limits_A E[X|\mathcal{H}]dP = \int\limits_A X dP$ for all $A \in \mathcal{H}$.

Now $\mathcal{P}_\mathcal{N}(X)$ is $\mathcal{H}$-measurable and

$$\int\limits_\Omega Y(X - \mathcal{P}_\mathcal{N}(X))dP = 0 \qquad \text{for all } Y \in \mathcal{N} \ .$$

In particular,

$$\int\limits_A (X - \mathcal{P}_\mathcal{N}(X))dP = 0 \qquad \text{for all } A \in \mathcal{H}$$

i.e.

$$\int\limits_A \mathcal{P}_\mathcal{N}(X)dP = \int\limits_A X dP \qquad \text{for all } A \in \mathcal{H} \ .$$

Hence, by uniqueness, $\mathcal{P}_\mathcal{N}(X) = E[X|\mathcal{H}]$. $\qquad\qquad\qquad\qquad\qquad\square$

From the general theory of Hilbert spaces we know that the solution $\widehat{X}_t$ of the problem (6.1.8) is given by the projection $\mathcal{P}_{\mathcal{K}_t}(X_t)$. Therefore Lemma 6.1.1 leads to the following useful result:

Theorem 6.1.2
$$\widehat{X}_t = \mathcal{P}_{\mathcal{K}_t}(X_t) = E[X_t|\mathcal{G}_t] \ .$$

This is the basis for the general Fujisaki-Kallianpur-Kunita equation of filtering theory. See for example Bensoussan (1992), Davis (1984) or Kallianpur (1980).

6.2 The 1-Dimensional Linear Filtering Problem

From now on we will concentrate on the linear case, which allows an explicit solution in terms of a stochastic differential equation for $\widehat{X}_t$ (*the Kalman-Bucy filter*):

In the *linear* filtering problem the system and observation equations have the form:

(linear system) $\qquad dX_t = F(t)X_t dt + C(t) dU_t;$

$$F(t) \in \mathbf{R}^{n \times n}, \ C(t) \in \mathbf{R}^{n \times p} \qquad (6.2.1)$$

(linear observations) $\quad dZ_t = G(t)X_t dt + D(t) dV_t;$

$$G(t) \in \mathbf{R}^{m \times n}, \ D(t) \in \mathbf{R}^{m \times r} \qquad (6.2.2)$$

To be able to focus on the main ideas in the solution of the filtering problem, we will first consider only the 1-dimensional case:

(linear system) $\qquad dX_t = F(t)X_t dt + C(t) dU_t; \ F(t), \ C(t) \in \mathbf{R} \qquad (6.2.3)$

(linear observations) $dZ_t = G(t)X_t dt + D(t) dV_t; \ G(t), \ D(t) \in \mathbf{R} \qquad (6.2.4)$

We assume (see (5.2.1)) that F, G, C, D are bounded on bounded intervals. Based on our interpretation (6.1.5) of Z_t we assume $Z_0 = 0$. We also assume that X_0 is normally distributed (and independent of $\{U_t\}, \{V_t\}$). Finally we assume that $D(t)$ is bounded away from 0 on bounded intervals.

The (important) extension to the several-dimensional case (6.2.1), (6.2.2) is technical, but does not require any essentially new ideas. Therefore we shall only state the result for this case (in the next section) after we have discussed the 1-dimensional situation. The reader is encouraged to work out the necessary modifications for the general case for himself or consult Bensoussan (1992), Davis (1977) or Kallianpur (1980) for a full treatment.

From now on we let X_t, Z_t be processes satisfying (6.2.3), (6.2.4). Here is an outline of the solution of the filtering problem in this case.

Step 1. Let $\mathcal{L} = \mathcal{L}(Z, t)$ be the closure in $L^2(P)$ of functions which are linear combinations of the form

$$c_0 + c_1 Z_{s_1}(\omega) + \cdots + c_k Z_{s_k}(\omega), \qquad \text{with } s_j \leq t, c_j \in \mathbf{R}.$$

Let

$$\mathcal{P}_{\mathcal{L}} \quad \text{denote the projection from } L^2(P) \text{ onto } \mathcal{L}.$$

Then, with $\mathcal{K}$ as in (6.1.9),

$$\widehat{X}_t = \mathcal{P}_{\mathcal{K}}(X_t) = E[X_t | \mathcal{G}_t] = \mathcal{P}_{\mathcal{L}}(X_t).$$

Thus, the best *Z-measurable* estimate of X_t coincides with the best *Z-linear* estimate of X_t.

Step 2. Replace Z_t by the *innovation process* N_t:

$$N_t = Z_t - \int_0^t (GX)_s^\wedge ds, \quad \text{where } (GX)_s^\wedge = \mathcal{P}_{\mathcal{L}(Z,s)}(G(s)X_s) = G(s)\widehat{X}_s.$$

Then

(i) N_t has *orthogonal increments*, i.e.
$E[(N_{t_1} - N_{s_1})(N_{t_2} - N_{s_2})] = 0$ for non-overlapping intervals $[s_1, t_1]$, $[s_2, t_2]$.

(ii) $\mathcal{L}(N, t) = \mathcal{L}(Z, t)$, so $\widehat{X}_t = \mathcal{P}_{\mathcal{L}(N,t)}(X_t)$.

Step 3. If we put

$$dR_t = \frac{1}{D(t)} dN_t \ ,$$

then R_t is a 1-dimensional Brownian motion. Moreover,

$$\mathcal{L}(N, t) = \mathcal{L}(R, t) \qquad \text{and}$$

$$\widehat{X}_t = \mathcal{P}_{\mathcal{L}(N,t)}(X_t) = \mathcal{P}_{\mathcal{L}(R,t)}(X_t) = E[X_t] + \int_0^t \frac{\partial}{\partial s} E[X_t R_s] dR_s \ .$$

Step 4. Find an expression for X_t by solving the (linear) stochastic differential equation

$$dX_t = F(t) X_t dt + C(t) dU_t \ .$$

Step 5. Substitute the formula for X_t from Step 4 into $E[X_t R_s]$ and use Step 3 to obtain a stochastic differential equation for $\widehat{X}_t$:

$$d\widehat{X}_t = \frac{\partial}{\partial s} E[X_t R_s]_{s=t} dR_t + \left(\int_0^t \frac{\partial^2}{\partial t \partial s} E[X_t R_s] dR_s \right) dt \quad \text{etc.}$$

Before we proceed to establish Steps 1–5, let us consider a simple, but motivating example:

Example 6.2.1 Suppose $X, W_1, W_2, \ldots$ are independent real random variables, $E[X] = E[W_j] = 0$ for all j, $E[X^2] = a^2$, $E[W_j^2] = m^2$ for all j. Put $Z_j = X + W_j$.

What is the best *linear* estimate $\widehat{X}$ of X based on $\{Z_j; j \leq k\}$? More precisely, let

$$\mathcal{L} = \mathcal{L}(Z, k) = \{c_1 Z_1 + \cdots + c_k Z_k; c_1, \ldots, c_k \in \mathbf{R}\} \ .$$

Then we want to find

$$\widehat{X}_k = \mathcal{P}_k(X) \ ,$$

where $\mathcal{P}_k$ denotes the projection into $\mathcal{L}(Z, k)$.

We use the Gram-Schmidt procedure to obtain random variables $A_1, A_2, \ldots$ such that

(i) $E[A_i A_j] = 0$ for $i \neq j$

(ii) $\mathcal{L}(A, k) = \mathcal{L}(Z, k)$ for all k.

Then

$$\widehat{X}_k = \sum_{j=1}^{k} \frac{E[XA_j]}{E[A_j^2]} A_j \qquad \text{for } k = 1, 2, \dots . \tag{6.2.5}$$

We obtain a recursive relation between $\widehat{X}_k$ and $\widehat{X}_{k-1}$ from this by observing that

$$A_j = Z_j - \widehat{X}_{j-1} , \tag{6.2.6}$$

which follows from

$$A_j = Z_j - \mathcal{P}_{j-1}(Z_j) = Z_j - \mathcal{P}_{j-1}(X) , \qquad \text{since } \mathcal{P}_{j-1}(W_j) = 0 .$$

By (6.2.6)

$$E[XA_j] = E[X(Z_j - \widehat{X}_{j-1})] = E[X(X - \widehat{X}_{j-1})] = E[(X - \widehat{X}_{j-1})^2]$$

and

$$E[A_j^2] = E[(X + W_j - \widehat{X}_{j-1})^2] = E[(X - \widehat{X}_{j-1})^2] + m^2 .$$

Hence

$$\widehat{X}_k = \widehat{X}_{k-1} + \frac{E[(X - \widehat{X}_{k-1})^2]}{E[(X - \widehat{X}_{k-1})^2] + m^2} (Z_k - \widehat{X}_{k-1}) . \tag{6.2.7}$$

If we introduce

$$\overline{Z}_k = \frac{1}{k} \sum_{j=1}^{k} Z_j ,$$

then this can be simplified to

$$\widehat{X}_k = \frac{a^2}{a^2 + \frac{1}{k} \cdot m^2} \overline{Z}_k . \tag{6.2.8}$$

(This can be seen as follows:
    Put

$$\alpha_k = \frac{a^2}{a^2 + \frac{1}{k} m^2} , \qquad U_k = \alpha_k \overline{Z}_k .$$

Then

(i) $U_k \in \mathcal{L}(Z, k)$
(ii) $X - U_k \perp \mathcal{L}(Z, k)$, since

$$E[(X - U_k)Z_i] = E[XZ_i] - \alpha_k E[\overline{Z}_k Z_i]$$
$$= E[X(X + W_i)] - \alpha_k \frac{1}{k} \sum_j E[Z_j Z_i]$$
$$= a^2 - \frac{1}{k} \alpha_k \sum_j E[(X + W_j)(X + W_i)] = a^2 - \frac{1}{k} \alpha_k [ka^2 + m^2] = 0 .)$$

The result can be interpreted as follows:

For large k we put $\widehat{X}_k \approx \overline{Z}_k$, while for small k the relation between a^2 and m^2 becomes more important. If $m^2 \gg a^2$, the observations are to a large extent neglected (for small k) and $\widehat{X}_k$ is put equal to its mean value, 0. See also Exercise 6.11.

This example gives the motivation for our approach:

We replace the process Z_t by an orthogonal increment process N_t (Step 2) in order to obtain a representation for $\widehat{X}_t$ analogous to (6.2.5). Such a representation is obtained in Step 3, after we have identified the best *linear* estimate with the best *measurable* estimate (Step 1) and established the connection between N_t and Brownian motion.

Step 1. Z-Linear and Z-Measurable Estimates

Lemma 6.2.2 *Let X, Z_s; $s \leq t$ be random variables in $L^2(P)$ and assume that*

$$(X, Z_{s_1}, Z_{s_2}, \ldots, Z_{s_n}) \in \mathbf{R}^{n+1}$$

has a normal distribution for all $s_1, s_2, \ldots, s_n \leq t$, $n \geq 1$. Then

$$\mathcal{P}_{\mathcal{L}}(X) = E[X|\mathcal{G}] = \mathcal{P}_{\mathcal{K}}(X) \, .$$

In other words, the best Z-linear estimate for X coincides with the best Z-measurable estimate in this case.

Proof. Put $\check{X} = \mathcal{P}_{\mathcal{L}}(X)$, $\widetilde{X} = X - \check{X}$. Then we claim that $\widetilde{X}$ is independent of $\mathcal{G}$: Recall that a random variable $(Y_1, \ldots, Y_k) \in \mathbf{R}^k$ is normal iff $c_1 Y_1 + \cdots + c_k Y_k$ is normal, for all choices of $c_1, \ldots, c_k \in \mathbf{R}$. And an L^2-limit of normal variables is again normal (Appendix A). Therefore

$$(\widetilde{X}, Z_{s_1}, \ldots, Z_{s_n}) \quad \text{is normal for all} \ \ s_1, \ldots, s_n \leq t \, .$$

Since $E[\widetilde{X} Z_{s_j}] = 0$, $\widetilde{X}$ and Z_{s_j} are uncorrelated, for $1 \leq j \leq n$. It follows (Appendix A) that

$$\widetilde{X} \text{ and } (Z_{s_1}, \ldots, Z_{s_n}) \text{ are independent} \, .$$

So $\widetilde{X}$ is independent from $\mathcal{G}$ as claimed. But then

$$E[\mathcal{X}_G(X - \check{X})] = E[\mathcal{X}_G\widetilde{X}] = E[\mathcal{X}_G] \cdot E[\widetilde{X}] = 0 \quad \text{for all } G \in \mathcal{G}$$

i.e. $\int\limits_G X dP = \int\limits_G \check{X} dP$. Since $\check{X}$ is $\mathcal{G}$-measurable, we conclude that
$\check{X} = E[X|\mathcal{G}]$.   □

There is a curious interpretation of this result: Suppose $X, \{Z_t\}_{t \in T}$ are $L^2(P)$-functions with given covariances. Among all possible distributions of

$$(X, Z_{t_1}, \ldots, Z_{t_n})$$

with these covariances, the *normal* distribution will be the "worst" w.r.t. estimation, in the following sense: For any distribution we have

$$E[(X - E[X|\mathcal{G}])^2] \leq E[(X - \mathcal{P}_{\mathcal{L}}(X))^2] \,,$$

with *equality* for the normal distribution, by Lemma 6.2.2. (Note that the quantity on the right hand side only depends on the covariances, not on the distribution we might choose to obtain these covariances). For a broad discussion of similar conclusions, based on an information theoretical game between nature and the observer, see Topsøe (1978).

Finally, to be able to apply Lemma 6.2.2 to our filtering problem, we need the following result:

Lemma 6.2.3

$$M_t = \begin{bmatrix} X_t \\ Z_t \end{bmatrix} \in \mathbf{R}^2 \quad \text{is a Gaussian process}.$$

Proof. We may regard M_t as the solution of a 2-dimensional linear stochastic differential equation of the form

$$dM_t = H(t)M_t dt + K(t)dB_t, M_0 = \begin{bmatrix} X_0 \\ 0 \end{bmatrix} ; \qquad (6.2.9)$$

where $H(t) \in \mathbf{R}^{2 \times 2}$, $K(t) \in \mathbf{R}^{2 \times 2}$ and B_t is 2-dimensional Brownian motion. Use Picard iteration to solve (6.2.9), i.e. put

$$M_t^{(n+1)} = M_0 + \int_0^t H(s)M_s^{(n)}ds + \int_0^t K(s)dB_s , \qquad n = 0, 1, 2, \ldots \ (6.2.10)$$

Then $M_t^{(n)}$ is Gaussian for all n and $M_t^{(n)} \to M_t$ in $L^2(P)$ (see the proof of Theorem 5.2.1) and therefore M_t is Gaussian (Theorem A.7).   □

Step 2. The Innovation Process

Before we introduce the innovation process we will establish a useful representation of the functions in the space

$$\mathcal{L}(Z, T) = \text{the closure in } L^2(P) \text{ of all linear combinations}$$

$$c_0 + c_1 Z_{t_1} + \cdots + c_k Z_{t_k}; \quad 0 \leq t_i \leq T, \; c_j \in \mathbf{R} \, .$$

If $f \in L^2[0,T]$, note that

$$E\left[\left(\int_0^T f(t)dZ_t\right)^2\right] = E\left[\left(\int_0^T f(t)G(t)X_t dt\right)^2\right] + E\left[\left(\int_0^T f(t)D(t)dV_t\right)^2\right]$$

$$+2E\left[\left(\int_0^T f(t)G(t)X_t dt\right)\left(\int_0^T f(t)D(t)dV_t\right)\right].$$

Since

$$E\left[\left(\int_0^T f(t)G(t)X_t dt\right)^2\right] \leq A_1 \cdot \int_0^T f(t)^2 dt \text{ by the Cauchy-Schwartz inequality,}$$

$$E\left[\left(\int_0^T f(t)D(t)dV_t\right)^2\right] = \int_0^T f(t)^2 D^2(t)dt \quad \text{by the Itô isometry}$$

and $\{X_t\}, \{V_t\}$ are independent, we conclude that

$$A_0 \int_0^T f^2(t)dt \leq E\left[\left(\int_0^T f(t)dZ_t\right)^2\right] \leq A_2 \int_0^T f^2(t)dt \, , \qquad (6.2.11)$$

for some constants A_0, A_1, A_2 not depending on f. We can now show

Lemma 6.2.4 $\mathcal{L}(Z,T) = \{c_0 + \int_0^T f(t)dZ_t; f \in L^2[0,T], c_0 \in \mathbf{R}\}.$

Proof. Denote the right hand side by $\mathcal{N}(Z,T)$. It is enough to show that

a) $\mathcal{N}(Z,T) \subset \mathcal{L}(Z,T)$
b) $\mathcal{N}(Z,T)$ contains all linear combinations of the form

$$c_0 + c_1 Z_{t_1} + \cdots + c_k Z_{t_k} \, ; \qquad 0 \leq t_i \leq T$$

c) $\mathcal{N}(Z,T)$ is closed in $L^2(P)$

a): This follows from the fact that if f is continuous then

$$\int_0^T f(t)dZ_t = \lim_{n \to \infty} \sum_j f(j \cdot 2^{-n}) \cdot (Z_{(j+1)2^{-n}} - Z_{j \cdot 2^{-n}}) \, .$$

b): Suppose $0 \leq t_1 < t_2 < \cdots < t_k \leq T$. We can write

$$\sum_{i=1}^{k} c_i Z_{t_i} = \sum_{j=0}^{k-1} c'_j \Delta Z_j = \sum_{j=0}^{k-1} \int_{t_j}^{t_{j+1}} c'_j dZ_t = \int_0^T \left(\sum_{j=0}^{k-1} c'_j \mathcal{X}_{[t_j, t_{j+1})}(t) \right) dZ_t \;,$$

where $\Delta Z_j = Z_{t_{j+1}} - Z_{t_j}$.

c): This follows from (6.2.11) and the fact that $L^2[0,T]$ is complete.   □

Now we define the **innovation process** N_t as follows:

$$N_t = Z_t - \int_0^t (GX)_s^\wedge ds, \text{ where } (GX)_s^\wedge = \mathcal{P}_{\mathcal{L}(Z,s)}(G(s)X_s) = G(s)\widehat{X}_s \;.$$

(6.2.12)

i.e.

$$dN_t = G(t)(X_t - \widehat{X}_t)dt + D(t)dV_t \;. \tag{6.2.13}$$

Lemma 6.2.5 (i) N_t *has orthogonal increments*

(ii) $E[N_t^2] = \int_0^t D^2(s)ds$

(iii) $\mathcal{L}(N,t) = \mathcal{L}(Z,t)$ *for all* $t \geq 0$

(iv) N_t *is a Gaussian process*

Proof. (i): If $s < t$ and $Y \in \mathcal{L}(Z,s)$ we have

$$E[(N_t - N_s)Y] = E\left[\left(\int_s^t G(r)(X_r - \widehat{X}_r)dr + \int_s^t D(r)dV_r \right)Y \right]$$

$$= \int_s^t G(r)E[(X_r - \widehat{X}_r)Y]dr + E\left[\left(\int_s^t DdV \right)Y \right] = 0 \;,$$

since $X_r - \widehat{X}_r \perp \mathcal{L}(Z,r) \supset \mathcal{L}(Z,s)$ for $r \geq s$ and V has independent increments.

(ii): By Itô's formula, with $g(t,x) = x^2$, we have

$$d(N_t^2) = 2N_t dN_t + \tfrac{1}{2}2(dN_t)^2 = 2N_t dN_t + D^2 dt \;.$$

So

$$E[N_t^2] = E\left[\int_0^t 2N_s dN_s \right] + \int_0^t D^2(s)ds \;.$$

Now

$$\int_0^t N_s dN_s = \lim_{\Delta t_j \to 0} \sum N_{t_j}[N_{t_{j+1}} - N_{t_j}] \;,$$

so since N has orthogonal increments we have

$$E\left[\int_0^t N_s dN_s\right] = 0 , \qquad \text{and (ii) follows .}$$

(iii): It is clear that $\mathcal{L}(N,t) \subset \mathcal{L}(Z,t)$ for all $t \geq 0$. To establish the opposite inclusion we use Lemma 6.2.4. So choose $f \in L^2[0,t]$ and let us see what functions can be obtained in the form

$$\int_0^t f(s)dN_s = \int_0^t f(s)dZ_s - \int_0^t f(r)G(r)\widehat{X}_r dr$$

$$= \int_0^t f(s)dZ_s - \int_0^t f(r)\left[\int_0^r g(r,s)dZ_s\right]dr - \int_0^t f(r)c(r)dr$$

$$= \int_0^t \left[f(s) - \int_s^t f(r)g(r,s)dr\right]dZ_s - \int_0^t f(r)c(r)dr ,$$

where we have used Lemma 6.2.2 and Lemma 6.2.4 to write, for each r,

$$(GX)_r^\wedge = c(r) + \int_0^r g(r,s)dZ_s \qquad \text{for some } g(r,\cdot) \in L^2[0,r], \ c(r) \in \mathbf{R} .$$

From the theory of Volterra integral equations (see e.g. Davis (1977), p. 125) there exists for all $h \in L^2[0,t]$ an $f \in L^2[0,t]$ such that

$$f(s) - \int_s^t f(r)g(r,s)dr = h(s).$$

So by choosing $h = \mathcal{X}_{[0,t_1]}$ where $0 \leq t_1 \leq t$, we obtain

$$\int_0^t f(r)c(r)dr + \int_0^t f(s)dN_s = \int_0^t \mathcal{X}_{[0,t_1]}(s)dZ_s = Z_{t_1} ,$$

which shows that $\mathcal{L}(N,t) \supset \mathcal{L}(Z,t)$.

(iv): $\widehat{X}_t$ is a limit (in $L^2(P)$) of linear combinations of the form

$$M = c_0 + c_1 Z_{s_1} + \cdots + c_k Z_{s_k} , \qquad \text{where } s_k \leq t .$$

Therefore

$$(\widehat{X}_{t_1}, \ldots, \widehat{X}_{t_m})$$

is a limit of m-dimensional random variables $(M^{(1)}, \ldots, M^{(m)})$ whose components $M^{(j)}$ are linear combinations of this form. $(M^{(1)}, \ldots, M^{(m)})$ has a normal distribution since $\{Z_t\}$ is Gaussian, and therefore the limit has. Hence $\{\widehat{X}_t\}$ is Gaussian. It follows that

$$N_t = Z_t - \int_0^t G(s)\widehat{X}_s ds$$

is Gaussian, by a similar argument.   □

Step 3. The Innovation Process and Brownian Motion

Let $N_t = Z_t - \int_0^t G(s)\widehat{X}_s ds$ be the innovation process defined in Step 2. Recall that we have assumed that $D(t)$ is bounded away from 0 on bounded intervals. Define the process $R_t(\omega)$ by

$$dR_t = \frac{1}{D(t)}dN_t(\omega) ; \qquad t \geq 0, \ R_0 = 0 . \tag{6.2.14}$$

Lemma 6.2.6 R_t *is a 1-dimensional Brownian motion.*

Proof. Observe that

(i) R_t has continuous paths
(ii) R_t has orthogonal increments (since N_t has)
(iii) R_t is Gaussian (since N_t is)
(iv) $E[R_t] = 0$ and $E[R_t R_s] = \min(s, t)$.

To prove the last assertion in (iv), note that by Itô's formula

$$d(R_t^2) = 2R_t dR_t + (dR_t)^2 = 2R_t dR_t + dt ,$$

so, since R_t has orthogonal increments,

$$E[R_t^2] = E[\int_0^t ds] = t .$$

Therefore, if $s < t$,

$$E[R_t R_s] = E[(R_t - R_s)R_s] + E[R_s^2] = E[R_s^2] = s .$$

Properties (i), (iii) and (iv) constitute one of the many characterizations of a 1-dimensional Brownian motion (see Simon (1979), Theorem 4.3).

(Alternatively, we could easily deduce that R_t has stationary, independent increments and therefore – by continuity – must be Brownian motion, by the result previously referred to in the beginning of Chapter 3. For a general characterization of Brownian motion see Corollary 8.4.5.) □

Since
$$\mathcal{L}(N, t) = \mathcal{L}(R, t)$$
we conclude that
$$\widehat{X}_t = \mathcal{P}_{\mathcal{L}(R,t)}(X_t) .$$
It turns out that the projection down to the space $\mathcal{L}(R, t)$ can be described very nicely: (compare with formula (6.2.5) in Example 6.2.1)

Lemma 6.2.7
$$\widehat{X}_t = E[X_t] + \int_0^t \frac{\partial}{\partial s} E[X_t R_s] dR_s . \tag{6.2.15}$$

Proof. From Lemma 6.2.4 we know that

$$\widehat{X}_t = c_0(t) + \int_0^t g(s) dR_s \qquad \text{for some } g \in L^2[0, t], \ c_0(t) \in \mathbf{R} .$$

Taking expectations we see that $c_0(t) = E[\widehat{X}_t] = E[X_t]$. We have

$$(X_t - \widehat{X}_t) \perp \int_0^t f(s) dR_s \qquad \text{for all } f \in L^2[0, t] .$$

Therefore

$$E\left[X_t \int_0^t f(s) dR_s\right] = E\left[\widehat{X}_t \int_0^t f(s) dR_s\right] = E\left[\int_0^t g(s) dR_s \int_0^t f(s) dR_s\right]$$

$$= E\left[\int_0^t g(s) f(s) ds\right] = \int_0^t g(s) f(s) ds , \qquad \text{for all } f \in L^2[0, t] ,$$

where we have used the Itô isometry. In particular, if we choose $f = \mathcal{X}_{[0,r]}$ for some $r \leq t$, we obtain

$$E[X_t R_r] = \int_0^r g(s) ds$$

or

$$g(r) = \frac{\partial}{\partial r} E[X_t R_r] , \qquad \text{as asserted .}$$

This completes Step 3.   □

Step 4. An Explicit Formula for X_t

This is easily obtained using Itô's formula, as in the examples in Chapter 5. The result is

$$X_t = \exp\left(\int_0^t F(s)ds\right)\left[X_0 + \int_0^t \exp\left(-\int_0^s F(u)du\right)C(s)dU_s\right]$$

$$= \exp\left(\int_0^t F(s)ds\right)X_0 + \int_0^t \exp\left(\int_s^t F(u)du\right)C(s)dU_s .$$

In particular, we note that $E[X_t] = E[X_0]\exp(\int_0^t F(s)ds)$.

More generally, if $0 \le r \le t$, (see Exercise 6.12)

$$X_t = \exp\left(\int_r^t F(s)ds\right)X_r + \int_r^t \exp\left(\int_s^t F(u)du\right)C(s)dU_s . \qquad (6.2.16)$$

Step 5. The Stochastic Differential Equation for $\widehat{X}_t$

We now combine the previous steps to obtain the solution of the filtering problem, i.e. a stochastic differential equation for $\widehat{X}_t$. Starting with the formula from Lemma 6.2.7

$$\widehat{X}_t = E[X_t] + \int_0^t f(s,t)dR_s ,$$

where

$$f(s,t) = \frac{\partial}{\partial s}E[X_t R_s] , \qquad (6.2.17)$$

we use that

$$R_s = \int_0^s \frac{G(r)}{D(r)}(X_r - \widehat{X}_r)dr + V_s \qquad \text{from (6.2.13) and (6.2.14))}$$

and obtain

$$E[X_t R_s] = \int\limits_0^s \frac{G(r)}{D(r)} E[X_t \widetilde{X}_r] dr \;,$$

where

$$\widetilde{X}_r = X_r - \widehat{X}_r \;. \tag{6.2.18}$$

Using formula (6.2.16) for X_t, we obtain

$$E[X_t \widetilde{X}_r] = \exp\left(\int\limits_r^t F(v) dv\right) E[X_r \widetilde{X}_r] = \exp\left(\int\limits_r^t F(v) dv\right) S(r) \;,$$

where

$$S(r) = E[(\widetilde{X}_r)^2] \;, \tag{6.2.19}$$

i.e. **the mean square error of the estimate at time** $r \geq 0$. Thus

$$E[X_t R_s] = \int\limits_0^s \frac{G(r)}{D(r)} \exp\left(\int\limits_r^t F(v) dv\right) S(r) dr$$

so that

$$f(s,t) = \frac{G(s)}{D(s)} \exp\left(\int\limits_s^t F(v) dv\right) S(s) \;. \tag{6.2.20}$$

We claim that $S(t)$ satisfies the (deterministic) differential equation

$$\frac{dS}{dt} = 2F(t)S(t) - \frac{G^2(t)}{D^2(t)} S^2(t) + C^2(t) \quad (\textit{The Riccati equation}) \;. \tag{6.2.21}$$

To prove (6.2.21) note that by the Pythagorean theorem, (6.2.15) and the Itô isometry

$$S(t) = E[(X_t - \widehat{X}_t)^2] = E[X_t^2] - 2E[X_t \widehat{X}_t] + E[\widehat{X}_t^2] = E[X_t^2] - E[\widehat{X}_t^2]$$

$$= T(t) - \int\limits_0^t f(s,t)^2 ds - E[X_t]^2 \;, \tag{6.2.22}$$

where

$$T(t) = E[X_t^2] \;. \tag{6.2.23}$$

Now by (6.2.16) and the Itô isometry we have

$$T(t) = \exp\left(2\int\limits_0^t F(s) ds\right) E[X_0^2] + \int\limits_0^t \exp\left(2\int\limits_s^t F(u) du\right) C^2(s) ds \;,$$

using that X_0 is independent of $\{U_s\}_{s \geq 0}$. So

$$\frac{dT}{dt} = 2F(t) \cdot \exp\left(2\int_0^t F(s)ds\right)E[X_0^2] + C^2(t)$$

$$+ \int_0^t 2F(t)\exp\left(2\int_s^t F(u)du\right)C^2(s)ds$$

i.e.

$$\frac{dT}{dt} = 2F(t)T(t) + C^2(t) \, . \tag{6.2.24}$$

Substituting in (6.2.22) we obtain, using Step 4,

$$\frac{dS}{dt} = \frac{dT}{dt} - f(t,t)^2 - \int_0^t 2f(s,t) \cdot \frac{\partial}{\partial t} f(s,t)ds - 2F(t)E[X_t]^2$$

$$= 2F(t)T(t) + C^2(t) - \frac{G^2(t)S^2(t)}{D^2(t)} - 2\int_0^t f^2(s,t)F(t)ds - 2F(t)E[X_t]^2$$

$$= 2F(t)S(t) + C^2(t) - \frac{G^2(t)S^2(t)}{D^2(t)} \, , \qquad \text{which is (6.2.21)} \, .$$

We are now ready for the stochastic differential equation for $\widehat{X}_t$:
From the formula

$$\widehat{X}_t = c_0(t) + \int_0^t f(s,t)dR_s \qquad \text{where } c_0(t) = E[X_t]$$

it follows that

$$d\widehat{X}_t = c_0'(t)dt + f(t,t)dR_t + \left(\int_0^t \frac{\partial}{\partial t} f(s,t)dR_s\right)dt \, , \tag{6.2.25}$$

since

$$\int_0^u \left(\int_0^t \frac{\partial}{\partial t} f(s,t)dR_s\right)dt = \int_0^u \left(\int_s^u \frac{\partial}{\partial t} f(s,t)dt\right)dR_s$$

$$= \int_0^u (f(s,u) - f(s,s))dR_s = \widehat{X}_u - c_0(u) - \int_0^u f(s,s)dR_s \, .$$

So

$$d\widehat{X}_t = c_0'(t)dt + \frac{G(t)S(t)}{D(t)}dR_t + \left(\int_0^t f(s,t)dR_s\right)F(t)dt$$

or

$$d\widehat{X}_t = c_0'(t)dt + F(t)\cdot(\widehat{X}_t - c_0(t))dt + \frac{G(t)S(t)}{D(t)}dR_t$$

$$= F(t)\widehat{X}_t dt + \frac{G(t)S(t)}{D(t)}dR_t \;, \tag{6.2.26}$$

since $c_0'(t) = F(t)c_0(t)$ (Step 4).

If we substitute

$$dR_t = \frac{1}{D(t)}[dZ_t - G(t)\widehat{X}_t dt]$$

we obtain

$$d\widehat{X}_t = (F(t) - \frac{G^2(t)S(t)}{D^2(t)})\widehat{X}_t dt + \frac{G(t)S(t)}{D^2(t)}dZ_t \;. \tag{6.2.27}$$

So the conclusion is:

Theorem 6.2.8 (The 1-dimensional Kalman-Bucy filter)
The solution $\widehat{X}_t = E[X_t|\mathcal{G}_t]$ *of the 1-dimensional linear filtering problem*

(linear system) $dX_t = F(t)X_t dt + C(t)dU_t; \; F(t), \; C(t) \in \mathbf{R}$ (6.2.3)

(linear observations) $dZ_t = G(t)X_t dt + D(t)dV_t; \; G(t), \; D(t) \in \mathbf{R}$ (6.2.4)

(with conditions as stated earlier) satisfies the stochastic differential equation

$$d\widehat{X}_t = \left(F(t) - \frac{G^2(t)S(t)}{D^2(t)}\right)\widehat{X}_t dt + \frac{G(t)S(t)}{D^2(t)}dZ_t \;; \quad \widehat{X}_0 = E[X_0] \tag{6.2.28}$$

where

$S(t) = E[(X_t - \widehat{X}_t)^2]$ *satisfies the (deterministic) Riccati equation*
$$\frac{dS}{dt} = 2F(t)S(t) - \frac{G^2(t)}{D^2(t)}S^2(t) + C^2(t), \; S(0) = E[(X_0 - E[X_0])^2] \;. \tag{6.2.29}$$

Example 6.2.9 (Noisy observations of a constant process)
Consider the simple case

(system) $dX_t = 0$, i.e. $X_t = X_0; \; E[X_0] = \widehat{X}_0, \; E[X_0^2] = a^2$

(observations) $dZ_t = X_t dt + mdV_t; \; Z_0 = 0$

(corresponding to

$$H_t = \frac{dZ}{dt} = X_t + mW_t, W_t = \text{ white noise}) \;.$$

First we solve the corresponding Riccati equation for

$$S(t) = E[(X_t - \widehat{X}_t)^2]:$$
$$\frac{dS}{dt} = -\frac{1}{m^2}S^2, \qquad S(0) = a^2$$

i.e.

$$S(t) = \frac{a^2m^2}{m^2 + a^2t}; \qquad t \geq 0.$$

This gives the following equation for $\widehat{X}_t$:

$$d\widehat{X}_t = -\frac{a^2}{m^2 + a^2t}\widehat{X}_t dt + \frac{a^2}{m^2 + a^2t}dZ_t; \qquad \widehat{X}_0 = E[X_0] = 0$$

or

$$d\left(\widehat{X}_t \exp\left(\int\limits_0^t \frac{a^2}{m^2 + a^2s}ds\right)\right) = \exp\left(\int\limits_0^t \frac{a^2}{m^2 + a^2s}ds\right)\frac{a^2}{m^2 + a^2t}dZ_t$$

which gives

$$\widehat{X}_t = \frac{m^2}{m^2 + a^2t}\widehat{X}_0 + \frac{a^2}{m^2 + a^2t}Z_t; \qquad t \geq 0. \qquad (6.2.30)$$

This is the continuous analogue of Example 6.2.1.

Example 6.2.10 (Noisy observations of a Brownian motion)
If we modify the preceding example slightly, so that

(system) $dX_t = cdU_t; \; E[X_0] = 0, \; E[X_0^2] = a^2, \; c$ constant
(observations) $dZ_t = X_t dt + mdV_t,$

the Riccati equation becomes

$$\frac{dS}{dt} = -\frac{1}{m^2}S^2 + c^2, S(0) = a^2$$

or

$$\frac{m^2 dS}{m^2c^2 - S^2} = dt, (S \neq mc).$$

This gives

$$\left|\frac{mc + s}{mc - s}\right| = K\exp\left(\frac{2ct}{m}\right); \qquad K = \left|\frac{mc + a^2}{mc - a^2}\right|.$$

Or

$$S(t) = \begin{cases} mc\dfrac{K\cdot\exp(\frac{2ct}{m})-1}{K\cdot\exp(\frac{2ct}{m})+1} \; ; & \text{if } S(0) < mc \\[3mm] mc \text{ (constant)} & \text{if } S(0) = mc \\[3mm] mc\dfrac{K\cdot\exp(\frac{2ct}{m})+1}{K\cdot\exp(\frac{2ct}{m})-1} & \text{if } S(0) > mc . \end{cases}$$

Thus in all cases the mean square error $S(t)$ tends to mc as $t \to \infty$.

For simplicity let us put $a = 0$, $m = c = 1$. Then

$$S(t) = \frac{\exp(2t) - 1}{\exp(2t) + 1} = \tanh(t) .$$

The equation for $\widehat{X}_t$ is

$$d\widehat{X}_t = -\tanh(t)\widehat{X}_t dt + \tanh(t)dZ_t , \qquad \widehat{X}_0 = 0$$

or

$$d(\cosh(t)\widehat{X}_t) = \sinh(t)dZ_t .$$

So

$$\widehat{X}_t = \frac{1}{\cosh(t)} \int\limits_0^t \sinh(s)dZ_s .$$

If we return to the interpretation of Z_t :

$$Z_t = \int\limits_0^t H_s ds ,$$

where H_s are the "original" observations (see (6.1.4)), we can write

$$\widehat{X}_t = \frac{1}{\cosh(t)} \int\limits_0^t \sinh(s)H_s ds , \qquad\qquad (6.2.31)$$

so $\widehat{X}_t$ is approximately (for large t) a weighted average of the observations H_s, with increasing emphasis on observations as time increases.

Remark. It is interesting to compare formula (6.2.31) with established formulas in forecasting. For example, the *exponentially weighted moving average* $\widetilde{X}_n$ suggested by C.C. Holt in 1958 is given by

$$\widetilde{X}_n = (1 - \alpha)^n Z_0 + \alpha \sum\limits_{k=1}^n (1 - \alpha)^{n-k} Z_k ,$$

where α is some constant; $0 \le \alpha \le 1$. See The Open University (1981), p. 16.

This may be written

$$\tilde{X}_n = \beta^{-n} Z_0 + (\beta - 1)\beta^{-n-1} \sum_{k=1}^{n} \beta^k Z_k \ ,$$

where $\beta = \frac{1}{1-\alpha}$ (assuming $\alpha < 1$), which is a discrete version of (6.2.31), or –
more precisely – of the formula corresponding to (6.2.31) in the general case
when $a \neq 0$ and m, c are not necessarily equal to 1.

Example 6.2.11 (Estimation of a parameter)
Suppose we want to estimate the value of a (constant) parameter θ, based
on observations Z_t satisfying the model

$$dZ_t = \theta M(t)dt + N(t)dB_t \ ,$$

where $M(t), N(t)$ are known functions. In this case the stochastic differential
equation for θ is of course

$$d\theta = 0 \ ,$$

so the Riccati equation for $S(t) = E[(\theta - \widehat{\theta}_t)^2]$ is

$$\frac{dS}{dt} = -\left(\frac{M(t)S(t)}{N(t)}\right)^2$$

which gives

$$S(t) = \left(S_0^{-1} + \int_0^t M(s)^2 N(s)^{-2} ds\right)^{-1}$$

and the Kalman-Bucy filter is

$$d\widehat{\theta}_t = \frac{M(t)S(t)}{N(t)^2}(dZ_t - M(t)\widehat{\theta}_t dt) \ .$$

This can be written

$$\left(S_0^{-1} + \int_0^t M(s)^2 N(s)^{-2} ds\right) d\widehat{\theta}_t + M(t)^2 N(t)^{-2} \widehat{\theta}_t dt = M(t)N(t)^{-2} dZ_t \ .$$

We recoqnize the left hand side a

$$d\left(\left(S_0^{-1} + \int_0^t M(s)^2 N(s)^{-2} ds\right)\widehat{\theta}_t\right)$$

so we obtain

$$\widehat{\theta}_t = \frac{\widehat{\theta}_0 S_0^{-1} + \int_0^t M(s)N(s)^{-2} dZ_s}{S_0^{-1} + \int_0^t M(s)^2 N(s)^{-2} ds} \ .$$

This estimate coincides with the maximum likelihood estimate in classical estimation theory if $S_0^{-1} = 0$. See Liptser and Shiryaev (1978).

For more information about estimates of drift parameters in diffusions and generalizations, see for example Aase (1982), Brown and Hewitt (1975) and Taraskin (1974).

Example 6.2.12 (Noisy observations of a population growth)
Consider a simple growth model (r constant)

$$dX_t = rX_tdt, E[X_0] = b > 0, \qquad E[(X_0 - b)^2] = a^2,$$

with observations

$$dZ_t = X_tdt + mdV_t; \qquad m \text{ constant}.$$

The corresponding Riccati equation

$$\frac{dS}{dt} = 2rS - \frac{1}{m^2}S^2, \qquad S(0) = a^2,$$

gives the logistic curve

$$S(t) = \frac{2rm^2}{1 + Ke^{-2rt}}; \qquad \text{where } K = \frac{2rm^2}{a^2} - 1.$$

So the equation for $\widehat{X}_t$ becomes

$$d\widehat{X}_t = \left(r - \frac{S}{m^2}\right)\widehat{X}_tdt + \frac{S}{m^2}dZ_t; \qquad \widehat{X}_0 = E[X_0] = b.$$

For simplicity let us assume that $a^2 = 2rm^2$, so that

$$S(t) = 2rm^2 \qquad \text{for all } t.$$

(In the general case $S(t) \to 2rm^2$ as $t \to \infty$, so this is not an unreasonable approximation for large t). Then we get

$$d(\exp(rt)\widehat{X}_t) = \exp(rt)2rdZ_t, \qquad \widehat{X}_0 = b$$

or

$$\widehat{X}_t = \exp(-rt)\left[\int_0^t 2r\exp(rs)dZ_s + b\right].$$

As in Example 6.2.10 this may be written

$$\widehat{X}_t = \exp(-rt)\left[\int_0^t 2r\exp(rs)H_sds + b\right], \qquad \text{if } Z_t = \int_0^t H_sds. \qquad (6.2.32)$$

For example, assume that $H_s = \beta$ (constant) for $0 \le s \le t$, i.e. that our observations (for some reason) give the same value β for all times $s \le t$. Then

$$\widehat{X}_t = 2\beta - (2\beta - b)\exp(-rt) \to 2\beta \qquad \text{as } t \to \infty .$$

If $H_s = \beta \cdot \exp(\alpha s)$, $s \ge 0$ (α constant), we get

$$\widehat{X}_t = \exp(-rt)\left[\frac{2r\beta}{r+\alpha}(\exp(r+\alpha)t - 1) + b\right]$$

$$\approx \frac{2r\beta}{r+\alpha}\exp\alpha t \qquad \text{for large } t .$$

Thus, only if $\alpha = r$, i.e. $H_s = \beta\exp(rs)$; $s \ge 0$, does the filter "believe" the observations in the long run. And only if $\alpha = r$ and $\beta = b$, i.e. $H_s = b\exp(rs)$; $s \ge 0$, does the filter "believe" the observations at all times.

Example 6.2.13 (Constant coefficients – general discussion)
Now consider the system

$$dX_t = FX_t dt + CdU_t ; \qquad F, C \text{ constants} \ne 0$$

with observations

$$dZ_t = GX_t dt + DdV_t ; \qquad G, D \text{ constants} \ne 0 .$$

The corresponding Riccati equation

$$S' = 2FS - \frac{G^2}{D^2}S^2 + C^2 , \qquad S(0) = a^2$$

has the solution

$$S(t) = \frac{\alpha_1 - K\alpha_2 \exp(\frac{(\alpha_2-\alpha_1)G^2 t}{D^2})}{1 - K\exp(\frac{(\alpha_2-\alpha_1)G^2 t}{D^2})} ,$$

where

$$\alpha_1 = G^{-2}(FD^2 - D\sqrt{F^2 D^2 + G^2 C^2})$$
$$\alpha_2 = G^{-2}(FD^2 + D\sqrt{F^2 D^2 + G^2 C^2})$$

and

$$K = \frac{a^2 - \alpha_1}{a^2 - \alpha_2} .$$

This gives the solution for $\widehat{X}_t$ of the form

$$\widehat{X}_t = \exp\left(\int_0^t H(s)ds\right)\widehat{X}_0 + \frac{G}{D^2}\int_0^t \exp\left(\int_s^t H(u)du\right)S(s)dZ_s ,$$

where

$$H(s) = F - \frac{G^2}{D^2}S(s) \ .$$

For large s we have $S(s) \approx \alpha_2$. This gives

$$\widehat{X}_t \approx \widehat{X}_0 \exp\left(\left(F - \frac{G^2\alpha_2}{D^2}\right)t\right) + \frac{G\alpha_2}{D^2}\int_0^t \exp\left(\left(F - \frac{G^2\alpha_2}{D^2}\right)(t-s)\right)dZ_s$$

$$= \widehat{X}_0 \exp(-\beta t) + \frac{G\alpha_2}{D^2}\exp(-\beta t)\int_0^t \exp(\beta s)dZ_s \qquad (6.2.33)$$

where $\beta = D^{-1}\sqrt{F^2D^2 + G^2C^2}$. So we get approximately the same behaviour as in the previous example.

6.3 The Multidimensional Linear Filtering Problem

Finally we formulate the solution of the n-dimensional linear filtering problem (6.2.1), (6.2.2):

Theorem 6.3.1 (The Multi-Dimensional Kalman-Bucy Filter)
The solution $\widehat{X}_t = E[X_t|\mathcal{G}_t]$ of the multi-dimensional linear filtering problem

(linear system) $dX_t = F(t)X_t dt + C(t)dU_t;$
$$F(t) \in \mathbf{R}^{n\times n}, \ C(t) \in \mathbf{R}^{n\times p} \qquad (6.3.1)$$

(linear observations) $dZ_t = G(t)X_t dt + D(t)dV_t;$
$$G(t) \in \mathbf{R}^{m\times n}, \ D(t) \in \mathbf{R}^{m\times r} \qquad (6.3.2)$$

satisfies the stochastic differential equation

$$d\widehat{X}_t = (F - SG^T(DD^T)^{-1}G)\widehat{X}_t dt + SG^T(DD^T)^{-1}dZ_t ; \quad \widehat{X}_0 = E[X_0]$$
$$(6.3.3)$$

where $S(t) := E[(X_t - \widehat{X}_t)(X_t - \widehat{X}_t)^T] \in \mathbf{R}^{n\times n}$ *satisfies the matrix Riccati equation*

$$\frac{dS}{dt} = FS + SF^T - SG^T(DD^T)^{-1}GS + CC^T ;$$
$$S(0) = E[(X_0 - E[X_0])(X_0 - E[X_0])^T] \ . \qquad (6.3.4)$$

The condition on $D(t) \in \mathbf{R}^{m\times r}$ is now that $D(t)D(t)^T$ is invertible for all t and that $(D(t)D(t)^T)^{-1}$ is bounded on every bounded t-interval.

A similar solution can be found for the more general situation (system) $dX_t = [F_0(t) + F_1(t)X_t + F_2(t)Z_t]dt + C(t)dU_t$ (6.3.5)

(observations) $dZ_t = [G_0(t) + G_1(t)X_t + G_2(t)Z_t]dt + D(t)dV_t$,   (6.3.6)

where $X_t \in \mathbf{R}^n$, $Z_t \in \mathbf{R}^m$ and $B_t = (U_t, V_t)$ is $n + m$-dimensional Brownian motion, with appropriate dimensions on the matrix coefficients. See Bensoussan (1992) and Kallianpur (1980), who also treat the non-linear case. An account of non-linear filtering theory is also given in Pardoux (1979) and Davis (1984).

For the solution of linear filtering problems governed by more general processes than Brownian motion (processes with orthogonal increments) see Davis (1977).

For various applications of filtering theory see Bucy and Joseph (1968), Jazwinski (1970), Gelb (1974), Maybeck (1979) and the references in these books.

Exercises

6.1.   **(Time-varying observations of a constant)**
Prove that if the (1-dimensional) system is

$$dX_t = 0, \ E[X_0] = 0 , \qquad E[X_0^2] = a^2$$

and the observation process is

$$dZ_t = G(t)X_t dt + dV_t , \qquad Z_0 = 0$$

then $S(t) = E[(X_t - \widehat{X}_t)^2]$ is given by

$$S(t) = \frac{1}{\frac{1}{S(0)} + \int_0^t G^2(s)ds} .$$ (6.3.7)

We say that we have *exact asymptotic estimation* if $S(t) \to 0$ as $t \to \infty$, i.e. if

$$\int_0^\infty G^2(s)ds = \infty .$$

Thus for

$$G(s) = \frac{1}{(1+s)^p} \qquad (p > 0 \text{ constant})$$

we have exact asymptotic estimation iff $p \leq \frac{1}{2}$.

6.2.   Consider the linear 1-dimensional filtering problem with no noise in the system:

(system) $dX_t = F(t)X_t dt$ (6.3.8)

(observations) $dZ_t = G(t)X_t dt + D(t)dV_t$ (6.3.9)

Put $S(t) = E[(X_t - \widehat{X}_t)^2]$ as usual and assume $S(0) > 0$.

a)   Show that

$$R(t) := \frac{1}{S(t)}$$

satisfies the *linear* differential equation

$$R'(t) = -2F(t)R(t) + \frac{G^2(t)}{D^2(t)} \; ; \qquad R(0) = \frac{1}{S(0)} \qquad (6.3.10)$$

b)   Use (6.3.10) to prove that for the filtering problem (6.3.8), (6.3.9) we have

$$\frac{1}{S(t)} = \frac{1}{S(0)} \exp\left(-2\int_0^t F(s)ds\right) + \int_0^t \exp\left(-2\int_s^t F(u)du\right) \frac{G^2(s)}{D^2(s)} ds \; .$$

$$(6.3.11)$$

6.3.   In Example 6.2.12 we found that

$$S(t) \to 2rm^2 \qquad \text{as } t \to \infty \; ,$$

so exact asymptotic estimation (Exercise 6.1) of X_t is not possible. However, prove that we can obtain exact asymptotic estimation of X_0, in the sense that

$$E[(X_0 - E[X_0|\mathcal{G}_t])^2] \to 0 \qquad \text{as } t \to \infty \; .$$

(Hint: Note that $X_0 = e^{-rt}X_t$ and therefore $E[X_0|\mathcal{G}_t] = e^{-rt}\widehat{X}_t$, so that

$$E[(X_0 - E[X_0|\mathcal{G}_t])^2] = e^{-2rt}S(t)) \; .$$

6.4.   Consider the multi-dimensional linear filtering problem with no noise in the system:

(system) $dX_t = F(t)X_t dt$;

                      $X_t \in \mathbf{R}^n$, $F(t) \in \mathbf{R}^{n \times n}$ (6.3.12)

(observations) $dZ_t = G(t)X_t dt + D(t)dV_t$;

                      $G(t) \in \mathbf{R}^{m \times n}$, $D(t) \in \mathbf{R}^{m \times r}$ (6.3.13)

Assume that $S(t)$ is nonsingular and define $R(t) = S(t)^{-1}$. Prove that $R(t)$ satisfies the *Lyapunov equation* (compare with Exercise 6.2)

$$R'(t) = -R(t)F(t) - F(t)^T R(t) + G(t)^T (D(t)D(t)^T)^{-1}G(t) \; . \quad (6.3.14)$$

(Hint: Note that since $S(t)S^{-1}(t) = I$ we have
$S'(t)S^{-1}(t) + S(t)(S^{-1})'(t) = 0$, which gives

$$(S^{-1})'(t) = -S^{-1}(t)S'(t)S^{-1}(t) \,.)$$

6.5. **(Prediction)**
In the *prediction problem* one seeks to estimate the value of the system X at a *future* time T based on the observations $\mathcal{G}_t$ up to the present time $t < T$.
Prove that in the linear setup (6.2.3), (6.2.4) the predicted value

$$E[X_T|\mathcal{G}_t] \,, \qquad T > t$$

is given by

$$E[X_T|\mathcal{G}_t] = \exp\left(\int\limits_t^T F(s)ds \right) \cdot \widehat{X}_t \,. \qquad (6.3.15)$$

(Hint: Use formula (6.2.16).)

6.6. **(Interpolation/smoothing)**
The *interpolation* or *smoothing problem* consists of estimating the value of the system X at a time $s < t$, given the observations up to time $t, \mathcal{G}_t$.
With notation as in (6.2.1), (6.2.2) one can show that $M_s := E[X_s|\mathcal{G}_t]$ satisfies the differential equation

$$\begin{cases} \frac{d}{ds}M_s = F(s)M_s + C(s)C^T(s)S^{-1}(s)(M_s - \widehat{X}_s) \,; & s < t \\ M_t = \widehat{X}_t \,. \end{cases}$$

$$(6.3.16)$$

(See Davis (1977, Theorem 4.4.4).)
Use this result to find $E[X_s|\mathcal{G}_t]$ in Example 6.2.9.

6.7. Consider the system

$$dX_t = \begin{bmatrix} dX_1(t) \\ dX_2(t) \end{bmatrix} = \begin{bmatrix} 0 \\ 0 \end{bmatrix} \,, \qquad E[X_0] = \begin{bmatrix} 0 \\ 0 \end{bmatrix}$$

with observations

$$\begin{bmatrix} dZ_1(t) \\ dZ_2(t) \end{bmatrix} = \begin{bmatrix} X_1 \\ X_1 + X_2 \end{bmatrix} dt + \begin{bmatrix} dV_1(t) \\ dV_2(t) \end{bmatrix} \,.$$

Apply (6.3.14) from Exercise 6.4 to prove that
$S(t) := E[(X_t - \widehat{X}_t)(X_t - \widehat{X}_t)^T]$ is given by

$$S(t)^{-1} = S^{-1}(0) + \begin{bmatrix} 2 & 1 \\ 1 & 1 \end{bmatrix} t$$

if $S(t)$ is invertible for all t. Then show that

$$d\widehat{X}_t = -S(t) \begin{bmatrix} 2 & 1 \\ 1 & 1 \end{bmatrix} \widehat{X}_t dt + S(t) \begin{bmatrix} 1 & 1 \\ 0 & 1 \end{bmatrix} dZ_t .$$

6.8. Transform the following Stratonovich equation

$$dX_t = b(t, X_t)dt + \sigma(t, X_t) \circ dB_t$$

into the corresponding Itô equation

$$dX_t = \widetilde{b}(t, X_t)dt + \sigma(t, X_t)dB_t$$

using (6.1.3):

a)

$$\begin{bmatrix} dX_1 \\ dX_2 \end{bmatrix} = \begin{bmatrix} 1 \\ X_2 + e^{2X_1} \end{bmatrix} dt + \begin{bmatrix} 0 \\ e^{X_1} \end{bmatrix} \circ dB_t \qquad (B_t \in \mathbf{R})$$

b)

$$\begin{bmatrix} dX_1 \\ dX_2 \end{bmatrix} = \begin{bmatrix} X_1 \\ X_2 \end{bmatrix} dt + \begin{bmatrix} X_2 \\ X_1 \end{bmatrix} \circ dB_t \qquad (B_t \in \mathbf{R})$$

6.9. Transform the following Itô equation

$$dX_t = b(t, X_t)dt + \sigma(t, X_t)dB_t$$

into the corresponding Stratonovich equation

$$dX_t = \widehat{b}(t, X_t)dt + \sigma(t, X_t) \circ dB_t ,$$

using (the converse of) (6.1.3):

a) $dX_t = -\frac{1}{2}X_t dt + KX_t dB_t$, where

$$K = \begin{bmatrix} 0 & -1 \\ 1 & 0 \end{bmatrix}, \quad X_t = \begin{bmatrix} X_1(t) \\ X_2(t) \end{bmatrix} \in \mathbf{R}^2 \quad \text{and} \quad B_t \in \mathbf{R}$$

(i.e. X_t is Brownian motion on the unit circle (Example 5.1.4)).

b) $\begin{bmatrix} dX_1 \\ dX_2 \end{bmatrix} = \begin{bmatrix} X_1 & -X_2 \\ X_2 & X_1 \end{bmatrix} \begin{bmatrix} dB_1 \\ dB_2 \end{bmatrix}.$

6.10. (On the support of an Itô diffusion)
The *support* of an Itô diffusion X in $\mathbf{R}^n$ starting at $x \in \mathbf{R}^n$ is the smallest closed set F with the property that

$$X_t(\omega) \in F \qquad \text{for all } t \geq 0, \text{ for a.a. } \omega .$$

In Example 5.1.4 we found that Brownian motion on the unit circle, X_t, satisfies the (Itô) stochastic differential equation

$$\begin{bmatrix} dX_1(t) \\ dX_2(t) \end{bmatrix} = -\frac{1}{2}\begin{bmatrix} X_1(t) \\ X_2(t) \end{bmatrix} dt + \begin{bmatrix} 0 & -1 \\ 1 & 0 \end{bmatrix}\begin{bmatrix} X_1(t) \\ X_2(t) \end{bmatrix} dB_t . \quad (6.3.17)$$

From this equation it is not at all apparent that its solution is situated on the same circle as the starting point. However, this can be detected by proceeding as follows: First transform the equation into its Stratonovich form, which in Exercise 6.9 is found to be

$$\begin{bmatrix} dX_1(t) \\ dX_2(t) \end{bmatrix} = \begin{bmatrix} 0 & -1 \\ 1 & 0 \end{bmatrix}\begin{bmatrix} X_1(t) \\ X_2(t) \end{bmatrix} \circ dB_t . \quad (6.3.18)$$

Then (formally) replace $\circ dB_t$ by $\phi'(t)dt$, where ϕ is some smooth (deterministic) function, $\phi(0) = 0$. This gives the deterministic equation

$$\begin{bmatrix} dX_1^{(\phi)}(t) \\ dX_2^{(\phi)}(t) \end{bmatrix} = \begin{bmatrix} 0 & -1 \\ 1 & 0 \end{bmatrix}\phi'(t)dt . \quad (6.3.19)$$

If $(X_1^{(\phi)}(0), X_2^{(\phi)}(0)) = (1,0)$ the solution of (6.3.19) is

$$\begin{bmatrix} X_1^{(\phi)}(t) \\ X_2^{(\phi)}(t) \end{bmatrix} = \begin{bmatrix} \cos\phi(t) \\ \sin\phi(t) \end{bmatrix} .$$

So for any smooth ϕ the corresponding solution $X^{(\phi)}(t)$ of (6.3.19) has its support on this unit circle. We can conclude that the original solution $X(t,\omega)$ is supported on the unit circle also, in virtue of the *Stroock-Varadhan support theorem*. This theorem says that, quite generally, the support of an Itô diffusion $X_t(\omega)$ coincides with the closure in $\mathbf{R}^n$ of $\{X^{(\phi)}(\cdot); \phi$ smooth$\}$, where $X^{(\phi)}(t)$ is obtained by replacing $\circ dB_t$ by $\phi'(t)dt$ in the same way as above. See e.g. Ikeda and Watanabe (1989, Th. VI. 8.1). (In this special case above the support could also have been found directly from (6.3.18)).
Use the procedure above to find the support of the process $X_t \in \mathbf{R}^2$ given by

$$dX_t = \frac{1}{2}X_tdt + \begin{bmatrix} 0 & 1 \\ 1 & 0 \end{bmatrix} X_t dB_t .$$

6.11. Consider Example 6.2.1, but now without the assumption that $E[X] = 0$. Show that

$$\widehat{X}_k = \frac{m^2}{ka^2 + m^2}E[X] + \frac{a^2}{a^2 + \frac{1}{k}m^2}\overline{Z}_k ; \qquad k = 1, 2, \ldots$$

where $a^2 = E[(X - E[X])^2]$. (Compare with (6.2.8).)

(Hint: Put $\xi = X - E[X]$, $\zeta_k = Z_k - E[X]$. Then apply (6.2.8) with X replaced by ξ and Z_k replaced by ζ_k.)

6.12. Prove formula (6.2.16).

(Hint: $\exp\left(-\int_r^s F(u)du\right)$ is an integrating factor for the stochastic differential equation (6.2.3).)

6.13. Consider the 1-dimensional linear filtering problem (6.2.3), (6.2.4). Find

$$E[\widehat{X}_t] \quad \text{and} \quad E[(\widehat{X}_t)^2] \, .$$

(Hint: Use Theorem 6.1.2 and use the definition of the mean square error $S(t)$.)

6.14. Let B_t be 1-dimensional Brownian motion.

a) Give an example of a process Z_t of the form

$$dZ_t = u(t,\omega)dt + dB_t$$

such that Z_t is a Brownian motion w.r.t. P and $u(t,\omega) \in \mathcal{V}$ is not identically 0.
(Hint: Choose Z_t to be the innovation process (6.2.13) in a linear filtering problem with $D(t) \equiv 1$.)

b) Show that the filtration $\{\mathcal{Z}_t\}_{t\geq 0}$ generated by a process Z_t as in a) must be strictly smaller than $\{\mathcal{F}_t\}_{t\geq 0}$, i.e. show that

$$\mathcal{Z}_t \subseteq \mathcal{F}_t \quad \text{for all } t \text{ and } \mathcal{Z}_t \neq \mathcal{F}_t \text{ for some } t \, .$$

(Hint: Use Exercise 4.12.)

6.15.* Suppose the state $X_t \in \mathbf{R}$ at time t is a geometric Brownian motion given by the equation

$$dX_t = \mu\, X_t dt + \sigma\, X_t dB_t \, ; \qquad X_0 = x > 0 \, . \qquad (6.3.20)$$

Here $\sigma \neq 0$ and x are known constants. The parameter μ is also constant, but we do not know its value, only its probability distribution, which is assumed to be normal with mean $\bar{\mu}$ and variance a^2. We assume that μ is independent of $\{B_s\}_{s\geq 0}$ and that $E[\mu^2] < \infty$.

We assume that we can observe the value of X_t for all t. Thus we have access to the information "(σ-algebra)" $\mathcal{M}_t$ generated by X_s; $s \leq t$. Let $\mathcal{N}_t$ be the σ-algebra generated by ξ_s, $s \leq t$, where

$$d\xi_t = \mu\, dt + \sigma\, dB_t \, ; \qquad \xi_0 = x. \qquad (6.3.21)$$

a) Prove that $\mathcal{M}_t = \mathcal{N}_t$.

b) Prove that

$$E[\mu|\mathcal{N}_t] = (\theta + \sigma^{-2}t)^{-1}(\bar{\mu}\,\theta + \sigma^{-2}\xi_t) \qquad (6.3.22)$$

where

$$\theta = E[(\mu - \bar{\mu})^2]^{-1}, \qquad \bar{\mu} = E[\mu]. \qquad (6.3.23)$$

c) Define

$$\widetilde{B}_t = \int_0^t \sigma^{-1}(\mu - E[\mu|\mathcal{M}_s])ds + B_t . \qquad (6.3.24)$$

Prove that $\widetilde{B}_t$ is a Brownian motion.

d) Prove that $\widetilde{B}_t$ is $\mathcal{M}_t$-measurable for all t. Hence

$$\widetilde{\mathcal{F}}_t \subseteq \mathcal{M}_t \qquad (6.3.25)$$

where $\widetilde{\mathcal{F}}_t$ is the σ-algebra generated by $\widetilde{B}_s$; $s \leq t$.

e) Prove that ξ_t is $\widetilde{\mathcal{F}}_t$-measurable for all t. Combined with d) and a) this gives that

$$\widetilde{\mathcal{F}}_t = \mathcal{M}_t = \mathcal{N}_t = \mathcal{F}_t .$$

f) Prove that

$$dX_t = E[\mu|\mathcal{M}_t]X_t dt + \sigma\, X_t d\widetilde{B}_t .$$

Note that in this representation of X_t all the coefficients are observable quantities.

6.16. Repeat Exercise 6.15, but this time with X_t being a mean-reverting Ornstein-Uhlenbeck process given by

$$dX_t = (\mu - \rho\, X_t)dt + \sigma\, dB_t ; \qquad X_0 = x \in \mathbf{R} .$$

Here $\rho, \sigma \neq 0$ and x are known constants, while μ is an unknown constant, as before. Conclude that X_t can be given a representation of the form

$$dX_t = (E[\mu|\mathcal{M}_t] - \rho\, X(t))dt + \sigma\, d\widetilde{B}_t .$$

Chapter 7
Diffusions: Basic Properties

7.1 The Markov Property

Suppose we want to describe the motion of a small particle suspended in a moving liquid, subject to random molecular bombardments. If $b(t,x) \in \mathbf{R}^3$ is the velocity of the fluid at the point x at time t, then a reasonable mathematical model for the position X_t of the particle at time t would be a stochastic differential equation of the form

$$\frac{dX_t}{dt} = b(t, X_t) + \sigma(t, X_t)W_t \ , \tag{7.1.1}$$

where $W_t \in \mathbf{R}^3$ denotes "white noise" and $\sigma(t,x) \in \mathbf{R}^{3\times3}$. The Itô interpretation of this equation is

$$dX_t = b(t, X_t)dt + \sigma(t, X_t)dB_t \ , \tag{7.1.2}$$

where B_t is 3-dimensional Brownian motion, and similarly (with a correction term added to b) for the Stratonovich interpretation (see (6.1.3)).

In a stochastic differential equation of the form

$$dX_t = b(t, X_t)dt + \sigma(t, X_t)dB_t \ , \tag{7.1.3}$$

where $X_t \in \mathbf{R}^n$, $b(t,x) \in \mathbf{R}^n$, $\sigma(t,x) \in \mathbf{R}^{n\times m}$ and B_t is m-dimensional Brownian motion, we will call b the *drift coefficient* and σ – or sometimes $\frac{1}{2}\sigma\sigma^T$ – the *diffusion coefficient* (see Theorem 7.3.3).

Thus the solution of a stochastic differential equation may be thought of as the mathematical description of the motion of a small particle in a moving fluid: Therefore such stochastic processes are called *(Itô) diffusions*.

In this chapter we establish some of the most basic properties and results about Itô diffusions:

7.1   The Markov property.

This will give us the necessary background for the applications in the remaining chapters.

Definition 7.1.1 *A (time-homogeneous)* Itô *diffusion is a stochastic process* $X_t(\omega) = X(t, \omega) \colon [0, \infty) \times \Omega \to \mathbf{R}^n$ *satisfying a stochastic differential equation of the form*

$$dX_t = b(X_t)dt + \sigma(X_t)dB_t , \qquad t \geq s ; \quad X_s = x \qquad (7.1.4)$$

where B_t is m-dimensional Brownian motion and $b \colon \mathbf{R}^n \to \mathbf{R}^n$, $\sigma \colon \mathbf{R}^n \to \mathbf{R}^{n \times m}$ satisfy the conditions in Theorem 5.2.1, which in this case simplify to:

$$|b(x) - b(y)| + |\sigma(x) - \sigma(y)| \leq D|x - y| ; \qquad x, y \in \mathbf{R}^n , \qquad (7.1.5)$$

i.e. that $b(\cdot)$ and $\sigma(\cdot)$ are Lipschitz continuous.

We will denote the (unique) solution of (7.1.4) by $X_t = X_t^{s,x}$; $t \geq s$. If $s = 0$ we write X_t^x for $X_t^{0,x}$. Note that we have assumed in (7.1.4) that b and σ do not depend on t but on x only. We shall see later (Chapters 10, 11) that the general case can be reduced to this situation. The resulting process $X_t(\omega)$ will have the property of being *time-homogeneous*, in the following sense:
    Note that

$$X_{s+h}^{s,x} = x + \int_s^{s+h} b(X_u^{s,x})du + \int_s^{s+h} \sigma(X_u^{s,x})dB_u$$

$$= x + \int_0^h b(X_{s+v}^{s,x})dv + \int_0^h \sigma(X_{s+v}^{s,x})d\widetilde{B}_v , \quad (u = s + v) \quad (7.1.6)$$

where $\widetilde{B}_v = B_{s+v} - B_s$; $v \geq 0$. (See Exercise 2.12). On the other hand of course

$$X_h^{0,x} = x + \int_0^h b(X_v^{0,x})dv + \int_0^h \sigma(X_v^{0,x})dB_v .$$

Since $\{\widetilde{B}_v\}_{v \geq 0}$ and $\{B_v\}_{v \geq 0}$ have the same P^0-distributions, it follows by weak uniqueness (Lemma 5.3.1) of the solution of the stochastic differential equation

$$dX_t = b(X_t)dt + \sigma(X_t)dB_t ; \qquad X_0 = x$$

that

$$\{X_{s+h}^{s,x}\}_{h \geq 0} \qquad \text{and} \qquad \{X_h^{0,x}\}_{h \geq 0}$$

have the same P^0-distributions, i.e. $\{X_t\}_{t\geq 0}$ is *time-homogeneous*.

We now introduce some notation which is common (and convenient) in the context of Markov processes:

From now on we let Q^x denote the probability law of a given (time-homogeneous) Itô diffusion $\{X_t\}_{t\geq 0}$ when its initial value is $X_0 = x \in \mathbf{R}^n$. The expectation with respect to Q^x is denoted by $E^x[\cdot]$. Hence we have

$$E^x[f_1(X_{t_1})\cdots f_k(X_{t_k})] = E[f_1(X^x_{t_1})\cdots f_k(X^x_{t_k})] \qquad (7.1.7)$$

for all bounded Borel functions $f_1,\ldots,f_k$ and all times $t_1,\ldots,t_k \geq 0$; $k = 1,2,\ldots$, where $E = E_P$ denotes the expectation with respect to the probability law $P = P^0$ for $\{B_t\}_{t\geq 0}$ when $B_0 = 0$ (see Section 2.2).

As before we let $\mathcal{F}_t^{(m)}$ be the σ-algebra generated by $\{B_r; r \leq t\}$. Similarly we let $\mathcal{M}_t$ be the σ-algebra generated by $\{X_r; r \leq t\}$. We have established earlier (see Theorem 5.2.1) that X_t is measurable with respect to $\mathcal{F}_t^{(m)}$, so $\mathcal{M}_t \subseteq \mathcal{F}_t^{(m)}$.

We now prove that X_t satisfies the important *Markov property*: The future behaviour of the process given what has happened up to time t is the same as the behaviour obtained when starting the process at X_t. The precise mathematical formulation of this is the following:

Theorem 7.1.2 (The Markov property for Itô diffusions)
Let f be a bounded Borel function from $\mathbf{R}^n$ to $\mathbf{R}$. Then, for $t,h \geq 0$

$$E^x[f(X_{t+h})|\mathcal{F}_t^{(m)}](\omega) = E^{X_t(\omega)}[f(X_h)] . \qquad (7.1.8)$$

(See Appendix B for definition and basic properties of conditional expectation). Here and in the following E^x denotes the expectation w.r.t. the probability measure Q^x. Thus $E^y[f(X_h)]$ means $E[f(X_h^y)]$, where E denotes the expectation w.r.t. the measure P^0. The right hand side means the function $E^y[f(X_h)]$ evaluated at $y = X_t(\omega)$.

Proof. Since, for $r \geq t$,

$$X_r(\omega) = X_t(\omega) + \int_t^r b(X_u)du + \int_t^r \sigma(X_u)dB_u ,$$

we have by uniqueness

$$X_r(\omega) = X_r^{t,X_t}(\omega) .$$

In other words, if we define

$$F(x,t,r,\omega) = X_r^{t,x}(\omega) \qquad \text{for } r \geq t ,$$

we have

$$X_r(\omega) = F(X_t,t,r,\omega); \ r \geq t . \qquad (7.1.9)$$

Note that $\omega \rightarrow F(x, t, r, \omega)$ is independent of $\mathcal{F}_t^{(m)}$. Using (7.1.9) we may rewrite (7.1.8) as

$$E[f(F(X_t, t, t + h, \omega)) | \mathcal{F}_t^{(m)}] = E[f(F(x, 0, h, \omega))]_{x=X_t} . \qquad (7.1.10)$$

Put $g(x, \omega) = f \circ F(x, t, t + h, \omega)$. Then $(x, \omega) \rightarrow g(x, \omega)$ is measurable. (See Exercise 7.6). Hence we can approximate g pointwise boundedly by functions on the form

$$\sum_{k=1}^{m} \phi_k(x) \psi_k(\omega) .$$

Using the properties of conditional expectation (see Appendix B) we get

$$
\begin{aligned}
E[g(X_t, \omega) | \mathcal{F}_t^{(m)}] &= E\left[\lim \sum \phi_k(X_t) \psi_k(\omega) | \mathcal{F}_t^{(m)}\right] \\
&= \lim \sum \phi_k(X_t) \cdot E[\psi_k(\omega) | \mathcal{F}_t^{(m)}] \\
&= \lim \sum E[\phi_k(y) \psi_k(\omega) | \mathcal{F}_t^{(m)}]_{y=X_t} \\
&= E[g(y, \omega) | \mathcal{F}_t^{(m)}]_{y=X_t} = E[g(y, \omega)]_{y=X_t} .
\end{aligned}
$$

Therefore, since $\{X_t\}$ is time-homogeneous,

$$
\begin{aligned}
E[f(F(X_t, t, t + h, \omega)) | \mathcal{F}_t^{(m)}] &= E[f(F(y, t, t + h, \omega))]_{y=X_t} \\
&= E[f(F(y, 0, h, \omega))]_{y=X_t}
\end{aligned}
$$

which is (7.1.10).   □

Remark. Theorem 7.1.2 states that X_t is a Markov process w.r.t. the family of σ-algebras $\{\mathcal{F}_t^{(m)}\}_{t \geq 0}$. Note that since $\mathcal{M}_t \subseteq \mathcal{F}_t^{(m)}$ this implies that X_t *is also a Markov process w.r.t. the σ-algebras* $\{\mathcal{M}_t\}_{t \geq 0}$. This follows from Theorem B.3 and Theorem B.2 c)(Appendix B):

$$
\begin{aligned}
E^x[f(X_{t+h}) | \mathcal{M}_t] &= E^x[E^x[f(X_{t+h}) | \mathcal{F}_t^{(m)}] | \mathcal{M}_t] \\
&= E^x[E^{X_t}[f(X_h)] | \mathcal{M}_t] = E^{X_t}[f(X_h)]
\end{aligned}
$$

since $E^{X_t}[f(X_h)]$ is $\mathcal{M}_t$-measurable.

7.2 The Strong Markov Property

Roughly, the strong Markov property states that a relation of the form (7.1.8) continues to hold if the time t is replaced by a random time $\tau(\omega)$ of a more general type called *stopping time* (or *Markov time*):

Definition 7.2.1 *Let* $\{\mathcal{N}_t\}$ *be an increasing family of σ-algebras (of subsets of Ω). A function $\tau\colon\Omega \to [0,\infty]$ is called a (strict) stopping time w.r.t. $\{\mathcal{N}_t\}$ if*

$$\{\omega; \tau(\omega) \le t\} \in \mathcal{N}_t , \qquad \text{for all } t \ge 0 .$$

In other words, it should be possible to decide whether or not $\tau \le t$ has occurred on the basis of the knowledge of $\mathcal{N}_t$.

Note that if $\tau(\omega) = t_0$ (constant) for all ω, then τ is trivially a stopping time w.r.t. any filtration, because in this case

$$\{\omega : \tau(\omega) \le t\} = \begin{cases} \Omega & \text{if } t_0 \le t \\ \emptyset & \text{if } t_0 > t \end{cases}$$

Example 7.2.2 Let $U \subset \mathbf{R}^n$ be open. Then the *first exit time*

$$\tau_U := \inf\{t > 0; X_t \notin U\}$$

is a stopping time w.r.t. $\{\mathcal{M}_t\}$, since

$$\{\omega; \tau_U \le t\} = \bigcap_m \bigcup_{\substack{r \in \mathbf{Q} \\ r < t}} \{\omega; X_r \notin K_m\} \in \mathcal{M}_t$$

where $\{K_m\}$ is an increasing sequence of closed sets such that $U = \bigcup_m K_m$.

More generally, if $H \subset \mathbf{R}^n$ is any set we define the *first exit time from H*, τ_H, as follows

$$\tau_H = \inf\{t > 0; X_t \notin H\} .$$

If we include the sets of measure 0 in $\mathcal{M}_t$ (which we do) then the family $\{\mathcal{M}_t\}$ is right-continuous i.e. $\mathcal{M}_t = \mathcal{M}_{t+}$, where $\mathcal{M}_{t+} = \bigcap_{s>t} \mathcal{M}_s$ (see Chung (1982, Theorem 2.3.4., p. 61)) and therefore τ_H is a stopping time for any Borel set H (see Dynkin (1965 II, 4.5.C.e.), p. 111)).

Definition 7.2.3 *Let τ be a stopping time w.r.t. $\{\mathcal{N}_t\}$ and let $\mathcal{N}_\infty$ be the smallest σ-algebra containing $\mathcal{N}_t$ for all $t \ge 0$. Then the σ-algebra $\mathcal{N}_\tau$ consists of all sets $N \in \mathcal{N}_\infty$ such that*

$$N \cap \{\tau \le t\} \in \mathcal{N}_t \qquad \text{for all } t \ge 0 .$$

In the case when $\mathcal{N}_t = \mathcal{M}_t$, an alternative and more intuitive description is:

$$\mathcal{M}_\tau = \text{ the σ-algebra generated by } \{X_{\min(s,\tau)}; s \ge 0\} . \tag{7.2.1}$$

(See Rao (1977, p. 2.15) or Stroock and Varadhan (1979, Lemma 1.3.3, p. 33).) Similarly, if $\mathcal{N}_t = \mathcal{F}_t^{(m)}$, we get

$$\mathcal{F}_\tau^{(m)} = \text{ the σ-algebra generated by } \{B_{s\wedge\tau}; s \ge 0\} .$$

Theorem 7.2.4 (The strong Markov property for Itô diffusions)
Let f be a bounded Borel function on $\mathbf{R}^n$, τ a stopping time w.r.t. $\mathcal{F}_t^{(m)}$, $\tau < \infty$ a.s. Then

$$E^x[f(X_{\tau+h})|\mathcal{F}_\tau^{(m)}] = E^{X_\tau}[f(X_h)] \qquad \text{for all } h \geq 0. \qquad (7.2.2)$$

Proof. We try to imitate the proof of the Markov property (Theorem 7.1.2). For a.a. ω we have that $X_r^{\tau,x}(\omega)$ satisfies

$$X_{\tau+h}^{\tau,x} = x + \int_\tau^{\tau+h} b(X_u^{\tau,x})du + \int_\tau^{\tau+h} \sigma(X_u^{\tau,x})dB_u.$$

By the strong Markov property for Brownian motion (Gihman and Skorohod (1972, p. 30)) the process

$$\widetilde{B}_v = B_{\tau+v} - B_\tau; \qquad v \geq 0$$

is again a Brownian motion and independent of $\mathcal{F}_\tau^{(m)}$. Therefore

$$X_{\tau+h}^{\tau,x} = x + \int_0^h b(X_{\tau+v}^{\tau,x})dv + \int_0^h \sigma(X_{\tau+v}^{\tau,x})d\widetilde{B}_v.$$

Hence $\{X_{\tau+h}^{\tau,x}\}_{h\geq 0}$ must coincide a.e. with the strongly unique (see (5.2.8)) solution Y_h of the equation

$$Y_h = x + \int_0^h b(Y_v)dv + \int_0^h \sigma(Y_v)d\widetilde{B}_v.$$

Since $\{Y_h\}_{h\geq 0}$ is independent of $\mathcal{F}_\tau^{(m)}$, $\{X_{\tau+h}^{\tau,x}\}$ must be independent also. Moreover, by weak uniqueness (Lemma 5.3.1) we conclude that

$$\{Y_h\}_{h\geq 0}, \text{ and hence } \{X_{\tau+h}^{\tau,x}\}_{h\geq 0}, \text{ has the same law as } \{X_h^{0,x}\}_{h\geq 0}. \qquad (7.2.3)$$

Put

$$F(x,t,r,\omega) = X_r^{t,x}(\omega) \qquad \text{for } r \geq t.$$

Then (7.2.2) can be written

$$E[f(F(x,0,\tau+h,\omega))|\mathcal{F}_\tau^{(m)}] = E[f(F(x,0,h,\omega))]_{x=X_\tau^{0,x}}.$$

Now, with $X_t = X_t^{0,x}$,

$$F(x, 0, \tau + h, \omega) = X_{\tau+h}(\omega) = x + \int_0^{\tau+h} b(X_s)ds + \int_0^{\tau+h} \sigma(X_s)dB_s$$

$$= x + \int_0^{\tau} b(X_s)ds + \int_0^{\tau} \sigma(X_s)dB_s + \int_{\tau}^{\tau+h} b(X_s)ds + \int_{\tau}^{\tau+h} \sigma(X_s)dB_s$$

$$= X_{\tau} + \int_{\tau}^{\tau+h} b(X_s)ds + \int_{\tau}^{\tau+h} \sigma(X_s)dB_s$$

$$= F(X_{\tau}, \tau, \tau + h, \omega) \ .$$

Hence (7.2.2) gets the form

$$E[f(F(X_{\tau}, \tau, \tau + h, \omega))|\mathcal{F}_{\tau}^{(m)}] = E[f(F(x, 0, h, \omega))]_{x=X_{\tau}} \ .$$

Put $g(x, t, r, \omega) = f(F(x, t, r, \omega))$. As in the proof of Theorem 7.1.2 we may assume that g has the form

$$g(x, t, r, \omega) = \sum_k \phi_k(x)\psi_k(t, r, \omega) \ .$$

Then, since $X_{\tau+h}^{\tau,x}$ is independent of $\mathcal{F}_{\tau}^{(m)}$ we get, using (7.2.3)

$$E[g(X_{\tau}, \tau, \tau + h, \omega)|\mathcal{F}_{\tau}^{(m)}] = \sum_k E[\phi_k(X_{\tau})\psi_k(\tau, \tau + h, \omega)|\mathcal{F}_{\tau}^{(m)}]$$

$$= \sum_k \phi_k(X_{\tau})E[\psi_k(\tau, \tau+h, \omega)|\mathcal{F}_{\tau}^{(m)}] = \sum_k E[\phi_k(x)\psi_k(\tau, \tau+h, \omega)|\mathcal{F}_{\tau}^{(m)}]_{x=X_{\tau}}$$

$$= E[g(x, \tau, \tau + h, \omega)|\mathcal{F}_{\tau}^{(m)}]_{x=X_{\tau}} = E[g(x, \tau, \tau + h, \omega)]_{x=X_{\tau}}$$

$$= E[f(X_{\tau+h}^{\tau,x})]_{x=X_{\tau}} = E[f(X_h^{0,x})]_{x=X_{\tau}} = E[f(F(x, 0, h, \omega))]_{x=X_{\tau}} \ .$$

$$\qquad\qquad\qquad\qquad\qquad\qquad\qquad\qquad\qquad\qquad\qquad\qquad\qquad\qquad \square$$

We now extend (7.2.2) to the following:

If $f_1, \cdots, f_k$ are bounded Borel functions on $\mathbf{R}^n$, τ an $\mathcal{F}_t^{(m)}$-stopping time, $\tau < \infty$ a.s. then

$$E^x[f_1(X_{\tau+h_1})f_2(X_{\tau+h_2}) \cdots f_k(X_{\tau+h_k})|\mathcal{F}_{\tau}^{(m)}] = E^{X_{\tau}}[f_1(X_{h_1}) \cdots f_k(X_{h_k})]$$
$$\tag{7.2.4}$$

for all $0 \le h_1 \le h_2 \le \cdots \le h_k$. This follows by induction: To illustrate the argument we prove it in the case $k = 2$:

$$E^x[f_1(X_{\tau+h_1})f_2(X_{\tau+h_2})|\mathcal{F}_{\tau}^{(m)}] = E^x[E^x[f_1(X_{\tau+h_1})f_2(X_{\tau+h_2})|\mathcal{F}_{\tau+h_1}]|\mathcal{F}_{\tau}^{(m)}]$$

$$= E^x[f_1(X_{\tau+h_1})E^x[f_2(X_{\tau+h_2})|\mathcal{F}_{\tau+h_1}]|\mathcal{F}_{\tau}^{(m)}]$$

$$= E^x[f_1(X_{\tau+h_1})E^{X_{\tau+h_1}}[f_2(X_{h_2-h_1})]|\mathcal{F}_{\tau}^{(m)}]$$

$$= E^{X_{\tau}}[f_1(X_{h_1})E^{X_{h_1}}[f_2(X_{h_2-h_1})]]$$

$$= E^{X_{\tau}}[f_1(X_{h_1})E^x[f_2(X_{h_2})|\mathcal{F}_{h_1}^{(m)}]] = E^{X_{\tau}}[f_1(X_{h_1})f_2(X_{h_2})] \ , \quad \text{as claimed} \ .$$

Next we proceed to formulate the general version we need: Let $\mathcal{H}$ be the set of all real $\mathcal{M}_\infty$-measurable functions. For $t \geq 0$ we define the *shift operator*

$$\theta_t \colon \mathcal{H} \to \mathcal{H}$$

as follows:

If $\eta = g_1(X_{t_1}) \cdots g_k(X_{t_k})$ (g_i Borel measurable, $t_i \geq 0$) we put

$$\theta_t \eta = g_1(X_{t_1+t}) \cdots g_k(X_{t_k+t}) \, .$$

Now extend in the natural way to all functions in $\mathcal{H}$ by taking limits of sums of such functions. Then it follows from (7.2.4) that

$$E^x[\theta_\tau \eta | \mathcal{F}_\tau^{(m)}] = E^{X_\tau}[\eta] \tag{7.2.5}$$

for all stopping times τ and all bounded $\eta \in \mathcal{H}$, where

$$(\theta_\tau \eta)(\omega) = (\theta_t \eta)(\omega) \qquad \text{if } \tau(\omega) = t \, .$$

Hitting distribution, harmonic measure and the mean value property

We will apply this to the following situation: Let $H \subset \mathbf{R}^n$ be measurable and let τ_H be the first exit time from H for an Itô diffusion X_t. Let α be another stopping time, g a bounded continuous function on $\mathbf{R}^n$ and put

$$\eta = g(X_{\tau_H}) \mathcal{X}_{\{\tau_H < \infty\}} \, , \qquad \tau_H^\alpha = \inf\{t > \alpha; X_t \notin H\} \, .$$

Then we have

$$\theta_\alpha \eta \cdot \mathcal{X}_{\{\alpha < \infty\}} = g(X_{\tau_H^\alpha}) \mathcal{X}_{\{\tau_H^\alpha < \infty\}} \, . \tag{7.2.6}$$

To prove (7.2.6) we approximate η by functions $\eta^{(k)}$; $k = 1, 2, \ldots$, of the form

$$\eta^{(k)} = \sum_j g(X_{t_j}) \mathcal{X}_{[t_j, t_{j+1})}(\tau_H) \, , \qquad t_j = j \cdot 2^{-k}, \ j = 0, 1, 2, \ldots$$

Now

$$\theta_t \mathcal{X}_{[t_j, t_{j+1})}(\tau_H) = \theta_t \mathcal{X}_{\{\forall r \in (0, t_j) X_r \in H \, \& \, \exists s \in [t_j, t_{j+1}) X_s \notin H\}}$$
$$= \mathcal{X}_{\{\forall r \in (0, t_j) X_{r+t} \in H \, \& \, \exists s \in [t_j, t_{j+1}) X_{s+t} \notin H\}}$$
$$= \mathcal{X}_{\{\forall u \in (t, t_j+t) X_u \in H \, \& \, \exists v \in [t_j+t, t_{j+1}+t) X_v \notin H\}} = \mathcal{X}_{[t_j+t, t_{j+1}+t)}(\tau_H^t) \, .$$

So we see that

$$\theta_t \eta = \lim_k \theta_t \eta^{(k)} = \lim_k \sum_j g(X_{t_j+t}) \mathcal{X}_{[t_j+t, t_{j+1}+t)}(\tau_H^t)$$

$$= g(X_{\tau_H^t}) \cdot \mathcal{X}_{\{\tau_H^t < \infty\}} \,, \quad \text{which is (7.2.6)} \,.$$

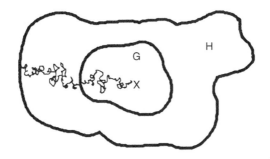

In particular, if $\alpha = \tau_G$ with $G \subset\subset H$ measurable, $\tau_H < \infty$ a.s. Q^x, then we have $\tau_H^\alpha = \tau_H$ and so

$$\theta_{\tau_G} g(X_{\tau_H}) = g(X_{\tau_H}) \,. \tag{7.2.7}$$

So if f is any bounded measurable function we obtain from (7.2.5) and (7.2.7):

$$E^x[f(X_{\tau_H})] = E^x[E^{X_{\tau_G}}[f(X_{\tau_H})]] = \int_{\partial G} E^y[f(X_{\tau_H})] \cdot Q^x[X_{\tau_G} \in dy] \tag{7.2.8}$$

for $x \in G$.

(Define $\mu_H^x(F) = Q^x(X_{\tau_H} \in F)$ and approximate f in $L^1(\mu_H^x)$ by continuous functions g satisfying (7.2.7)). In other words, the expected value of f at X_{τ_H} when starting at $x \in G$ can be obtained by integrating the expected value when starting at $y \in \partial G$ with respect to the *hitting distribution* ("harmonic measure") of X on ∂G. This can be restated as follows:

Define **the harmonic measure of X on ∂G**, μ_G^x, by

$$\mu_G^x(F) = Q^x[X_{\tau_G} \in F] \qquad \text{for } F \subset \partial G, \, x \in G \,.$$

Then the function

$$\phi(x) = E^x[f(X_{\tau_H})]$$

satisfies **the mean value property**:

$$\phi(x) = \int_{\partial G} \phi(y) d\mu_G^x(y) \,, \qquad \text{for all } x \in G \tag{7.2.9}$$

for all Borel sets $G \subset\subset H$.

This is an important ingredient in our solution of the generalized Dirichlet problem in Chapter 9.

7.3 The Generator of an Itô Diffusion

It is fundamental for many applications that we can associate a second order partial differential operator A to an Itô diffusion X_t. The basic connection between A and X_t is that A is the *generator* of the process X_t:

Definition 7.3.1 *Let $\{X_t\}$ be a (time-homogeneous) Itô diffusion in $\mathbf{R}^n$. The (infinitesimal) generator A of X_t is defined by*

$$Af(x) = \lim_{t \downarrow 0} \frac{E^x[f(X_t)] - f(x)}{t} \; ; \qquad x \in \mathbf{R}^n \; .$$

The set of functions $f: \mathbf{R}^n \to \mathbf{R}$ such that the limit exists at x is denoted by $\mathcal{D}_A(x)$, while $\mathcal{D}_A$ denotes the set of functions for which the limit exists for all $x \in \mathbf{R}^n$.

To find the relation between A and the coefficients b, σ in the stochastic differential equation (7.1.4) defining X_t we need the following result, which is useful in many connections:

Lemma 7.3.2 *Let $Y_t = Y_t^x$ be an Itô process in $\mathbf{R}^n$ of the form*

$$Y_t^x(\omega) = x + \int_0^t u(s, \omega)ds + \int_0^t v(s, \omega)dB_s(\omega)$$

where B is m-dimensional. Let $f \in C_0^2(\mathbf{R}^n)$, i.e. $f \in C^2(\mathbf{R}^n)$ and f has compact support, and let τ be a stopping time with respect to $\{\mathcal{F}_t^{(m)}\}$, and assume that $E^x[\tau] < \infty$. Assume that $u(t, \omega)$ and $v(t, \omega)$ are bounded on the set of (t, ω) such that $Y(t, \omega)$ belongs to the support of f. Then

$$E^x[f(Y_\tau)] =$$

$$f(x) + E^x\left[\int_0^\tau \left(\sum_i u_i(s, \omega) \frac{\partial f}{\partial x_i}(Y_s) + \frac{1}{2} \sum_{i,j} (vv^T)_{i,j}(s, \omega) \frac{\partial^2 f}{\partial x_i \partial x_j}(Y_s) \right) ds \right],$$

where E^x is the expectation w.r.t. the natural probability law R^x for Y_t starting at x:

$$R^x[Y_{t_1} \in F_1, \dots, Y_{t_k} \in F_k] = P^0[Y_{t_1}^x \in F_1, \dots, Y_{t_k}^x \in F_k] \;, \qquad F_i \;\; Borel \; sets \; .$$

Proof. Put $Z = f(Y)$ and apply Itô's formula (To simplify the notation we suppress the index t and let $Y_1, \dots, Y_n$ and $B_1, \dots, B_m$ denote the coordinates of Y and B, respectively)

$$dZ = \sum_i \frac{\partial f}{\partial x_i}(Y)dY_i + \frac{1}{2}\sum_{i,j}\frac{\partial^2 f}{\partial x_i \partial x_j}(Y)dY_i dY_j$$

$$= \sum_i u_i \frac{\partial f}{\partial x_i}dt + \frac{1}{2}\sum_{i,j}\frac{\partial^2 f}{\partial x_i \partial x_j}(vdB)_i(vdB)_j + \sum_i \frac{\partial f}{\partial x_i}(vdB)_i .$$

Since

$$(vdB)_i \cdot (vdB)_j = \left(\sum_k v_{ik}dB_k\right)\left(\sum_n v_{jn}dB_n\right)$$

$$= \left(\sum_k v_{ik}v_{jk}\right)dt = (vv^T)_{ij}dt ,$$

this gives

$$f(Y_t) = f(Y_0) + \int_0^t \left(\sum_i u_i \frac{\partial f}{\partial x_i} + \frac{1}{2}\sum_{i,j}(vv^T)_{ij}\frac{\partial^2 f}{\partial x_i \partial x_j}\right)ds$$

$$+ \sum_{i,k}\int_0^t v_{ik}\frac{\partial f}{\partial x_i}dB_k . \tag{7.3.1}$$

Hence

$$E^x[f(Y_\tau)] = f(x) + E^x\left[\int_0^\tau \left(\sum_i u_i \frac{\partial f}{\partial x_i}(Y) + \frac{1}{2}\sum_{i,j}(vv^T)_{i,j}\frac{\partial^2 f}{\partial x_i \partial x_j}(Y)\right)ds\right]$$

$$+ \sum_{i,k}E^x\left[\int_0^\tau v_{ik}\frac{\partial f}{\partial x_i}(Y)dB_k\right] . \tag{7.3.2}$$

If g is a bounded Borel function, $|g| \leq M$ say, then for all integers k we have

$$E^x\left[\int_0^{\tau \wedge k} g(Y_s)dB_s\right] = E^x\left[\int_0^k \mathcal{X}_{\{s<\tau\}}g(Y_s)dB_s\right] = 0 ,$$

since $g(Y_s)$ and $\mathcal{X}_{\{s<\tau\}}$ are both $\mathcal{F}_s^{(m)}$-measurable. Moreover

$$E^x\left[\left(\int_0^\tau g(Y_s)dB_s - \int_0^{\tau \wedge k} g(Y_s)dB_s\right)^2\right] = E^x\left[\int_{\tau \wedge k}^\tau g^2(Y_s)ds\right]$$

$$\leq M^2 E^x[\tau - \tau \wedge k] \to 0 .$$

Therefore

$$0 = \lim_{k \to \infty} E^x \left[\int_0^{\tau \wedge k} g(Y_s) dB_s \right] = E^x [\int_0^\tau g(Y_s) dB_s] .$$

Combining this with (7.3.2) we get Lemma 7.3.2.   □

This gives immediately the formula for the generator A of an Itô diffusion:

Theorem 7.3.3 *Let* X_t *be the Itô diffusion*

$$dX_t = b(X_t)dt + \sigma(X_t)dB_t .$$

If $f \in C_0^2(\mathbf{R}^n)$ *then* $f \in \mathcal{D}_A$ *and*

$$Af(x) = \sum_i b_i(x)\frac{\partial f}{\partial x_i} + \frac{1}{2}\sum_{i,j}(\sigma\sigma^T)_{i,j}(x)\frac{\partial^2 f}{\partial x_i \partial x_j} . \qquad (7.3.3)$$

Proof. This follows from Lemma 7.3.2 (with $\tau = t$) and the definition of A.

□

Example 7.3.4 The n-dimensional Brownian motion is of course the solution of the stochastic differential equation

$$dX_t = dB_t ,$$

i.e. we have $b = 0$ and $\sigma = I_n$, the n-dimensional identity matrix. So the generator of B_t is

$$Af = \frac{1}{2}\sum_i \frac{\partial^2 f}{\partial x_i^2} ; \qquad f = f(x_1, \ldots, x_n) \in C_0^2(\mathbf{R}^n)$$

i.e. $A = \frac{1}{2}\Delta$, where Δ is the Laplace operator.

Example 7.3.5 (The graph of Brownian motion) Let B denote 1-dimensional Brownian motion and let $X = \begin{pmatrix} X_1 \\ X_2 \end{pmatrix}$ be the solution of the stochastic differential equation

$$\begin{cases} dX_1 = dt ; & X_1(0) = t_0 \\ dX_2 = dB; & X_2(0) = x_0 \end{cases}$$

i.e.

$$dX = bdt + \sigma dB ; \qquad X(0) = \begin{pmatrix} t_0 \\ X_0 \end{pmatrix} ,$$

with $b = \begin{pmatrix} 1 \\ 0 \end{pmatrix}$ and $\sigma = \begin{pmatrix} 0 \\ 1 \end{pmatrix}$. In other words, X may be regarded as the graph of Brownian motion. The generator A of X is given by the *heat operator*

$$Af = \frac{\partial f}{\partial t} + \frac{1}{2}\frac{\partial^2 f}{\partial x^2} \; ; \qquad f = f(t, x) \in C_0^2(\mathbf{R}^n) \; .$$

From now on we will, unless otherwise stated, let $A = A_X$ denote the generator of the Itô diffusion X_t. We let $L = L_X$ denote the differential operator given by the right hand side of (7.3.3). From Theorem 7.3.3 we know that A_X and L_X coincide on $C_0^2(\mathbf{R}^n)$.

7.4 The Dynkin Formula

If we combine (7.3.2) and (7.3.3) we get:

Theorem 7.4.1 (Dynkin's formula)
Let $f \in C_0^2(\mathbf{R}^n)$. *Suppose* τ *is a stopping time,* $E^x[\tau] < \infty$. *Then*

$$E^x[f(X_\tau)] = f(x) + E^x\left[\int_0^\tau Af(X_s)ds\right] . \tag{7.4.1}$$

Remarks.

(i)   Note that if τ is the first exit time of a bounded set, $E^x[\tau] < \infty$, then (7.4.1) holds for any function $f \in C^2$.

(ii) For a more general version of Theorem 7.4.1 see Dynkin (1965 I), p. 133.

Example 7.4.2 Consider n-dimensional Brownian motion $B = (B_1, \ldots, B_n)$ starting at $a = (a_1, \ldots, a_n) \in \mathbf{R}^n (n \geq 1)$ and assume $|a| < R$. What is the expected value of the first exit time τ_K of B from the ball

$$K = K_R = \{x \in \mathbf{R}^n; |x| < R\} \; ?$$

Choose an integer k and apply Dynkin's formula with $X = B$, $\tau = \sigma_k = \min(k, \tau_K)$, and $f \in C_0^2$ such that $f(x) = |x|^2$ for $|x| \leq R$:

$$E^a[f(B_{\sigma_k})] = f(a) + E^a\left[\int_0^{\sigma_k} \frac{1}{2}\Delta f(B_s)ds\right]$$

$$= |a|^2 + E^a\left[\int_0^{\sigma_k} n \cdot ds\right] = |a|^2 + n \cdot E^a[\sigma_k] \; .$$

Hence $E^a[\sigma_k] \leq \frac{1}{n}(R^2 - |a|^2)$ for all k. So letting $k \to \infty$ we conclude that $\tau_K = \lim \sigma_k < \infty$ a.s. and

$$E^a[\tau_K] = \frac{1}{n}(R^2 - |a|^2) . \tag{7.4.2}$$

Next we assume that $n \geq 2$ and $|b| > R$. What is the probability that B starting at b ever hits K?

Let α_k be the first exit time from the annulus

$$A_k = \{x; R < |x| < 2^k R\} ; \qquad k = 1, 2, \ldots$$

and put

$$T_K = \inf\{t > 0; B_t \in K\} .$$

Let $f = f_{n,k}$ be a C^2 function with compact support such that, if $R \leq |x| \leq 2^k R$,

$$f(x) = \begin{cases} -\log |x| & \text{when } n = 2 \\ |x|^{2-n} & \text{when } n > 2 . \end{cases}$$

Then, since $\Delta f = 0$ in A_k, we have by Dynkin's formula

$$E^b[f(B_{\alpha_k})] = f(b) \quad \text{for all } k . \tag{7.4.3}$$

Put

$$p_k = P^b[|B_{\alpha_k}| = R] , \quad q_k = P^b[|B_{\alpha_k}| = 2^k R] .$$

Let us now consider the two cases $n = 2$ and $n > 2$ separately:

$\boldsymbol{n = 2.}$ Then we get from (7.4.3)

$$-\log R \cdot p_k - (\log R + k \cdot \log 2)q_k = -\log |b| \qquad \text{for all } k . \tag{7.4.4}$$

This implies that $q_k \to 0$ as $k \to \infty$, so that

$$P^b[T_K < \infty] = 1 , \tag{7.4.5}$$

i.e. Brownian motion is *recurrent* in $\mathbf{R}^2$. (See Port and Stone (1979)).

$\boldsymbol{n > 2.}$ In this case (7.4.3) gives

$$p_k \cdot R^{2-n} + q_k \cdot (2^k R)^{2-n} = |b|^{2-n} .$$

Since $0 \leq q_k \leq 1$ we get by letting $k \to \infty$

$$\lim_{k \to \infty} p_k = P^b[T_K < \infty] = \left(\frac{|b|}{R}\right)^{2-n} ,$$

i.e. Brownian motion is *transient* in $\mathbf{R}^n$ for $n > 2$.

7.5 The Characteristic Operator

We now introduce an operator which is closely related to the generator A, but is more suitable in many situations, for example in the solution of the Dirichlet problem.

Definition 7.5.1 *Let $\{X_t\}$ be an Itô diffusion. The* characteristic operator *$\mathcal{A} = \mathcal{A}_X$ of $\{X_t\}$ is defined by*

$$\mathcal{A}f(x) = \lim_{U \downarrow x} \frac{E^x[f(X_{\tau_U})] - f(x)}{E^x[\tau_U]} , \qquad (7.5.1)$$

where the $U's$ are open sets U_k decreasing to the point x, in the sense that $U_{k+1} \subset U_k$ and $\bigcap_k U_k = \{x\}$, and $\tau_U = \inf\{t > 0; X_t \notin U\}$ is the first exit time from U for X_t. The set of functions f such that the limit (7.5.1) exists for all $x \in \mathbf{R}^n$ (and all $\{U_k\}$) is denoted by $\mathcal{D}_\mathcal{A}$. If $E^x[\tau_U] = \infty$ for all open $U \ni x$, we define $\mathcal{A}f(x) = 0$.

It turns out that $\mathcal{D}_A \subseteq \mathcal{D}_\mathcal{A}$ always and that

$$Af = \mathcal{A}f \qquad \text{for all } f \in \mathcal{D}_A .$$

(See Dynkin (1965 I, p. 143).)

We will only need that $\mathcal{A}_X$ and L_X coincide on C^2. To obtain this we first clarify a property of exit times.

Definition 7.5.2 *A point $x \in \mathbf{R}^n$ is called a* trap *for $\{X_t\}$ if*

$$Q^x(\{X_t = x \ for \ all \ t\}) = 1 .$$

In other words, x is trap if and only if $\tau_{\{x\}} = \infty$ a.s. Q^x. For example, if $b(x_0) = \sigma(x_0) = 0$, then x_0 is a trap for X_t (by strong uniqueness of X_t).

Lemma 7.5.3 *If x is not a trap for X_t, then there exists an open set $U \ni x$ such that*

$$E^x[\tau_U] < \infty .$$

Proof. See Lemma 5.5 p. 139 in Dynkin (1965 I).

Theorem 7.5.4 *Let $f \in C^2$. Then $f \in \mathcal{D}_\mathcal{A}$ and*

$$\mathcal{A}f = \sum_i b_i \frac{\partial f}{\partial x_i} + \frac{1}{2} \sum_{i,j} (\sigma \sigma^T)_{ij} \frac{\partial^2 f}{\partial x_i \partial x_j} . \qquad (7.5.2)$$

Proof. As before we let L denote the operator defined by the right hand side of (7.5.2). If x is a trap for $\{X_t\}$ then $\mathcal{A}f(x) = 0$. Choose a bounded open set V such that $x \in V$. Modify f to f_0 outside V such that $f_0 \in C_0^2(\mathbf{R}^n)$. Then

$f_0 \in \mathcal{D}_A(x)$ and $0 = Af_0(x) = Lf_0(x) = Lf(x)$. Hence $\mathcal{A}f(x) = Lf(x) = 0$ in this case. If x is not a trap, choose a bounded open set $U \ni x$ such that $E^x[\tau_U] < \infty$. Then by Dynkin's formula (Theorem 7.4.1) (and the following Remark (i)), writing $\tau_U = \tau$

$$\left| \frac{E^x[f(X_\tau)] - f(x)}{E^x[\tau]} - Lf(x) \right| = \frac{|E^x[\int_0^\tau \{(Lf)(X_s) - Lf(x)\}ds]|}{E^x[\tau]}$$

$$\leq \sup_{y \in U} |Lf(x) - Lf(y)| \to 0 \qquad \text{as } U \downarrow x,$$

since Lf is a continuous function. $\square$

Remark. We have now obtained that an Itô diffusion is a continuous, strong Markov process such that the domain of definition of its characteristic operator includes C^2. Thus an Itô diffusion is a *diffusion* in the sense of Dynkin (1965 I).

Example 7.5.5 (Brownian motion on the unit circle) The characteristic operator of the process $Y = \begin{pmatrix} Y_1 \\ Y_2 \end{pmatrix}$ from Example 5.1.4 satisfying the stochastic differential equations (5.1.13), i.e.

$$\begin{cases} dY_1 = -\frac{1}{2}Y_1 dt - Y_2 dB \\ \\ dY_2 = -\frac{1}{2}Y_2 dt + Y_1 dB \end{cases}$$

is

$$\mathcal{A}f(y_1, y_2) = \frac{1}{2}\left[y_2^2 \frac{\partial^2 f}{\partial y_1^2} - 2y_1 y_2 \frac{\partial^2 f}{\partial y_1 \partial y_2} + y_1^2 \frac{\partial^2 f}{\partial y_2^2} - y_1 \frac{\partial f}{\partial y_1} - y_2 \frac{\partial f}{\partial y_2} \right].$$

This is because $dY = -\frac{1}{2}Y dt + KY dB$, where

$$K = \begin{pmatrix} 0 & -1 \\ 1 & 0 \end{pmatrix}$$

so that

$$dY = b(Y)dt + \sigma(Y)dB$$

with

$$b(y_1, y_2) = \begin{pmatrix} -\frac{1}{2}y_1 \\ -\frac{1}{2}y_2 \end{pmatrix}, \qquad \sigma(y_1, y_2) = \begin{pmatrix} -y_2 \\ y_1 \end{pmatrix}$$

and

$$a = \frac{1}{2}\sigma\sigma^T = \frac{1}{2}\begin{pmatrix} y_2^2 & -y_1 y_2 \\ -y_1 y_2 & y_1^2 \end{pmatrix}.$$

Example 7.5.6 Let D be an open subset of $\mathbf{R}^n$ such that $\tau_D < \infty$ a.s. Q^x for all x. Let ϕ be a bounded, measurable function on ∂D and define

$$\widetilde{\phi}(x) = E^x[\phi(X_{\tau_D})]$$

($\widetilde{\phi}$ is called the *X-harmonic extension of* ϕ). Then if U is open, $x \in U \subset\subset D$, we have by (7.2.8) that

$$E^x[\widetilde{\phi}(X_{\tau_U})] = E^x[E^{X_{\tau_U}}[\phi(X_{\tau_D})]] = E^x[\phi(X_{\tau_D})] = \widetilde{\phi}(x) \ .$$

So $\widetilde{\phi} \in \mathcal{D}_A$ and

$$\mathcal{A}\widetilde{\phi} = 0 \quad \text{in } D \ ,$$

in spite of the fact that in general $\widetilde{\phi}$ need not even be continuous in D (See Example 9.2.1).

Exercises

7.1.* Find the generator of the following Itô diffusions:

a) $dX_t = \mu X_t dt + \sigma dB_t$ (The Ornstein-Uhlenbeck process) ($B_t \in \mathbf{R}$; μ, σ constants).

b) $dX_t = rX_t dt + \alpha X_t dB_t$ (The geometric Brownian motion) ($B_t \in \mathbf{R}$; r, α constants).

c) $dY_t = r\, dt + \alpha Y_t dB_t$ ($B_t \in \mathbf{R}$; r, α constants)

d) $dY_t = \begin{bmatrix} dt \\ dX_t \end{bmatrix}$ where X_t is as in a)

e) $\begin{bmatrix} dX_1 \\ dX_2 \end{bmatrix} = \begin{bmatrix} 1 \\ X_2 \end{bmatrix} dt + \begin{bmatrix} 0 \\ e^{X_1} \end{bmatrix} dB_t$ ($B_t \in \mathbf{R}$)

f) $\begin{bmatrix} dX_1 \\ dX_2 \end{bmatrix} = \begin{bmatrix} 1 \\ 0 \end{bmatrix} dt + \begin{bmatrix} 1 & 0 \\ 0 & X_1 \end{bmatrix} \begin{bmatrix} dB_1 \\ dB_2 \end{bmatrix}$

g) $X(t) = (X_1, X_2, \cdots, X_n)$, where

$$dX_k(t) = r_k X_k dt + X_k \cdot \sum_{j=1}^{n} \alpha_{kj} dB_j \ ; \qquad 1 \le k \le n$$

$((B_1, \cdots, B_n)$ is Brownian motion in $\mathbf{R}^n$, r_k and α_{kj} are constants).

7.2.* Find an Itô diffusion (i.e. write down the stochastic differential equation for it) whose generator is the following:

a) $Af(x) = f'(x) + f''(x)$; $f \in C_0^2(\mathbf{R})$

b) $Af(t,x) = \frac{\partial f}{\partial t} + cx\frac{\partial f}{\partial x} + \frac{1}{2}\alpha^2 x^2 \frac{\partial^2 f}{\partial x^2}$; $f \in C_0^2(\mathbf{R}^2)$,
where c, α are constants.

c) $Af(x_1,x_2) = 2x_2\frac{\partial f}{\partial x_1} + \ln(1 + x_1^2 + x_2^2)\frac{\partial f}{\partial x_2}$
$+ \frac{1}{2}(1 + x_1^2)\frac{\partial^2 f}{\partial x_1^2} + x_1\frac{\partial^2 f}{\partial x_1 \partial x_2} + \frac{1}{2}\cdot\frac{\partial^2 f}{\partial x_2^2}$; $f \in C_0^2(\mathbf{R}^2)$.

7.3. Let B_t be Brownian motion on $\mathbf{R}, B_0 = 0$ and define

$$X_t = X_t^x = x \cdot e^{ct+\alpha B_t} ,$$

where c, α are constants. Prove directly from the definition that X_t is a Markov process.

7.4.* Let B_t^x be 1-dimensional Brownian motion starting at $x \in \mathbf{R}^+$. Put

$$\tau = \inf\{t > 0; B_t^x = 0\} .$$

a)   Prove that $\tau < \infty$ a.s. P^x for all $x > 0$. (Hint: See Example 7.4.2, second part).

b)   Prove that $E^x[\tau] = \infty$ for all $x > 0$. (Hint: See Example 7.4.2, first part).

7.5. Let the functions b, σ satisfy condition (5.2.1) of Theorem 5.2.1, with a constant C independent of t, i.e.

$$|b(t,x)| + |\sigma(t,x)| \le C(1 + |x|) \qquad \text{for all } x \in \mathbf{R}^n \text{ and all } t \ge 0 .$$

Let X_t be a solution of

$$dX_t = b(t, X_t)dt + \sigma(t, X_t)dB_t .$$

Show that

$$E[|X_t|^2] \le (1 + E[|X_0|^2])e^{Kt} - 1$$

for some constant K independent of t.
(Hint: Use Dynkin's formula with $f(x) = |x|^2$ and $\tau = t \wedge \tau_R$, where $\tau_R = \inf\{t > 0; |X_t| \ge R\}$, and let $R \to \infty$ to achieve the inequality

$$E[|X_t|^2] \le E[|X_0|^2] + K \cdot \int_0^t (1 + E[|X_s|^2])ds ,$$

which is of the form (5.2.9).)

7.6. Let $g(x,\omega) = f \circ F(x, t, t+h, \omega)$ be as in the proof of Theorem 7.1.2. Assume that f is continuous.

a)   Prove that the map $x \to g(x, \cdot)$ is continuous from $\mathbf{R}^n$ into $L^2(P)$ by using (5.2.9).

For simplicity assume that $n = 1$ in the following.

b)   Use a) to prove that $(x, \omega) \to g(x, \omega)$ is measurable. (Hint: For each $m = 1, 2, \ldots$ put $\xi_k = \xi_k^{(m)} = k \cdot 2^{-m}$, $k = 1, 2, \ldots$ Then

$$g^{(m)}(x, \cdot) := \sum_k g(\xi_k, \cdot) \cdot \mathcal{X}_{\{\xi_k \leq x < \xi_{k+1}\}}$$

converges to $g(x, \cdot)$ in $L^2(P)$ for each x. Deduce that $g^{(m)} \to g$ in $L^2(dm_R \times dP)$ for all R, where dm_R is Lebesgue measure on $\{|x| \leq R\}$. So a subsequence of $g^{(m)}(x, \omega)$ converges to $g(x, \omega)$ for a.a. (x, ω).)

7.7.   Let B_t be Brownian motion on $\mathbf{R}^n$ starting at $x \in \mathbf{R}^n$ and let $D \subset \mathbf{R}^n$ be an open ball centered at x.

a)   Use Exercise 2.15 to prove that the harmonic measure μ_D^x of B_t is rotation invariant (about x) on the sphere ∂D. Conclude that μ_D^x coincides with normalized surface measure σ on ∂D.

b)   Let ϕ be a bounded measurable function on a bounded open set $W \subset \mathbf{R}^n$ and define

$$u(x) = E^x[\phi(B_{\tau_W})] \qquad \text{for } x \in W .$$

Prove that u satisfies the classical mean value property:

$$u(x) = \int_{\partial D} u(y) d\sigma(y) \tag{7.5.3}$$

for all balls D centered at x with $\overline{D} \subset W$.

c)   Let W be as in b) and let $w : W \to \mathbf{R}$ be *harmonic* in W, i.e.

$$\Delta w := \sum_{i=1}^n \frac{\partial^2 w}{\partial x_i^2} = 0 \qquad \text{in } W. \tag{7.5.4}$$

Prove that w satisfies the classical mean value property (7.5.3).

Remark. For a converse of this see e.g. Øksendal and Stroock (1982) and the references therein.

7.8.   Let $\{\mathcal{N}_t\}$ be a right-continuous family of σ-algebras of subsets of Ω, containing all sets of measure zero.

    a)   Let τ_1, τ_2 be stopping times (w.r.t. $\mathcal{N}_t$). Prove that $\tau_1 \wedge \tau_2$ and $\tau_1 \vee \tau_2$ are stopping times.

    b)   If $\{\tau_n\}$ is a decreasing family of stopping times prove that $\tau := \lim_n \tau_n$ is a stopping time.

    c)   If X_t is an Itô diffusion in $\mathbf{R}^n$ and $F \subset \mathbf{R}^n$ is closed, prove that τ_F is a stopping time w.r.t. $\mathcal{M}_t$. (Hint: Consider open sets decreasing to F).

7.9.   Let X_t be a geometric Brownian motion, i.e.

$$dX_t = rX_t dt + \alpha X_t dB_t , \qquad X_0 = x > 0$$

where $B_t \in \mathbf{R}$; r, α are constants.

    a)   Find the generator A of X_t and compute $Af(x)$ when $f(x) = x^\gamma$; $x > 0$, γ constant.

    b)   If $r < \frac{1}{2}\alpha^2$ then $X_t \to 0$ as $t \to \infty$, a.s. Q^x (Example 5.1.1). But what is the probability p that X_t, when starting from $x < R$, ever hits the value R ? Use Dynkin's formula with $f(x) = x^{\gamma_1}$, $\gamma_1 = 1 - \frac{2r}{\alpha^2}$, to prove that

$$p = \left(\frac{x}{R}\right)^{\gamma_1} .$$

    c)   If $r > \frac{1}{2}\alpha^2$ then $X_t \to \infty$ as $t \to \infty$, a.s. Q^x. Put

$$\tau = \inf\{t > 0; X_t \geq R\} .$$

Use Dynkin's formula with $f(x) = \ln x$, $x > 0$ to prove that

$$E^x[\tau] = \frac{\ln \frac{R}{x}}{r - \frac{1}{2}\alpha^2} .$$

(Hint: First consider exit times from (ρ, R), $\rho > 0$ and then let $\rho \to 0$. You need estimates for

$$(1 - p(\rho)) \ln \rho ,$$

where

$$p(\rho) = Q^x[X_t \text{ reaches the value } R \text{ before } \rho] ,$$

which you can get from the calculations in a), b).)

7.10.   Let X_t be the geometric Brownian motion

$$dX_t = rX_t dt + \alpha X_t dB_t .$$

Find $E^x[X_T|\mathcal{F}_t]$ for $t \leq T$ by
a) using the Markov property
and
b) writing $X_t = x\, e^{rt} M_t$, where

$$M_t = \exp(\alpha B_t - \tfrac{1}{2}\alpha^2 t) \qquad \text{is a martingale .}$$

7.11.   Let X_t be an Itô diffusion in $\mathbf{R}^n$ and let $f: \mathbf{R}^n \to \mathbf{R}$ be a function
such that

$$E^x\left[\int_0^\infty |f(X_t)| dt \right] < \infty \qquad \text{for all } x \in \mathbf{R}^n .$$

Let τ be a stopping time. Use the strong Markov property to prove
that

$$E^x\left[\int_\tau^\infty f(X_t) dt \right] = E^x[g(X_\tau)] ,$$

where

$$g(y) = E^y\left[\int_0^\infty f(X_t) dt \right] .$$

7.12.   **(Local martingales)**
An $\mathcal{N}_t$-adapted stochastic process $Z(t) \in \mathbf{R}^n$ is called a *local martingale* with respect to the given filtration $\{\mathcal{N}_t\}$ if there exists an
increasing sequence of $\mathcal{N}_t$-stopping times τ_k such that

$$\tau_k \to \infty \quad \text{a.s. as } k \to \infty$$

and

$$Z(t \wedge \tau_k) \qquad \text{is an } \mathcal{N}_t\text{-martingale for all } k .$$

a)   Show that if $Z(t)$ is a local martingale and there exists a constant
$T \leq \infty$ such that the family $\{Z(\tau)\}_{\tau \leq T}$ is uniformly integrable
(Appendix C) then $\{Z(t)\}_{t \leq T}$ is a martingale.

b)   In particular, if $Z(t)$ is a local martingale and there exists a
constant $K < \infty$ such that

$$E[Z^2(\tau)] \leq K$$

for all stopping times $\tau \leq T$, then $\{Z(t)\}_{t \leq T}$ is a martingale.

c)   Show that if $Z(t)$ is a *lower bounded* local martingale, then $Z(t)$
is a supermartingale (Appendix C).

d)   Let $\phi \in \mathcal{W}(0, T)$. Show that

$$Z(t) := \int_0^t \phi(s, w) dB(s) ; \qquad 0 \le t \le T$$

is a local martingale.

7.13.   a)   Let $B_t \in \mathbf{R}^2$, $B_0 = x \ne 0$. Fix $0 < \epsilon < R < \infty$ and define

$$X_t = \ln |B_{t \wedge \tau}| ; \qquad t \ge 0$$

where
$$\tau = \inf \{ t > 0; |B_t| \le \epsilon \quad \text{or} \quad |B_t| \ge R \} .$$

Prove that X_t is an $\mathcal{F}_{t \wedge \tau}$-martingale. (Hint: Use Exercise 4.8.)
Deduce that $\ln |B_t|$ is a local martingale (Exercise 7.12).

b)   Let $B_t \in \mathbf{R}^n$ for $n \ge 3$, $B_0 = x \ne 0$. Fix $\epsilon > 0$, $R < \infty$ and
define
$$Y_t = |B_{t \wedge \tau}|^{2-n} ; \qquad t \ge 0$$

where
$$\tau = \inf \{ t > 0; |B_t| \le \epsilon \quad \text{or} \quad |B_t| \ge R \} .$$

Prove that Y_t is an $\mathcal{F}_{t \wedge \tau}$-martingale.
Deduce that $|B_t|^{2-n}$ is a local martingale.

7.14.   **(Doob's h-transform)**
Let B_t be n-dimensional Brownian motion, $D \subset \mathbf{R}^n$ a bounded open
set and $h > 0$ a harmonic function on D (i.e. $\Delta h = 0$ in D). Let X_t
be the solution of the stochastic differential equation

$$dX_t = \nabla (\ln h)(X_t) dt + dB_t$$

More precisely, choose an increasing sequence $\{D_k\}$ of open subsets of
D such that $\overline{D}_k \subset D$ and $\bigcup_{k=1}^{\infty} D_k = D$. Then for each k the equation
above can be solved (strongly) for $t < \tau_{D_k}$. This gives in a natural
way a solution for $t < \tau := \lim_{k \to \infty} \tau_{D_k}$.

a)   Show that the generator A of X_t satisfies

$$Af = \frac{\Delta(hf)}{2h} \qquad \text{for } f \in C_0^2(D) .$$

In particular, if $f = \frac{1}{h}$ then $\mathcal{A}f = 0$.

b)   Use a) to show that if there exists $x_0 \in \partial D$ such that

$$\lim_{x \to y \in \partial D} h(x) = \begin{cases} 0 & \text{if } y \ne x_0 \\ \infty & \text{if } y = x_0 \end{cases}$$

(i.e. h is a *kernel function*), then

$$\lim_{t \to \tau} X_t = x_0 \text{ a.s.}$$

(Hint: Consider $E^x[f(X_T)]$ for suitable stopping times T and with $f = \frac{1}{h}$)

In other words, we have imposed a drift on B_t which causes the process to exit from D at the point x_0 only. This can also be formulated as follows: X_t is obtained by *conditioning B_t to exit from D at x_0*. See Doob (1984).

7.15.* Let B_t be 1-dimensional and define

$$F(\omega) = (B_T(\omega) - K)^+$$

where $K > 0$, $T > 0$ are constants.
By the Itô representation theorem (Theorem 4.3.3) we know that there exists $\phi \in \mathcal{V}(0, T)$ such that

$$F(\omega) = E[F] + \int_0^T \phi(t, \omega) dB_t \ .$$

How do we find ϕ explicitly? This problem is of interest in mathematical finance, where ϕ may be regarded as the replicating portfolio for the contingent claim F (see Chapter 12). Using the Clark-Ocone formula (see Karatzas and Ocone (1991), Øksendal (1996) or Aase et al (2000)) one can deduce that

$$\phi(t, \omega) = E[\mathcal{X}_{[K,\infty)}(B_T)|\mathcal{F}_t] \ ; \qquad t < T \ . \tag{7.5.5}$$

Use (7.5.5) and the Markov property of Brownian motion to prove that for $t < T$ we have

$$\phi(t, \omega) = \frac{1}{\sqrt{2\pi(T-t)}} \int_K^\infty \exp\left(-\frac{(x - B_t(\omega))^2}{2(T-t)}\right) dx \ . \tag{7.5.6}$$

7.16. Let B_t be 1-dimensional and let $f \colon \mathbf{R} \to \mathbf{R}$ be a bounded function. Prove that if $t < T$ then

$$E^x[f(B_T)|\mathcal{F}_t] = \frac{1}{\sqrt{2\pi(T-t)}} \int_{\mathbf{R}} f(x) \exp\left(-\frac{(x - B_t(\omega))^2}{2(T-t)}\right) dx \ .$$

$$\tag{7.5.7}$$

(Compare with (7.5.6).)

7.17. Let B_t be 1-dimensional and put

$$X_t = (x^{1/3} + \tfrac{1}{3}B_t)^3 \ ; \qquad t \geq 0 \ .$$

Then we have seen in Exercise 4.15 that X_t is a solution of the stochastic differential equation

$$dX_t = \tfrac{1}{3}X_t^{1/3}dt + X_t^{2/3}dB_t ; \qquad X_0 = x . \qquad (7.5.8)$$

Define

$$\tau = \inf\{t > 0; X_t = 0\}$$

and put

$$Y_t = \begin{cases} X_t & \text{for } t \le \tau \\ 0 & \text{for } t > \tau . \end{cases}$$

Prove that Y_t is also a (strong) solution of (7.5.8). Why does not this contradict the uniqueness assertion of Theorem 5.2.1?
(Hint: Verify that

$$Y_t = x + \int_0^t \tfrac{1}{3}Y_s^{1/3}ds + \int_0^t Y_s^{2/3}dB_s$$

for all t by splitting the integrals as follows:

$$\int_0^t = \int_0^{t\wedge\tau} + \int_{t\wedge\tau}^t \ . \ )$$

7.18.*   a)   Let

$$dX_t = b(X_t)dt + \sigma(X_t)dB_t ; \qquad X_0 = x$$

be a 1-dimensional Itô diffusion with characteristic operator $\mathcal{A}$. Let $f \in C^2(\mathbf{R})$ be a solution of the differential equation

$$\mathcal{A}f(x) = b(x)f'(x) + \tfrac{1}{2}\sigma^2(x)f''(x) = 0 ; \qquad x \in \mathbf{R} . \qquad (7.5.9)$$

Let $(a,b) \subset \mathbf{R}$ be an open interval such that $x \in (a,b)$ and put

$$\tau = \inf\{t > 0; X_t \notin (a,b)\} .$$

Assume that $\tau < \infty$ a.s. Q^x and define

$$p = P^x[X_\tau = b] .$$

Use Dynkin's formula to prove that

$$p = \frac{f(x) - f(a)}{f(b) - f(a)} . \qquad (7.5.10)$$

In other words, the *harmonic measure* $\mu_{(a,b)}^x$ of X on $\partial(a,b) = \{a,b\}$ is given by

$$\mu^x_{(a,b)}(b) = \frac{f(x) - f(a)}{f(b) - f(a)}, \quad \mu^x_{(a,b)}(a) = \frac{f(b) - f(x)}{f(b) - f(a)} . \quad (7.5.11)$$

b)   Now specialize to the process

$$X_t = x + B_t ; \qquad t \geq 0 .$$

Prove that

$$p = \frac{x - a}{b - a} . \qquad (7.5.12)$$

c)   Find p if

$$X_t = x + \mu t + \sigma B_t ; \qquad t \geq 0$$

where $\mu, \sigma \in \mathbf{R}$ are nonzero constants.

7.19.   Let B^x_t be 1-dimensional Brownian motion starting at $x > 0$. Define

$$\tau = \tau(x, \omega) = \inf\{t > 0; \, B^x_t(\omega) = 0\} .$$

From Exercise 7.4 we know that

$$\tau < \infty \quad \text{a.s. } P^x \text{ and } E^x[\tau] = \infty .$$

What is the distribution of the random variable $\tau(\omega)$?

a)   To answer this, first find the *Laplace transform*

$$g(\lambda) := E^x[e^{-\lambda \tau}] \qquad \text{for } \lambda > 0 .$$

(Hint: Let $M_t = \exp(-\sqrt{2\lambda}\, B_t - \lambda t)$. Then

$$\{M_{t \wedge \tau}\}_{t \geq 0} \quad \text{is a bounded martingale. })$$

[Solution: $g(\lambda) = \exp(-\sqrt{2\lambda}\, x) .$]

b)   To find the density $f(t)$ of τ it suffices to find $f(t) = f(t, x)$ such that

$$\int\limits_0^\infty e^{-\lambda t} f(t) dt = \exp(-\sqrt{2\lambda}\, x) \qquad \text{for all } \lambda > 0$$

i.e. to find the *inverse* Laplace transform of $g(\lambda)$. Verify that

$$f(t, x) = \frac{x}{\sqrt{2\pi t^3}} \exp\left(-\frac{x^2}{2t}\right); \qquad t > 0 .$$

7.20.   **(Population growth in a stochastic, crowded environment (II))**
As an alternative to the model in Exercise 5.15 consider the equation

$$dX_t = r\, X_t(K - X_t)dt + \alpha\, X_t(K - X_t)dB_t \ ; \qquad X_0 = x \geq 0.$$

This equation does not satisfy the conditions for existence and uniqueness in Theorem 5.2.1. However, we can still prove that a unique strong solution exists by proceeding as follows:

a) For $n = 1, 2, \ldots$ define

$$b_n(y) = \begin{cases} y(K - y) & \text{if} \quad 0 \leq y \leq n \\ n(K - n) & \text{if} \quad y > n \end{cases}$$

and

$$\sigma_n(y) = \begin{cases} \alpha\, y(K - y) & \text{if} \quad 0 \leq y \leq n \\ \alpha\, n(K - n) & \text{if} \quad y > n \end{cases}$$

and let $X_t = X_t^{(n)}$ be the unique solution of

$$dX_t = b_n(X_t)dt + \sigma_n(X_t)dB_t \ ; \qquad X_0 = x.$$

Define
$$\tau_n = \inf\{t > 0; X_t^{(n)} = n\}.$$

Show that
$$X_t^{(n)} = X_t^{(n+1)} \qquad \text{for all} \ \ t < \tau_n$$

and use this to find a unique strong solution X_t for $t < \tau_\infty$:
$\lim\limits_{n\to\infty} \tau_n.$

b) Prove that $\tau_\infty = \infty$ a.s.
c) Prove that
(i) $X_0 = 0 \Rightarrow X_t = 0$ for all t
(ii) $X_0 = K \Rightarrow X_t = K$ for all t
(iii) $0 < X_0 < K \Rightarrow 0 < X_t < K$ for all t
(iv) $X_0 > K \Rightarrow X_t > K$ for all t.

For a discussion of optimal harvesting from this population model see Lungu and Øksendal (1997).

Chapter 8
Other Topics in Diffusion Theory

In this chapter we study some other important topics in diffusion theory and related areas. Some of these topics are not strictly necessary for the remaining chapters, but they are all central in the theory of stochastic analysis and essential for further applications. The following topics will be treated:

8.1 Kolmogorov's backward equation. The resolvent.
8.2 The Feynman-Kac formula. Killing.
8.3 The martingale problem.
8.4 When is an Itô process a diffusion?
8.5 Random time change.
8.6 The Girsanov formula.

8.1 Kolmogorov's Backward Equation. The Resolvent

In the following we let X_t be an Itô diffusion in $\mathbf{R}^n$ with generator A. If we choose $f \in C_0^2(\mathbf{R}^n)$ and $\tau = t$ in Dynkin's formula (7.4.1) we see that

$$u(t, x) = E^x[f(X_t)]$$

is differentiable with respect to t and

$$\frac{\partial u}{\partial t} = E^x[Af(X_t)] . \tag{8.1.1}$$

It turns out that the right hand side of (8.1.1) can be expressed in terms of u also:

Theorem 8.1.1 (Kolmogorov's backward equation)
Let $f \in C_0^2(\mathbf{R}^n)$.

a) *Define*
$$u(t, x) = E^x[f(X_t)] . \tag{8.1.2}$$

Then $u(t, \cdot) \in \mathcal{D}_A$ for each t and

$$\frac{\partial u}{\partial t} = Au , \qquad t > 0, \ x \in \mathbf{R}^n \tag{8.1.3}$$

$$u(0, x) = f(x) ; \qquad x \in \mathbf{R}^n \tag{8.1.4}$$

where the right hand side is to be interpreted as A applied to the function $x \to u(t, x)$.

b) *Moreover, if $w(t, x) \in C^{1,2}(\mathbf{R} \times \mathbf{R}^n)$ is a bounded function satisfying (8.1.3), (8.1.4) then $w(t, x) = u(t, x)$, given by (8.1.2).*

Proof. a) Let $g(x) = u(t, x)$. Then since $t \to u(t, x)$ is differentiable we have

$$\frac{E^x[g(X_r)] - g(x)}{r} = \frac{1}{r} \cdot E^x[E^{X_r}[f(X_t)] - E^x[f(X_t)]]$$

$$= \frac{1}{r} \cdot E^x[E^x[f(X_{t+r})|\mathcal{F}_r] - E^x[f(X_t)|\mathcal{F}_r]]$$

$$= \frac{1}{r} \cdot E^x[f(X_{t+r}) - f(X_t)]$$

$$= \frac{u(t + r, x) - u(t, x)}{r} \to \frac{\partial u}{\partial t} \qquad \text{as } r \downarrow 0 .$$

Hence

$$Au = \lim_{r \downarrow 0} \frac{E^x[g(X_r)] - g(x)}{r} \qquad \text{exists and } \frac{\partial u}{\partial t} = Au, \text{ as asserted .}$$

Conversely, to prove the uniqueness statement in b) assume that a function $w(t, x) \in C^{1,2}(\mathbf{R} \times \mathbf{R}^n)$ satisfies (8.1.3)–(8.1.4). Then

$$\widetilde{A}w := -\frac{\partial w}{\partial t} + Aw = 0 \qquad \text{for } t > 0, \ x \in \mathbf{R}^n \tag{8.1.5}$$

and

$$w(0, x) = f(x) , \qquad x \in \mathbf{R}^n . \tag{8.1.6}$$

Fix $(s, x) \in \mathbf{R} \times \mathbf{R}^n$. Define the process Y_t in $\mathbf{R}^{n+1}$ by $Y_t = (s - t, X_t^{0,x})$, $t \geq 0$. Then Y_t has generator $\widetilde{A}$ and so by (8.1.5) and Dynkin's formula we have, for all $t \geq 0$,

$$E^{s,x}[w(Y_{t \wedge \tau_R})] = w(s, x) + E^{s,x}\left[\int_0^{t \wedge \tau_R} \widetilde{A}w(Y_r)dr \right] = w(s, x) ,$$

where $\tau_R = \inf\{t > 0; |X_t| \geq R\}$.

Letting $R \to \infty$ we get

$$w(s, x) = E^{s,x}[w(Y_t)] ; \qquad \forall t \geq 0 .$$

In particular, choosing $t = s$ we get

$$w(s,x) = E^{s,x}[w(Y_s)] = E[w(0, X_s^{0,x})] = E[f(X_s^{0,x})] = E^x[f(X_s)] \,. \qquad \Box$$

Remark. If we introduce the operator $Q_t \colon f \to E^\bullet[f(X_t)]$ then we have $u(t,x) = (Q_t f)(x)$ and we may rewrite (8.1.1) and (8.1.3) as follows:

$$\frac{d}{dt}(Q_t f) = Q_t(Af) \,; \qquad f \in C_0^2(\mathbf{R}^n) \tag{8.1.1}'$$

$$\frac{d}{dt}(Q_t f) = A(Q_t f) \,; \qquad f \in C_0^2(\mathbf{R}^n) \,. \tag{8.1.3}'$$

Thus the equivalence of (8.1.1) and (8.1.3) amounts to saying that the operators Q_t and A commute, in some sense. Arguing formally, it is tempting to say that the solution of (8.1.1)$'$ and (8.1.3)$'$ is

$$Q_t = e^{tA}$$

and therefore $Q_t A = A Q_t$. However, this argument would require a further explanation, because in general A is an unbounded operator.

It is an important fact that if a positive multiple of the identity is subtracted from A then the operator A always has an inverse. This inverse can be expressed explicitly in terms of the diffusion X_t:

Definition 8.1.2 *For $\alpha > 0$ and $g \in C_b(\mathbf{R}^n)$ we define the resolvent operator R_α by*

$$R_\alpha g(x) = E^x \left[\int_0^\infty e^{-\alpha t} g(X_t) dt \right] \,. \tag{8.1.7}$$

Lemma 8.1.3 $R_\alpha g$ *is a bounded continuous function.*

Proof. Since $R_\alpha g(x) = \int_0^\infty e^{-\alpha t} E^x[g(X_t)] dt$, we see that Lemma 8.1.3 is a direct consequence of the next result:

Lemma 8.1.4 *Let g be a lower bounded, measurable function on $\mathbf{R}^n$ and define, for fixed $t \geq 0$*

$$u(x) = E^x[g(X_t)] \,.$$

a) *If g is lower semicontinuous, then u is lower semicontinuous.*
b) *If g is bounded and continuous, then u is continuous. In other words, any Itô diffusion X_t is Feller-continuous.*

Proof. By (5.2.10) we have

$$E[|X_t^x - X_t^y|^2] \leq |y - x|^2 C(t) \,,$$

where $C(t)$ does not depend on x and y. Let $\{y_n\}$ be a sequence of points converging to x. Then

$$X_t^{y_n} \to X_t^x \qquad \text{in } L^2(\Omega, P) \text{ as } n \to \infty .$$

So, by taking a subsequence $\{z_n\}$ of $\{y_n\}$ we obtain that

$$X_t^{z_n}(\omega) \to X_t^x(\omega) \qquad \text{for a.a. } \omega \in \Omega .$$

a) If g is lower bounded and lower semicontinuous, then by the Fatou lemma

$$u(x) = E[g(X_t^x)] \le E[\varliminf_{n \to \infty} g(X_t^{z_n})] \le \varliminf_{n \to \infty} E[g(X_t^{z_n})] = \varliminf_{n \to \infty} u(z_n) .$$

Therefore every sequence $\{y_n\}$ converging to x has a subsequence $\{z_n\}$ such that $u(x) \le \varliminf_{n \to \infty} u(z_n)$. That proves that u is lower semicontinuous.

b) If g is bounded and continuous, the result in a) can be applied both to g and $-g$. Hence both u and $-u$ are lower semicontinuous and we conclude that u is continuous.   □

We now prove that R_α and $\alpha - A$ are inverse operators:

Theorem 8.1.5 a)   *If $f \in C_0^2(\mathbf{R}^n)$ then $R_\alpha(\alpha - A)f = f$ for all $\alpha > 0$.*
b)   *If $g \in C_b(\mathbf{R}^n)$ then $R_\alpha g \in \mathcal{D}_A$ and $(\alpha - A)R_\alpha g = g$ for all $\alpha > 0$.*

Proof. a) If $f \in C_0^2(\mathbf{R}^n)$ then by Dynkin's formula

$$R_\alpha(\alpha - A)f(x) = (\alpha R_\alpha f - R_\alpha A f)(x)$$

$$= \alpha \int_0^\infty e^{-\alpha t} E^x[f(X_t)]dt - \int_0^\infty e^{-\alpha t} E^x[Af(X_t)]dt$$

$$= \Big|_0^\infty -e^{-\alpha t} E^x[f(X_t)] + \int_0^\infty e^{-\alpha t} \frac{d}{dt} E^x[f(X_t)]dt - \int_0^\infty e^{-\alpha t} E^x[Af(X_t)]dt$$

$$= E^x[f(X_0)] = f(x) .$$

b) If $g \in C_b(\mathbf{R}^n)$ then by the Markov property

$$E^x[R_\alpha g(X_t)] = E^x[E^{X_t}[\int_0^\infty e^{-\alpha s} g(X_s)ds]]$$

$$= E^x[E^x[\theta_t(\int_0^\infty e^{-\alpha s} g(X_s)ds)|\mathcal{F}_t]] = E^x[E^x[\int_0^\infty e^{-\alpha s} g(X_{t+s})ds|\mathcal{F}_t]]$$

$$= E^x[\int_0^\infty e^{-\alpha s} g(X_{t+s})ds] = \int_0^\infty e^{-\alpha s} E^x[g(X_{t+s})]ds .$$

Integration by parts gives

$$E^x[R_\alpha g(X_t)] = \alpha \int_0^\infty e^{-\alpha s} \int_t^{t+s} E^x[g(X_v)]dv\, ds\;.$$

This identity implies that $R_\alpha g \in \mathcal{D}_A$ and

$$A(R_\alpha g) = \alpha R_\alpha g - g\;.\qquad\qquad\qquad\square$$

8.2 The Feynman-Kac Formula. Killing

With a little harder work we can obtain the following useful generalization of Kolmogorov's backward equation:

Theorem 8.2.1 (The Feynman-Kac formula)
Let $f \in C_0^2(\mathbf{R}^n)$ and $q \in C(\mathbf{R}^n)$. Assume that q is lower bounded.

a)   *Put*

$$v(t,x) = E^x\left[\exp\left(-\int_0^t q(X_s)ds\right)f(X_t)\right]\;.\qquad(8.2.1)$$

      Then

$$\frac{\partial v}{\partial t} = Av - qv\;;\qquad t > 0,\; x \in \mathbf{R}^n\qquad(8.2.2)$$

$$v(0,x) = f(x)\;;\qquad x \in \mathbf{R}^n\qquad(8.2.3)$$

b)   *Moreover, if $w(t,x) \in C^{1,2}(\mathbf{R} \times \mathbf{R}^n)$ is bounded on $K \times \mathbf{R}^n$ for each compact $K \subset \mathbf{R}$ and w solves (8.2.2), (8.2.3), then $w(t,x) = v(t,x)$, given by (8.2.1).*

Proof. a)   Let $Y_t = f(X_t), Z_t = \exp(-\int_0^t q(X_s)ds)$. Then dY_t is given by (7.3.1) and

$$dZ_t = -Z_t q(X_t)dt\;.$$

So
$$d(Y_t Z_t) = Y_t dZ_t + Z_t dY_t\;,\qquad \text{since } dZ_t \cdot dY_t = 0\;.$$

Note that since $Y_t Z_t$ is an Itô process it follows from Lemma 7.3.2 that $v(t,x) = E^x[Y_t Z_t]$ is differentiable w.r.t. t.
Therefore, with $v(t,x)$ as in (8.2.1) we get

$$\frac{1}{r}(E^x[v(t,X_r)] - v(t,x)) = \frac{1}{r}E^x[E^{X_r}[Z_t f(X_t)] - E^x[Z_t f(X_t)]]$$

$$= \frac{1}{r} E^x [E^x [f(X_{t+r}) \exp \Big(- \int_0^t q(X_{s+r}) ds \Big) |\mathcal{F}_r] - E^x [Z_t f(X_t)|\mathcal{F}_r]]$$

$$= \frac{1}{r} E^x [Z_{t+r} \cdot \exp \Big(\int_0^r q(X_s) ds \Big) f(X_{t+r}) - Z_t f(X_t)]$$

$$= \frac{1}{r} E^x [f(X_{t+r}) Z_{t+r} - f(X_t) Z_t]$$

$$+ \frac{1}{r} E^x \Big[f(X_{t+r}) Z_{t+r} \cdot \Big(\exp \Big(\int_0^r q(X_s) ds \Big) - 1 \Big) \Big]$$

$$\rightarrow \frac{\partial}{\partial t} v(t, x) + q(x) v(t, x) \qquad \text{as } r \rightarrow 0 ,$$

because

$$\frac{1}{r} f(X_{t+r}) Z_{t+r} \Big(\exp \Big(\int_0^r q(X_s) ds \Big) - 1 \Big) \rightarrow f(X_t) Z_t q(X_0)$$

pointwise boundedly. That completes the proof of a).

b) Assume that $w(t, x) \in C^{1,2}(\mathbf{R} \times \mathbf{R}^n)$ satisfies (8.2.2) and (8.2.3) and that $w(t, x)$ is bounded on $K \times \mathbf{R}^n$ for each compact $K \subset \mathbf{R}$. Then

$$\widehat{A} w(t, x) := -\frac{\partial w}{\partial t} + A w - q w = 0 \qquad \text{for } t > 0, \, x \in \mathbf{R}^n \qquad (8.2.4)$$

and

$$w(0, x) = f(x) ; \qquad x \in \mathbf{R}^n . \qquad (8.2.5)$$

Fix $(s, x, z) \in \mathbf{R} \times \mathbf{R}^n \times \mathbf{R}$ and define $Z_t = z + \int_0^t q(X_s) ds$ and $H_t = (s - t, X_t^{0,x}, Z_t)$. Then H_t is an Itô diffusion with generator

$$A_H \phi(s, x, z) = -\frac{\partial \phi}{\partial s} + A \phi + q(x) \frac{\partial \phi}{\partial z} ; \qquad \phi \in C_0^2(\mathbf{R} \times \mathbf{R}^n \times \mathbf{R}) .$$

Hence by (8.2.4) and Dynkin's formula we have, for all $t \geq 0$, $R > 0$ and with $\phi(s, x, z) = \exp(-z) w(s, x)$:

$$E^{s,x,z}[\phi(H_{t \wedge \tau_R})] = \phi(s, x, z) + E^{s,x,z} \Big[\int_0^{t \wedge \tau_R} A_H \phi(H_r) dr \Big] ,$$

where $\tau_R = \inf\{t > 0; |H_t| \geq R\}$.

Note that with this choice of ϕ we have by (8.2.4)

$$A_H \phi(s, x, z) = \exp(-z)\left[-\frac{\partial w}{\partial s} + Aw - q(x)w \right] = 0 \ .$$

Hence

$$w(s, x) = \phi(s, x, 0) = E^{s,x,0}[\phi(H_{t \wedge \tau_R})]$$

$$= E^x\left[\exp\left(-\int_0^{t \wedge \tau_R} q(X_r)dr \right) w(s - t \wedge \tau_R, X_{t \wedge \tau_R}) \right]$$

$$\to E^x\left[\exp\left(-\int_0^t q(X_r)dr \right) w(s - t, X_t) \right] \qquad \text{as } R \to \infty \ ,$$

since $w(r, x)$ is bounded for $(r, x) \in K \times \mathbf{R}^n$. In particular, choosing $t = s$ we get

$$w(s, x) = E^x\left[\exp\left(-\int_0^s q(X_r)dr \right) w(0, X_s^{0,x}) \right] = v(s, x) \ , \quad \text{as claimed} \ .$$

$\square$

Remark. (About killing a diffusion)

In Theorem 7.3.3 we have seen that the generator of an Itô diffusion X_t given by

$$dX_t = b(X_t)dt + \sigma(X_t)dB_t \tag{8.2.6}$$

is a partial differential operator L of the form

$$Lf = \sum a_{ij}\frac{\partial^2 f}{\partial x_i \partial x_j} + \sum b_i \frac{\partial f}{\partial x_i} \tag{8.2.7}$$

where $[a_{ij}] = \frac{1}{2}\sigma\sigma^T$, $b = [b_i]$. It is natural to ask if one can also find processes whose generator has the form

$$Lf = \sum a_{ij}\frac{\partial^2 f}{\partial x_i \partial x_j} + \sum b_i \frac{\partial f}{\partial x_i} - cf \ , \tag{8.2.8}$$

where $c(x)$ is a bounded and continuous function.

If $c(x) \geq 0$ the answer is yes and a process $\widetilde{X}_t$ with generator (8.2.8) is obtained by *killing* X_t at a certain *(killing) time* ζ. By this we mean that there exists a random time ζ such that if we put

$$\widetilde{X}_t = X_t \qquad \text{if } t < \zeta \tag{8.2.9}$$

and leave $\widetilde{X}_t$ undefined if $t \geq \zeta$ (alternatively, put $\widetilde{X}_t = \partial$ if $t \geq \zeta$, where $\partial \notin \mathbf{R}^n$ is some "coffin" state), then $\widetilde{X}_t$ is also a strong Markov process and

$$E^x[f(\widetilde{X}_t)] := E^x[f(X_t) \cdot \mathcal{X}_{[0,\zeta)}(t)] = E^x\left[f(X_t) \cdot e^{-\int_0^t c(X_s)ds}\right] \quad (8.2.10)$$

for all bounded continuous functions f on $\mathbf{R}^n$.

Let $v(t, x)$ denote the right hand side of (8.2.10) with $f \in C_0^2(\mathbf{R}^n)$. Then

$$\lim_{t \to 0} \frac{E^x[f(\widetilde{X}_t)] - f(x)}{t} = \frac{\partial}{\partial t}v(t, x)_{t=0} = (Av - cv)_{t=0} = Af(x) - c(x)f(x) \;,$$

by the Feynman-Kac formula.

So the generator of $\widetilde{X}_t$ is (8.2.8), as required. The function $c(x)$ can be interpreted as the *killing rate*:

$$c(x) = \lim_{t \downarrow 0} \frac{1}{t}Q^x[X_0 \text{ is killed in the time interval } (0, t]] \;.$$

Thus by applying such a killing procedure we can come from the special case $c = 0$ in (8.2.7) to the general case (8.2.8) with $c(x) \geq 0$. Therefore, for many purposes it is enough to consider the equation (8.2.7).

If the function $c(x) \geq 0$ is given, an explicit construction of the killing time ζ such that (8.2.10) holds can be found in Karlin and Taylor (1981), p. 314. For a more general discussion see Blumenthal and Getoor (1968), Chap. III.

8.3 The Martingale Problem

If $dX_t = b(X_t)dt + \sigma(X_t)dB_t$ is an Itô diffusion in $\mathbf{R}^n$ with generator A and if $f \in C_0^2(\mathbf{R}^n)$ then by (7,3.1)

$$f(X_t) = f(x) + \int_0^t Af(X_s)ds + \int_0^t \nabla f^T(X_s)\sigma(X_s)dB_s \;. \quad (8.3.1)$$

Define

$$M_t = f(X_t) - \int_0^t Af(X_r)dr \quad (= f(x) + \int_0^t \nabla f^T(X_r)\sigma(X_r)dB_r) \;. \quad (8.3.2)$$

Then, since Itô integrals are martingales (w.r.t. the σ-algebras $\{\mathcal{F}_t^{(m)}\}$) we have for $s > t$

$$E^x[M_s|\mathcal{F}_t^{(m)}] = M_t \;.$$

It follows that

$$E^x[M_s|\mathcal{M}_t] = E^x[E^x[M_s|\mathcal{F}_t^{(m)}]|\mathcal{M}_t] = E^x[M_t|\mathcal{M}_t] = M_t \;,$$

since M_t is $\mathcal{M}_t$-measurable. We have proved:

Theorem 8.3.1 *If X_t is an Itô diffusion in $\mathbf{R}^n$ with generator A, then for all $f \in C_0^2(\mathbf{R}^n)$ the process*

$$M_t = f(X_t) - \int_0^t Af(X_r)dr$$

is a martingale w.r.t. $\{\mathcal{M}_t\}$.

If we identify each $\omega \in \Omega$ with the function

$$\omega_t = \omega(t) = X_t^x(\omega)$$

we see that the probability space $(\Omega, \mathcal{M}, Q^x)$ is identified with

$$((\mathbf{R}^n)^{[0,\infty)}, \mathcal{B}, \widetilde{Q}^x)$$

where $\mathcal{B}$ is the Borel σ-algebra on $(\mathbf{R}^n)^{[0,\infty)}$ (see Chapter 2). Thus, regarding the law of X_t^x as a probability measure $\widetilde{Q}^x$ on $\mathcal{B}$ we can formulate Theorem 8.3.1 as follows:

Theorem 8.3.1'. *If $\widetilde{Q}^x$ is the probability measure on $\mathcal{B}$ induced by the law Q^x of an Itô diffusion X_t, then for all $f \in C_0^2(\mathbf{R}^n)$ the process*

$$M_t = f(X_t) - \int_0^t Af(X_r)dr \; (= f(\omega_t) - \int_0^t Af(\omega_r)dr) \; ; \quad \omega \in (\mathbf{R}^n)^{[0,\infty)} \quad (8.3.3)$$

is a $\widetilde{Q}^x$-martingale w.r.t. the Borel σ-algebras $\mathcal{B}_t$ of $(\mathbf{R}^n)^{[0,t]}$, $t \geq 0$. In other words, the measure $\widetilde{Q}^x$ solves the martingale problem for the differential operator A, in the following sense:

Definition 8.3.2 *Let L be a semi-elliptic differential operator of the form*

$$L = \sum b_i \frac{\partial}{\partial x_i} + \sum_{i,j} a_{ij} \frac{\partial^2}{\partial x_i \partial x_j}$$

where the coefficients b_i, a_{ij} are locally bounded Borel measurable functions on $\mathbf{R}^n$. Then we say that a probability measure $\widetilde{P}^x$ on $((\mathbf{R}^n)^{[0,\infty)}, \mathcal{B})$ solves the martingale problem for L (starting at x) if the process

$$M_t = f(\omega_t) - \int_0^t Lf(\omega_r)dr \, , M_0 = f(x) \quad a.s. \; \widetilde{P}^x$$

is a $\widetilde{P}^x$ martingale w.r.t. $\mathcal{B}_t$, for all $f \in C_0^2(\mathbf{R}^n)$. The martingale problem is called well posed *if there is a unique measure $\widetilde{P}^x$ solving the martingale problem.*

The argument of Theorem 8.3.1 actually proves that $\widetilde{Q}^x$ solves the martingale problem for A whenever X_t is a weak solution of the stochastic differential equation

$$dX_t = b(X_t)dt + \sigma(X_t)dB_t \; . \tag{8.3.4}$$

Conversely, it can be proved that if $\widetilde{P}^x$ solves the martingale problem for

$$L = \sum b_i \frac{\partial}{\partial x_i} + \tfrac{1}{2} \sum (\sigma\sigma^T)_{ij} \frac{\partial^2}{\partial x_i \partial x_j} \tag{8.3.5}$$

starting at x, for all $x \in \mathbf{R}^n$, then there exists a weak solution X_t of the stochastic differential equation (8.3.4). Moreover, this weak solution X_t is a Markov process if and only if the martingale problem for L is well posed. (See Stroock and Varadhan (1979) or Rogers and Williams (1987)). Therefore, if the coefficients b, σ of (8.3.4) satisfy the conditions (5.2.1), (5.2.2) of Theorem 5.2.1, we conclude that

$$\widetilde{Q}^x \text{ is the } \textit{unique} \text{ solution of the martingale problem}$$
$$\text{for the operator } L \text{ given by (8.3.5) .} \tag{8.3.6}$$

Lipschitz-continuity of the coefficients of L is not *necessary* for the uniqueness of the martingale problem. For example, one of the spectacular results of Stroock and Varadhan (1979) is that

$$L = \sum b_i \frac{\partial}{\partial x_i} + \sum a_{ij} \frac{\partial^2}{\partial x_i \partial x_j}$$

has a unique solution of the martingale problem if $[a_{ij}]$ is everywhere positive definite, $a_{ij}(x)$ is continuous, $b(x)$ is measurable and there exists a constant D such that

$$|b(x)| + |a(x)|^{\frac{1}{2}} \le D(1 + |x|) \qquad \text{for all } x \in \mathbf{R}^n \; .$$

8.4 When is an Itô Process a Diffusion?

The Itô formula gives that if we apply a C^2 function $\phi \colon U \subset \mathbf{R}^n \to \mathbf{R}^n$ to an Itô process X_t the result $\phi(X_t)$ is another Itô process. A natural question is: If X_t is an Itô diffusion will $\phi(X_t)$ be an Itô diffusion too? The answer is no in general, but it may be yes in some cases:

Example 8.4.1 (The Bessel process) Let $n \ge 2$. In Example 4.2.2 we found that the process

$$R_t(\omega) = |B(t, \omega)| = (B_1(t, \omega)^2 + \cdots + B_n(t, \omega)^2)^{\frac{1}{2}}$$

satisfies the equation

$$dR_t = \sum_{i=1}^{n} \frac{B_i dB_i}{R_t} + \frac{n-1}{2R_t} dt .$$ (8.4.1)

However, as it stands this is *not* a stochastic differential equation of the form (5.2.3), so it is not apparent from (8.4.1) that R is an Itô diffusion. But this will follow if we can show that

$$Y_t := \int_0^t \sum_{i=1}^{n} \frac{B_i}{|B|} dB_i$$

coincides in law with (i.e. has the same finite-dimensional distributions as) 1-dimensional Brownian motion $\tilde{B}_t$. For then (8.4.1) can be written

$$dR_t = \frac{n-1}{2R_t} dt + d\tilde{B}$$

which is of the form (5.2.3), thus showing by weak uniqueness (Lemma 5.3.1) that R_t is an Itô diffusion with generator

$$Af(x) = \tfrac{1}{2} f''(x) + \frac{n-1}{2x} f'(x)$$

as claimed in Example 4.2.2. One way of seeing that the process Y_t coincides in law with 1-dimensional Brownian motion $\tilde{B}_t$ is to apply the following result:

Theorem 8.4.2 *An Itô process*

$$dY_t = v dB_t ; \qquad Y_0 = 0 \;\; with \;\; v(t, \omega) \in \mathcal{V}_{\mathcal{H}}^{n \times m}$$

coincides (in law) with n-dimensional Brownian motion if and only if

$$vv^T(t, \omega) = I_n \qquad for \; a.a. \; (t, \omega) \; w.r.t. \; dt \times dP$$ (8.4.2)

where I_n is the n-dimensional identity matrix.

Note that in the example above we have

$$Y_t = \int_0^t v dB$$

with

$$v = \left[\frac{B_1}{|B|}, \ldots, \frac{B_n}{|B|} \right], \qquad B = \begin{pmatrix} B_1 \\ \vdots \\ B_n \end{pmatrix}$$

and since $vv^T = 1$, we get that Y_t is a 1-dimensional Brownian motion, as required.

Theorem 8.4.2 is a special case of the following result, which gives a necessary and sufficient condition for an Itô process to coincide in law with a given diffusion: (We use the symbol $\simeq$ for "coincides in law with").

Theorem 8.4.3 *Let X_t be an Itô diffusion given by*

$$dX_t = b(X_t)dt + \sigma(X_t)dB_t , \quad b \in \mathbf{R}^n , \quad \sigma \in \mathbf{R}^{n \times m}, \quad X_0 = x ,$$

and let Y_t be an Itô process given by

$$dY_t = u(t,\omega)dt + v(t,\omega)dB_t , \quad u \in \mathbf{R}^n , \quad v \in \mathbf{R}^{n \times m}, \quad Y_0 = x .$$

Then $X_t \simeq Y_t$ if and only if

$$E^x[u(t,\cdot)|\mathcal{N}_t] = b(Y_t^x) \qquad and \ vv^T(t,\omega) = \sigma\sigma^T(Y_t^x) \tag{8.4.3}$$

for a.a. (t,ω) w.r.t. $dt \times dP$, where $\mathcal{N}_t$ is the σ-algebra generated by Y_s; $s \leq t$.

Proof. Assume that (8.4.3) holds. Let

$$A = \sum b_i \frac{\partial}{\partial x_i} + \tfrac{1}{2} \sum_{i,j} (\sigma\sigma^T)_{ij} \frac{\partial^2}{\partial x_i \partial x_j}$$

be the generator of X_t and define, for $f \in C_0^2(\mathbf{R}^n)$,

$$Hf(t,\omega) = \sum_i u_i(t,\omega) \frac{\partial f}{\partial x_i}(Y_t) + \tfrac{1}{2} \sum_{i,j} (vv^T)_{ij}(t,\omega) \frac{\partial^2 f}{\partial x_i \partial x_j}(Y_t) .$$

Then by Itô's formula (see (7.3.1)) we have, for $s > t$,

$$E^x[f(Y_s)|\mathcal{N}_t] = f(Y_t) + E^x\left[\int_t^s Hf(r,\omega)dr|\mathcal{N}_t\right] + E^x\left[\int_t^s \nabla f^T v dB_r|\mathcal{N}_t\right]$$

$$= f(Y_t) + E^x\left[\int_t^s E^x[Hf(r,\omega)|\mathcal{N}_r]dr|\mathcal{N}_t\right]$$

$$= f(Y_t) + E^x\left[\int_t^s Af(Y_r)dr|\mathcal{N}_t\right] \qquad \text{by (8.4.3) ,} \tag{8.4.4}$$

where E^x denotes expectation w.r.t. the law R^x of Y_t (see Lemma 7.3.2). Therefore, if we define

$$M_t = f(Y_t) - \int_0^t Af(Y_r)dr \qquad (8.4.5)$$

then, for $s > t$,

$$E^x[M_s|\mathcal{N}_t] = f(Y_t) + E^x\left[\int_t^s Af(Y_r)dr|\mathcal{N}_t\right] - E^x\left[\int_0^s Af(Y_r)dr|\mathcal{N}_t\right]$$

$$= f(Y_t) - E^x\left[\int_0^t Af(Y_r)dr|\mathcal{N}_t\right] = M_t .$$

Hence M_t is a martingale w.r.t. the σ-algebras $\mathcal{N}_t$ and the law R^x. By uniqueness of the solution of the martingale problem (see (8.3.6)) we conclude that $X_t \simeq Y_t$.

Conversely, assume that $X_t \simeq Y_t$. Choose $f \in C_0^2$. By Itô's formula (7.3.1) we have, for a.a. (t,ω) w.r.t. $dt \times dP$,

$$\lim_{h \downarrow 0} \frac{1}{h}(E^x[f(Y_{t+h})|\mathcal{N}_t] - f(Y_t))$$

$$= \lim_{h \downarrow 0} \frac{1}{h}\left(\int_t^{t+h} E^x\left[\sum_i u_i(s,\omega)\frac{\partial f}{\partial x_i}(Y_s)\right.\right.$$

$$\left.\left. + \frac{1}{2}\sum_{i,j}(vv^T)_{ij}(s,\omega)\frac{\partial^2 f}{\partial x_i \partial x_j}(Y_s)|\mathcal{N}_t\right]ds\right) \qquad (8.4.6)$$

$$= \sum_i E^x[u_i(t,\omega)|\mathcal{N}_t]\frac{\partial f}{\partial x_i}(Y_t) + \frac{1}{2}\sum_{i,j}E^x[(vv^T)_{ij}(t,\omega)|\mathcal{N}_t]\frac{\partial^2 f}{\partial x_i \partial x_j}(Y_t) . (8.4.7)$$

On the other hand, since $X_t \simeq Y_t$ we know that Y_t is a Markov process. Therefore (8.4.6) coincides with

$$\lim_{h \downarrow 0} \frac{1}{h}(E^{Y_t}[f(Y_h)] - E^{Y_t}[f(Y_0)])$$

$$= \sum_i E^{Y_t}\left[u_i(0,\omega)\frac{\partial f}{\partial x_i}(Y_0)\right] + \frac{1}{2}\sum_{i,j}E^{Y_t}\left[(vv^T)_{ij}(0,\omega)\frac{\partial^2 f}{\partial x_i \partial x_j}(Y_0)\right]$$

$$= \sum_i E^{Y_t}[u_i(0,\omega)]\frac{\partial f}{\partial x_i}(Y_t) + \frac{1}{2}\sum_{i,j}E^{Y_t}[(vv^T)_{ij}(0,\omega)]\frac{\partial^2 f}{\partial x_i \partial x_j}(Y_t) . \qquad (8.4.8)$$

Comparing (8.4.7) and (8.4.8) we conclude that

$$E^x[u(t,\omega)|\mathcal{N}_t] = E^{Y_t}[u(0,\omega)] \quad \text{and} \quad E^x[vv^T(t,\omega)|\mathcal{N}_t] = E^{Y_t}[vv^T(0,\omega)]$$

$$(8.4.9)$$

for a.a. (t, ω).

On the other hand, since the generator of Y_t coincides with the generator A of X_t we get from (8.4.8) that

$$E^{Y_t}[u(0, \omega)] = b(Y_t) \quad \text{and} \quad E^{Y_t}[vv^T(0, \omega)] = \sigma\sigma^T(Y_t) \quad \text{for a.a. } (t, \omega) . \tag{8.4.10}$$

Combining (8.4.9) and (8.4.10) we conclude that

$$E^x[u|\mathcal{N}_t] = b(Y_t) \quad \text{and} \quad E^x[vv^T|\mathcal{N}_t] = \sigma\sigma^T(Y_t) \quad \text{for a.a. } (t, \omega) . \tag{8.4.11}$$

From this we obtain (8.4.3) by using that in fact $vv^T(t, \cdot)$ is always $\mathcal{N}_t$-measurable, in the following sense:

Lemma 8.4.4 Let $dY_t = u(t, \omega)dt + v(t, \omega)dB_t, Y_0 = x$ be as in Theorem 8.4.3. Then there exists an $\mathcal{N}_t$-adapted process $W(t, \omega)$ such that

$$vv^T(t, \omega) = W(t, \omega) \qquad \text{for a.a. } (t, \omega) .$$

Proof. By Itô's formula we have (if $Y_i(t, \omega)$ denotes component number i of $Y(t, \omega)$)

$$Y_i Y_j(t, \omega) = x_i x_j + \int_0^t Y_i dY_j(s) + \int_0^t Y_j dY_i(s) + \int_0^t (vv^T)_{ij}(s, \omega)ds .$$

Therefore, if we put

$$H_{ij}(t, \omega) = Y_i Y_j(t, \omega) - x_i x_j - \int_0^t Y_i dY_j - \int_0^t Y_j dY_i , \qquad 1 \le i, \ j \le n$$

then H_{ij} is $\mathcal{N}_t$-adapted and

$$H_{ij}(t, \omega) = \int_0^t (vv^T)_{ij}(s, \omega)ds .$$

Therefore

$$(vv^T)_{ij}(t, \omega) = \lim_{r \downarrow 0} \frac{H(t, \omega) - H(t - r, \omega)}{r}$$

for a.a. t. This shows Lemma 8.4.4 and the proof of Theorem 8.4.3 is complete.
□

Remarks. 1) One may ask if also $u(t, \cdot)$ must be $\mathcal{N}_t$-measurable. However, the following example shows that this fails even in the case when $v = n = 1$:

Let B_1, B_2 be two independent 1-dimensional Brownian motions and define

$$dY_t = B_1(t)dt + dB_2(t) .$$

Then we may regard Y_t as noisy observations of the process $B_1(t)$. So by Example 6.2.10 we have that

$$E[(B_1(t,\omega) - \widehat{B}_1(t,\omega))^2] = \tanh(t) ,$$

where $\widehat{B}_1(t,\omega) = E[B_1(t)|\mathcal{N}_t]$ is the Kalman-Bucy filter. In particular, $B_1(t,\omega)$ cannot be $\mathcal{N}_t$-measurable.

2) The process $v(t,\omega)$ need not be $\mathcal{N}_t$-adapted either: Let B_t be 1-dimensional Brownian motion and define

$$dY_t = \text{sign}(B_t)dB_t \tag{8.4.12}$$

where

$$\text{sign}(z) = \begin{cases} 1 & \text{if } z > 0 \\ -1 & \text{if } z \leq 0 . \end{cases}$$

Tanaka's formula says that

$$|B_t| = |B_0| + \int_0^t \text{sign}(B_s)dB_s + L_t \tag{8.4.13}$$

where $L_t = L_t(\omega)$ is *local time* of B_t at 0, a non-decreasing process which only increases when $B_t = 0$ (see Exercise 4.10). Therefore the σ-algebra $\mathcal{N}_t$ generated by $\{Y_s; s \leq t\}$ is contained in the σ-algebra $\mathcal{H}_t$ generated by $\{|B_s|; s \leq t\}$. It follows that $v(t,\omega) = \text{sign}(B_t)$ cannot be $\mathcal{N}_t$-adapted.

Corollary 8.4.5 (How to recognize a Brownian motion)
Let

$$dY_t = u(t,\omega)dt + v(t,\omega)dB_t$$

be an Itô process in $\mathbf{R}^n$. *Then* Y_t *is a Brownian motion if and only if*

$$E^x[u(t,\cdot)|\mathcal{N}_t] = 0 \quad and \quad vv^T(t,\omega) = I_n \tag{8.4.14}$$

for a.a. (t,ω).

Remark. Using Theorem 8.4.3 one may now proceed to investigate when the image $Y_t = \phi(X_t)$ of an Itô diffusion X_t by a C^2-function ϕ coincides in law with an Itô diffusion Z_t. Applying the criterion (8.4.3) one obtains the following result:

$$\phi(X_t) \sim Z_t \quad \text{if and only if}$$
$$A[f o \phi] = \widehat{A}[f] o \phi \tag{8.4.15}$$

for all second order polynomials $f(x_1, \ldots, x_n) = \sum a_i x_i + \sum c_{ij} x_i x_j$ (and hence for all $f \in C_0^2$) where A and $\widehat{A}$ are the generators of X_t and Z_t respectively. (Here o denotes function composition: $(f \circ \phi)(x) = f(\phi(x))$.) For generalizations of this result, see Csink and Øksendal (1983), and Csink, Fitzsimmons and Øksendal (1990).

8.5 Random Time Change

Let $c(t, \omega) \geq 0$ be an $\mathcal{F}_t$-adapted process. Define

$$\beta_t = \beta(t, \omega) = \int_0^t c(s, \omega) ds . \tag{8.5.1}$$

We will say that β_t is a (random) *time change* with *time change rate* $c(t, \omega)$.

Note that $\beta(t, \omega)$ is also $\mathcal{F}_t$-adapted and for each ω the map $t \to \beta_t(\omega)$ is non-decreasing. Define $\alpha_t = \alpha(t, \omega)$ by

$$\alpha_t = \inf\{s; \beta_s > t\} . \tag{8.5.2}$$

Then α_t is a *right*-inverse of β_t, for each ω :

$$\beta(\alpha(t, \omega), \omega) = t \qquad \text{for all } t \geq 0 . \tag{8.5.3}$$

Moreover, $t \to \alpha_t(\omega)$ is right-continuous.

If $c(s, \omega) > 0$ for a.a. (s, ω) then $t \to \beta_t(\omega)$ is strictly increasing, $t \to \alpha_t(\omega)$ is continuous and α_t is also a *left*-inverse of β_t:

$$\alpha(\beta(t, \omega), \omega) = t \qquad \text{for all } t \geq 0 . \tag{8.5.4}$$

In general $\omega \to \alpha(t, \omega)$ is an $\{\mathcal{F}_s\}$-stopping time for each t, since

$$\{\omega; \alpha(t, \omega) < s\} = \{\omega; t < \beta(s, \omega)\} \in \mathcal{F}_s . \tag{8.5.5}$$

We now ask the question: Suppose X_t is an Itô diffusion and Y_t an Itô process as in Theorem 8.4.3. When does there exist a time change β_t such that $Y_{\alpha_t} \simeq X_t$? (Note that α_t is only defined up to time β_∞. If $\beta_\infty < \infty$ we interpret $Y_{\alpha_t} \simeq X_t$ to mean that Y_{α_t} has the same law as X_t up to time β_∞).

Here is a partial answer (see Øksendal (1990)):

Theorem 8.5.1 *Let X_t, Y_t be as in Theorem 8.4.3 and let β_t be a time change with right inverse α_t as in (8.5.1), (8.5.2) above. Assume that*

$$u(t, \omega) = c(t, \omega) b(Y_t) \quad \text{and} \quad vv^T(t, \omega) = c(t, \omega) \cdot \sigma\sigma^T(Y_t) \tag{8.5.6}$$

for a.a. t, ω. Then

$$Y_{\alpha_t} \simeq X_t .$$

This result allows us to recognize time changes of Brownian motion:

Theorem 8.5.2 *Let* $dY_t = v(t,\omega)dB_t$, $v \in \mathbf{R}^{n \times m}$, $B_t \in \mathbf{R}^m$ *be an Itô integral in* $\mathbf{R}^n$, $Y_0 = 0$ *and assume that*

$$vv^T(t,\omega) = c(t,\omega)I_n \qquad (8.5.7)$$

for some process $c(t,\omega) \geq 0$. *Let* α_t, β_t *be as in (8.5.1), (8.5.2). Then*

$$Y_{\alpha_t} \quad \text{is an } n\text{-dimensional Brownian motion} .$$

Corollary 8.5.3 *Let* $dY_t = \sum\limits_{i=1}^{n} v_i(t,\omega)dB_i(t,\omega)$, $Y_0 = 0$, *where* $B = (B_1, \ldots, B_n)$ *is a Brownian motion in* $\mathbf{R}^n$. *Then*

$$\widehat{B}_t := Y_{\alpha_t} \quad \text{is a 1-dimensional Brownian motion} ,$$

where α_t *is defined by (8.5.2) and*

$$\beta_s = \int\limits_0^s \left\{ \sum\limits_{i=1}^{n} v_i^2(r,\omega) \right\} dr . \qquad (8.5.8)$$

Corollary 8.5.4 *Let* Y_t, β_s *be as in Corollary 8.5.3. Assume that*

$$\sum\limits_{i=1}^{n} v_i^2(r,\omega) > 0 \quad \text{for a.a. } (r,\omega) . \qquad (8.5.9)$$

Then there exists a Brownian motion $\widehat{B}_t$ *such that*

$$Y_t = \widehat{B}_{\beta_t} . \qquad (8.5.10)$$

Proof. Let

$$\widehat{B}_t = Y_{\alpha_t} \qquad (8.5.11)$$

be the Brownian motion from Corollary 8.5.3. By (8.5.9) β_t is strictly increasing and hence (8.5.4) holds, So choosing $t = \beta_s$ in (8.5.11) we get (8.5.10). $\qquad \square$

Corollary 8.5.5 *Let* $c(t,\omega) \geq 0$ *be given and define*

$$Y_t = \int\limits_0^t \sqrt{c(s,\omega)} \, dB_s ,$$

where B_s *is an* n-*dimensional Brownian motion. Then*

$$Y_{\alpha_t} \quad \text{is also an } n\text{-dimensional Brownian motion} .$$

We now use this to prove that a time change of an Itô integral is again an Itô integral, but driven by a different Brownian motion $\widetilde{B}_t$. First we construct $\widetilde{B}_t$:

Lemma 8.5.6 *Suppose* $s \to \alpha(s,\omega)$ *is continuous,* $\alpha(0,\omega) = 0$ *for a.a.* ω. *Fix* $t > 0$ *such that* $\beta_t < \infty$ *a.s. and assume that* $E[\alpha_t] < \infty$. *For* $k = 1, 2, \ldots$ *put*

$$t_j = \begin{cases} j \cdot 2^{-k} & \text{if } j \cdot 2^{-k} \leq \alpha_t \\ t & \text{if } j \cdot 2^{-k} > \alpha_t \end{cases}$$

and choose r_j *such that* $\alpha_{r_j} = t_j$. *Suppose* $f(s,\omega) \geq 0$ *is* $\mathcal{F}_s$-*adapted, bounded and* s-*continuous for a.a.* ω. *Then*

$$\lim_{k \to \infty} \sum_j f(\alpha_j, \omega) \Delta B_{\alpha_j} = \int_0^{\alpha_t} f(s,\omega) dB_s \quad a.s. \, , \qquad (8.5.12)$$

where $\alpha_j = \alpha_{r_j}$, $\Delta B_{\alpha_j} = B_{\alpha_{j+1}} - B_{\alpha_j}$ *and the limit is in* $L^2(\Omega, P)$.

Proof. For all k we have

$$E\left[\left(\sum_j f(\alpha_j, \omega) \Delta B_{\alpha_j} - \int_0^{\alpha_t} f(s,\omega) dB_s\right)^2\right]$$

$$= E\left[\left(\sum_j \int_{\alpha_j}^{\alpha_{j+1}} (f(\alpha_j, \omega) - f(s,\omega) dB_s\right)^2\right]$$

$$= \sum_j E\left[\left(\int_{\alpha_j}^{\alpha_{j+1}} (f(\alpha_j, \omega) - f(s,\omega)) dB_s\right)^2\right]$$

$$= \sum_j E\left[\int_{\alpha_j}^{\alpha_{j+1}} (f(\alpha_j, \omega) - f(s,\omega))^2 ds\right] = E\left[\int_0^{\alpha_t} (f - f_k)^2 ds\right],$$

where $f_k(s,\omega) = \sum_j f(t_j, \omega) \mathcal{X}_{[t_j, t_{j+1})}(s)$ is the elementary approximation to f. (See Corollary 3.1.8). This implies (8.5.12).   □

We now use this to establish a general time change formula for Itô integrals. An alternative proof in the case $n = m = 1$ can be found in McKean (1969, §2.8).

Theorem 8.5.7 (Time change formula for Itô integrals)
Suppose $c(s,\omega)$ *and* $\alpha(s,\omega)$ *are* s-*continuous,* $\alpha(0,\omega) = 0$ *for a.a.* ω *and that* $E[\alpha_t] < \infty$. *Let* B_s *be an* m-*dimensional Brownian motion and let* $v(s,\omega) \in \mathcal{V}_{\mathcal{H}}^{n \times m}$ *be bounded and* s-*continuous. Define*

$$\tilde{B}_t = \lim_{k \to \infty} \sum_j \sqrt{c(\alpha_j, \omega)} \, \Delta B_{\alpha_j} = \int_0^{\alpha_t} \sqrt{c(s, \omega)} \, dB_s \, . \tag{8.5.13}$$

Then $\tilde{B}_t$ is an (m-dimensional) $\mathcal{F}_{\alpha_t}^{(m)}$-Brownian motion (i.e. $\tilde{B}_t$ is a Brownian motion and $\tilde{B}_t$ is a martingale w.r.t. $\mathcal{F}_{\alpha_t}^{(m)}$) and

$$\int_0^{\alpha_t} v(s, \omega) dB_s = \int_0^t v(\alpha_r, \omega) \sqrt{\alpha_r'(\omega)} \, d\tilde{B}_r \qquad a.s. \ P \, , \tag{8.5.14}$$

where $\alpha_r'(\omega)$ is the derivative of $\alpha(r, \omega)$ w.r.t. r, so that

$$\alpha_r'(\omega) = \frac{1}{c(\alpha_r, \omega)} \qquad for \ a.a. \ r \geq 0, \ a.a. \ \omega \in \Omega \, . \tag{8.5.15}$$

Proof. The existence of the limit in (8.5.13) and the second identity in (8.5.13) follow by applying Lemma 8.5.6 to the function

$$f(s, \omega) = \sqrt{c(s, \omega)} \, .$$

Then by Corollary 8.5.5 we have that $\tilde{B}_t$ is an $\mathcal{F}_{\alpha_t}^{(m)}$-Brownian motion. It remains to prove (8.5.14):

$$\begin{aligned}
\int_0^{\alpha_t} v(s, \omega) dB_s &= \lim_{k \to \infty} \sum_j v(\alpha_j, \omega) \Delta B_{\alpha_j} \\
&= \lim_{k \to \infty} \sum_j v(\alpha_j, \omega) \sqrt{\frac{1}{c(\alpha_j, \omega)}} \sqrt{c(\alpha_j, \omega)} \, \Delta B_{\alpha_j} \\
&= \lim_{k \to \infty} \sum_j v(\alpha_j, \omega) \sqrt{\frac{1}{c(\alpha_j, \omega)}} \, \Delta \tilde{B}_j \\
&= \int_0^t v(\alpha_r, \omega) \sqrt{\frac{1}{c(\alpha_r, \omega)}} \, d\tilde{B}_r
\end{aligned}$$

and the proof is complete. $\qquad\qquad\qquad\qquad\qquad\qquad\qquad\qquad\qquad\square$

Example 8.5.8 (Brownian motion on the unit sphere in $\mathbf{R}^n$; $n > 2$)
In Examples 5.1.4 and 7.5.5 we constructed Brownian motion on the unit circle. It is not obvious how to extend the method used there to obtain Brownian motion on the unit sphere S of $\mathbf{R}^n$; $n \geq 3$. However, we may proceed as follows: Apply the function $\phi: \mathbf{R}^n \setminus \{0\} \to S$ defined by

$$\phi(x) = x \cdot |x|^{-1} \, ; \qquad x \in \mathbf{R}^n \setminus \{0\}$$

to n-dimensional Brownian motion $B = (B_1, \ldots, B_n)$. The result is a stochastic integral $Y = (Y_1, \ldots, Y_n) = \phi(B)$ which by Itô's formula is given by

$$dY_i = \frac{|B|^2 - B_i^2}{|B|^3} dB_i - \sum_{j \neq i} \frac{B_j B_i}{|B|^3} dB_j - \frac{n-1}{2} \cdot \frac{B_i}{|B|^3} dt \; ; \quad i = 1, 2, \ldots, n \; .$$

$$(8.5.16)$$

Hence

$$dY = \frac{1}{|B|} \cdot \sigma(Y) dB + \frac{1}{|B|^2} b(Y) dt \; ,$$

where

$$\sigma = [\sigma_{ij}] \in \mathbf{R}^{n \times n} \; , \quad \text{with } \sigma_{ij}(Y) = \delta_{ij} - Y_i Y_j ; \; 1 \leq i, \; j \leq n$$

and

$$b(y) = -\frac{n-1}{2} \cdot \begin{pmatrix} y_1 \\ \vdots \\ y_n \end{pmatrix} \in \mathbf{R}^n \; , \quad (y_1, \ldots, y_n \text{ are the coordinates of } y \in \mathbf{R}^n) \; .$$

Now perform the following time change: Define

$$Z_t(\omega) = Y_{\alpha(t,\omega)}(\omega)$$

where

$$\alpha_t = \beta_t^{-1} \; , \quad \beta(t, \omega) = \int_0^t \frac{1}{|B|^2} ds \; .$$

Then Z is again an Itô process and by Theorem 8.5.7

$$dZ = \sigma(Z) d\widetilde{B} + b(Z) dt \; .$$

Hence Z is a diffusion with characteristic operator

$$\mathcal{A}f(y) = \tfrac{1}{2} \left(\Delta f(y) - \sum_{i,j} y_i y_j \frac{\partial^2 f}{\partial y_i \partial y_j} \right) - \frac{n-1}{2} \cdot \sum_i y_i \frac{\partial f}{\partial y_i} \; ; \quad |y| = 1 \; .$$

$$(8.5.17)$$

Thus, $\phi(B) = \frac{B}{|B|}$ is – after a suitable change of time scale – equal to a diffusion Z living on the unit sphere S of $\mathbf{R}^n$. Note that Z is invariant under orthogonal transformations in $\mathbf{R}^n$ (since B is). It is reasonable to call Z *Brownian motion on the unit sphere S*. For other constructions see Itô and McKean (1965, p. 269 (§7.15)) and Stroock (1971).

More generally, given a Riemannian manifold M with metric tensor $g = [g_{ij}]$ one may define a *Brownian motion on M* as a diffusion on M whose characteristic operator $\mathcal{A}$ in local coordinates x_i is given by $\frac{1}{2}$ times the Laplace-Beltrami operator (here $[g^{ij}] = [g_{ij}]^{-1}$)

$$\Delta_M = \frac{1}{\sqrt{\det(g)}} \cdot \sum_i \frac{\partial}{\partial x_i} \left(\sqrt{\det(g)} \sum_j g^{ij} \frac{\partial}{\partial x_j} \right) . \qquad (8.5.18)$$

See for example Meyer (1966, p. 256–270), McKean (1969, §4.3). The subject of stochastic differential equations on manifolds is also treated in Ikeda and Watanabe (1989), Emery (1989) and Elworthy (1982).

Example 8.5.9 (Harmonic and analytic functions)
Let $B = (B_1, B_2)$ be 2-dimensional Brownian motion. Let us investigate what happens if we apply a C^2 function

$$\phi(x_1, x_2) = (u(x_1, x_2), v(x_1, x_2))$$

to B:
Put $Y = (Y_1, Y_2) = \phi(B_1, B_2)$ and apply Itô's formula:

$$dY_1 = u_1'(B_1, B_2)dB_1 + u_2'(B_1, B_2)dB_2 + \tfrac{1}{2}[u_{11}''(B_1, B_2) + u_{22}''(B_1, B_2)]dt$$

and

$$dY_2 = v_1'(B_1, B_2)dB_1 + v_2'(B_1, B_2)dB_2 + \tfrac{1}{2}[v_{11}''(B_1, B_2) + v_{22}''(B_1, B_2)]dt ,$$

where $u_1' = \frac{\partial u}{\partial x_1}$ etc. So

$$dY = b(B_1, B_2)dt + \sigma(B_1, B_2)dB ,$$

with $b = \tfrac{1}{2} \begin{pmatrix} \Delta u \\ \Delta v \end{pmatrix}$, $\sigma = \begin{pmatrix} u_1' & u_2' \\ v_1' & v_2' \end{pmatrix} = D_\phi$ (the derivative of ϕ).

So $Y = \phi(B_1, B_2)$ is a martingale if (and, in fact, only if) ϕ is harmonic, i.e. $\Delta\phi = 0$. If ϕ is harmonic, we get by Corollary 8.5.3 that

$$\phi(B_1, B_2) = (\widetilde{B}_{\beta_1}^{(1)}, \widetilde{B}_{\beta_2}^{(2)})$$

where $\widetilde{B}^{(1)}$ and $\widetilde{B}^{(2)}$ are two (not necessarily independent) versions of 1-dimensional Brownian motion, and

$$\beta_1(t, \omega) = \int_0^t |\nabla u|^2(B_1, B_2)ds , \quad \beta_2(t, \omega) = \int_0^t |\nabla v|^2(B_1, B_2)ds .$$

Since

$$\sigma\sigma^T = \begin{pmatrix} |\nabla u|^2 & \nabla u \cdot \nabla v \\ \nabla u \cdot \nabla v & |\nabla v|^2 \end{pmatrix}$$

we see that if (in addition to $\Delta u = \Delta v = 0$)

$$|\nabla u|^2 = |\nabla v|^2 \quad \text{and} \quad \nabla u \cdot \nabla v = 0 \qquad (8.5.19)$$

then

$$Y_t = Y_0 + \int\limits_0^t \sigma dB$$

with

$$\sigma \sigma^T = |\nabla u|^2 (B_1, B_2) I_2 \ , \quad Y_0 = \phi(B_1(0), B_2(0)) \ .$$

Therefore, if we let

$$\beta_t = \beta(t, \omega) = \int\limits_0^t |\nabla u|^2 (B_1, B_2) ds \ , \qquad \alpha_t = \beta_t^{-1}$$

we obtain by Theorem 8.5.2 that Y_{α_t} is a *2-dimensional Brownian motion*. Conditions (8.5.19) – in addition to $\Delta u = \Delta v = 0$ – are easily seen to be equivalent to requiring that the function $\phi(x + iy) = \phi(x, y)$ regarded as a complex function is either *analytic* or *conjugate analytic*.

Thus we have proved a theorem of P. Lévy that $\phi(B_1, B_2)$ is – after a change of time scale – again Brownian motion in the plane if and only if ϕ is either analytic or conjugate analytic. For extensions of this result see Bernard, Campbell and Davie (1979), Csink and Øksendal (1983) and Csink, Fitzsimmons and Øksendal (1990).

8.6 The Girsanov Theorem

We end this chapter by discussing a result, the Girsanov theorem, which is fundamental in the general theory of stochastic analysis. It is also very important in many applications, for example in economics (see Chapter 12).

Basically the Girsanov theorem says that if we change the *drift* coefficient of a given Itô process (with a nondegenerate diffusion coefficient), then the law of the process will not change dramatically. In fact, the law of the new process will be absolutely continuous w.r.t. the law of the original process and we can compute explicitly the Radon-Nikodym derivative.

We now proceed to make this precise. First we state (without proof) the useful Lévy characterization of Brownian motion. A proof can be found in e.g. Ikeda & Watanabe (1989), Theorem II.6.1, or in Karatzas & Shreve (1991), Theorem 3.3.16.

Theorem 8.6.1 (The Lévy characterization of Brownian motion)
Let $X(t) = (X_1(t), \ldots, X_n(t))$ be a continuous stochastic process on a probability space $(\Omega, \mathcal{H}, Q)$ with values in $\mathbf{R}^n$. Then the following, a) and b), are equivalent

a) $X(t)$ *is a Brownian motion w.r.t. Q, i.e. the law of $X(t)$ w.r.t. Q is the same as the law of an n-dimensional Brownian motion.*

b) (i) $X(t) = (X_1(t), \ldots, X_n(t))$ *is a martingale w.r.t. Q (and w.r.t. its own filtration) and*

(ii) $X_i(t)X_j(t) - \delta_{ij}t$ *is a martingale w.r.t. Q (and w.r.t. its own filtration) for all $i, j \in \{1, 2, \ldots, n\}$.*

Remark. In this Theorem one may replace condition (ii) by the condition

(ii)′ The *cross-variation processes* $\langle X_i, X_j \rangle_t$ satisfy the identity

$$\langle X_i, X_j \rangle_t(\omega) = \delta_{ij}t \qquad a.s., \ 1 \leq i, \ j \leq n \qquad (8.6.1)$$

where

$$\langle X_i, X_j \rangle_t = \tfrac{1}{4}[\langle X_i + X_j, X_i + X_j \rangle_t - \langle X_i - X_j, X_i - X_j \rangle_t], \quad (8.6.2)$$

$\langle Y, Y \rangle_t$ being the quadratic variation process. (See Exercise 4.7.)

Next we need an auxiliary result about conditional expectation:

Lemma 8.6.2 (Bayes' rule) *Let μ and ν be two probability measures on a measurable space $(\Omega, \mathcal{G})$ such that $d\nu(\omega) = f(\omega)d\mu(\omega)$ for some $f \in L^1(\mu)$. Let X be a random variable on $(\Omega, \mathcal{G})$ such that*

$$E_\nu[|X|] = \int_\Omega |X(\omega)|f(\omega)d\mu(\omega) < \infty .$$

Let $\mathcal{H}$ be a σ-algebra, $\mathcal{H} \subset \mathcal{G}$. Then

$$E_\nu[X|\mathcal{H}] \cdot E_\mu[f|\mathcal{H}] = E_\mu[fX|\mathcal{H}] \quad a.s. \qquad (8.6.3)$$

Proof. By the definition of conditional expectation (Appendix B) we have that if $H \in \mathcal{H}$ then

$$\int_H E_\nu[X|\mathcal{H}]fd\mu = \int_H E_\nu[X|\mathcal{H}]d\nu = \int_H Xd\nu$$

$$= \int_H Xfd\mu = \int_H E_\mu[fX|\mathcal{H}]d\mu \qquad (8.6.4)$$

On the other hand, by Theorem B.3 (Appendix B) we have

$$\int_H E_\nu[X|\mathcal{H}]fd\mu = E_\mu[E_\nu[X|\mathcal{H}]f \cdot \mathcal{X}_H] = E_\mu[E_\mu[E_\nu[X|\mathcal{H}]f \cdot \mathcal{X}_H|\mathcal{H}]]$$

$$= E_\mu[\mathcal{X}_H E_\nu[X|\mathcal{H}] \cdot E_\mu[f|\mathcal{H}]] = \int_H E_\nu[X|\mathcal{H}] \cdot E_\mu[f|\mathcal{H}]d\mu . \qquad (8.6.5)$$

Combining (8.6.4) and (8.6.5) we get

$$\int_H E_\nu[X|\mathcal{H}] \cdot E_\mu[f|\mathcal{H}]d\mu = \int_H E_\mu[fX|\mathcal{H}]d\mu \ .$$

Since this holds for all $H \in \mathcal{H}$, (8.6.3) follows.   □

Before stating the Girsanov theorem we make some general remarks about absolute continuity of measures:

Let $(\Omega, \mathcal{F}, \{\mathcal{F}\}_{t\geq 0}, P)$ be a filtered probability space (i.e. $(\Omega, \mathcal{F}, P)$ is a probability space and $\{\mathcal{F}\}_{t\geq 0}$ is a filtration on $(\Omega, \mathcal{F})$). Fix $T > 0$ and let Q be another probability measure on $\mathcal{F}_T$. We say that Q is *absolutely continuous* w.r.t. $P|_{\mathcal{F}_T}$ (the restriction of P to $\mathcal{F}_T$) and write $Q \ll P$ if

$$P(H) = 0 \Rightarrow Q(H) = 0 \qquad \text{for all } H \in \mathcal{F}_T.$$

By the Radon-Nikodym theorem this occurs if and only if there exists an $\mathcal{F}_T$-measurable random variable $Z_T(\omega) \geq 0$ such that

$$dQ(\omega) = Z_T(\omega)dP(\omega) \qquad \text{on } \mathcal{F}_T.$$

In this case we write

$$\frac{dQ}{dP} = Z_T \qquad \text{on } \mathcal{F}_T$$

and we call Z_T the *Radon-Nikodym derivative* of Q with respect to P.

The following observation may be regarded as a weak partial converse of the Girsanov theorem:

Lemma 8.6.3 *Suppose $Q \ll P|_{\mathcal{F}_T}$ with $\frac{dQ}{dP} = Z_T$ on $\mathcal{F}_T$. Then $Q|_{\mathcal{F}_t} \ll P|_{\mathcal{F}_t}$ for all $t \in [0, T]$ and if we define*

$$Z_t := \frac{d(Q|_{\mathcal{F}_t})}{d(P|_{\mathcal{F}_t})}$$

then Z_t is a martingale w.r.t. $\mathcal{F}_t$ and P.

Proof. Since $Q \ll P$ on $\mathcal{F}_T$ and $\mathcal{F}_t \subset \mathcal{F}_T$ it is obvious that $Q \ll P$ on $\mathcal{F}_t$. Choose $F \in \mathcal{F}_t$. Then

$$E_P[\mathcal{X}_F \cdot E_P[Z_T|\mathcal{F}_t]] = E_P[E_P[\mathcal{X}_F \cdot Z_T|\mathcal{F}_t]]$$
$$= E_P[\mathcal{X}_F \cdot Z_T] = E_Q[\mathcal{X}_F] = E_P[\mathcal{X}_F \cdot Z_t].$$

Since this holds for all $F \in \mathcal{F}_t$ we conclude that

$$E_P[Z_T|\mathcal{F}_t] = Z_t \quad \text{a.s.} \quad P|_{\mathcal{F}_t}.$$

                                                                                    □

We can now prove the first version of the Girsanov formula:

Theorem 8.6.4 (The Girsanov theorem I)
Let $Y(t) \in \mathbf{R}^n$ be an Itô process of the form

$$dY(t) = a(t, \omega)dt + dB(t) ; \qquad t \leq T, \ Y_0 = 0 .$$

where $T \leq \infty$ is a given constant and $B(t)$ is n-dimensional Brownian motion. Put

$$M_t = \exp\left(-\int_0^t a(s, \omega)dB_s - \tfrac{1}{2}\int_0^t a^2(s, \omega)ds \right) ; \qquad 0 \leq t \leq T . \quad (8.6.6)$$

Assume that M_t is a martingale with respect to $\mathcal{F}_t^{(n)}$ and P. Define the measure Q on $\mathcal{F}_T^{(n)}$ by

$$dQ(\omega) = M_T(\omega)dP(\omega) . \qquad (8.6.7)$$

Then Q is a probability measure on $\mathcal{F}_T^{(n)}$ and $Y(t)$ is an n-dimensional Brownian motion w.r.t. Q, for $0 \leq t \leq T$.

Remarks.

(1) The transformation $P \to Q$ given by (8.6.7) is called the *Girsanov transformation of measures*.
(2) As pointed out in Exercise 4.4 the following *Novikov condition* (8.6.8) is sufficient to guarantee that $\{M_t\}_{t \leq T}$ is a martingale (w.r.t. $\mathcal{F}_t^{(n)}$ and P):

$$E\left[\exp\left(\tfrac{1}{2}\int_0^T a^2(s, \omega)ds \right) \right] < \infty \qquad (8.6.8)$$

where $E = E_P$ is the expectation w.r.t. P.
(3) Note that since M_t is a martingale we actually have that

$$M_T dP = M_t dP \qquad \text{on } \mathcal{F}_t^{(n)}; \ t \leq T . \qquad (8.6.9)$$

To see this, let f be a bounded $\mathcal{F}_t^{(n)}$-measurable function. Then by Theorem B.3 we have

$$\int_\Omega f(\omega)M_T(\omega)dP(\omega) = E[fM_T] = E[E[fM_T|\mathcal{F}_t]]$$

$$= E[fE[M_T|\mathcal{F}_t]] = E[fM_t] = \int_\Omega f(\omega)M_t(\omega)dP(\omega) .$$

Proof of Theorem 8.6.4. Since M_t is a martingale we have

$$Q(\Omega) = E_Q[1] = E_P[M_T] = 1.$$

Hence Q is a probability measure. For simplicity we assume that $a(s, \omega)$ is bounded. In view of Theorem 8.6.1 we have to verify that

(i) $Y(t) = (Y_1(t), \ldots, Y_n(t))$ is a martingale w.r.t. Q (8.6.10)

and

(ii) $Y_i(t)Y_j(t) - \delta_{ij}t$ is a martingale w.r.t. Q,

               for all $i, j \in \{1, 2, \ldots, n\}$.   (8.6.11)

To verify (i) we put $K(t) = M_t Y(t)$ and use Itô's formula to get (see Exercises 4.3, 4.4)

$$
\begin{aligned}
dK_i(t) &= M_t dY_i(t) + Y_i(t)dM_t + dY_i(t)dM_t \\
&= M_t(a_i(t)dt + dB_i(t)) + Y_i(t)M_t\left(\sum_{k=1}^{n} -a_k(t)dB_k(t)\right) \\
&\quad + (dB_i(t))\left(-M_t\sum_{k=1}^{n} a_k(t)dB_k(t)\right) \\
&= M_t(dB_i(t) - Y_i(t)\sum_{k=1}^{n} a_k(t)dB_k(t)) = M_t\,\gamma^{(i)}(t)dB(t) \quad (8.6.12)
\end{aligned}
$$

where $\gamma^{(i)}(t) = (\gamma_1^{(i)}(t), \ldots, \gamma_n^{(i)}(t))$, with

$$
\gamma_j^{(i)}(t) = \begin{cases} -Y_i(t)a_j(t) & \text{for } j \neq i \\ 1 - Y_i(t)a_i(t) & \text{for } j = i . \end{cases}
$$

Hence $K_i(t)$ is a martingale w.r.t. P, so by Lemma 8.6.2 we get, for $t > s$,

$$
\begin{aligned}
E_Q[Y_i(t)|\mathcal{F}_s] &= \frac{E[M_t Y_i(t)|\mathcal{F}_s]}{E[M_t|\mathcal{F}_s]} = \frac{E[K_i(t)|\mathcal{F}_s]}{M_s} \\
&= \frac{K_i(s)}{M_s} = Y_i(s) ,
\end{aligned}
$$

which shows that $Y_i(t)$ is a martingale w.r.t. Q. This proves (i). The proof of (ii) is similar and is left to the reader. $\square$

Remark. Theorem 8.6.4 states that for all Borel sets $F_1, \ldots, F_k \subset \mathbf{R}^n$ and all $t_1, t_2, \ldots, t_k \leq T$, $k = 1, 2, \ldots$ we have

$$
Q[Y(t_1) \in F_1, \ldots, Y(t_k) \in F_k] = P[B(t_1) \in F_1, \ldots, B(t_k) \in F_k] \quad (8.6.13)
$$

An equivalent way of expressing (8.6.7) is to say that $Q \ll P$ (Q is absolutely continuous w.r.t. P) with Radon-Nikodym derivative

$$
\frac{dQ}{dP} = M_T \quad \text{on } \mathcal{F}_T^{(n)} .   (8.6.14)
$$

Note that $M_T(\omega) > 0$ a.s., so we also have that $P \ll Q$. Hence the two measures Q and P are equivalent. Therefore we get from (8.6.13)

$$P[Y(t_1) \in F_1, \ldots, Y(t_k) \in F_k] > 0$$
$$\Longleftrightarrow Q[Y(t_1) \in F_1, \ldots, Y(t_k) \in F_k] > 0$$
$$\Longleftrightarrow P[B(t_1) \in F_1, \ldots, B(t_k) \in F_k] > 0; \quad t_1, \ldots, t_k \in [0, T] \quad (8.6.15)$$

Example 8.6.5 Suppose $Y(t) \in \mathbf{R}^n$ is given by

$$dY(t) = g(t)dt + dB(t), \qquad 0 \le t \le T$$

where $g : [0, T] \rightarrow \mathbf{R}^n$ is a continuous *deterministic* function. Then the Novikov condition (8.6.8) holds trivially and $Y(t)$ is a Brownian motion w.r.t. Q, where

$$dQ(\omega) = \exp\left(-\int_0^T g(s)dB(s) - \tfrac{1}{2}\int_0^T g^2(s)ds \right) dP(\omega) \qquad \text{on } \mathcal{F}_T^{(n)}.$$

Theorem 8.6.6 (The Girsanov theorem II)
Let $Y(t) \in \mathbf{R}^n$ be an Itô process of the form

$$dY(t) = \beta(t, \omega)dt + \theta(t, \omega)dB(t) ; \qquad t \le T \qquad (8.6.16)$$

where $B(t) \in \mathbf{R}^m$, $\beta(t, \omega) \in \mathbf{R}^n$ and $\theta(t, \omega) \in \mathbf{R}^{n \times m}$. Suppose there exist processes $u(t, \omega) \in \mathcal{W}_{\mathcal{H}}^m$ and $\alpha(t, \omega) \in \mathcal{W}_{\mathcal{H}}^n$ such that

$$\theta(t, \omega)u(t, \omega) = \beta(t, \omega) - \alpha(t, \omega) \qquad (8.6.17)$$

Put

$$M_t = \exp\left(-\int_0^t u(s, \omega)dB_s - \tfrac{1}{2}\int_0^t u^2(s, \omega)ds \right) ; \qquad t \le T \qquad (8.6.18)$$

and

$$dQ(\omega) = M_T(\omega)dP(\omega) \qquad \text{on } \mathcal{F}_T^{(m)} . \qquad (8.6.19)$$

Assume that M_t is a martingale (w.r.t. $\mathcal{F}_t^{(n)}$ and P). Then Q is a probability measure on $\mathcal{F}_T^{(m)}$, the process

$$\widehat{B}(t) := \int_0^t u(s, \omega)ds + B(t) ; \qquad t \le T \qquad (8.6.20)$$

is a Brownian motion w.r.t. Q and in terms of $\widehat{B}(t)$ the process $Y(t)$ has the stochastic integral representation

$$dY(t) = \alpha(t, \omega)dt + \theta(t, \omega)d\widehat{B}(t) . \tag{8.6.21}$$

Remark. As in the Remark following Theorem 8.6.4 we note that the following Novikov condition is sufficient to guarantee that M_t is a martingale:

$$E\left[\exp\left(\tfrac{1}{2}\int_0^T u^2(s, \omega)ds\right)\right] < \infty . \tag{8.6.22}$$

Proof. It follows from Theorem 8.6.4 that Q is a probability measure on $\mathcal{F}_T^{(m)}$ and $\widehat{B}(t)$ is a Brownian motion w.r.t. Q. So, substituting (8.6.20) in (8.6.16) we get, by (8.6.17),

$$
\begin{aligned}
dY(t) &= \beta(t, \omega)dt + \theta(t, \omega)(d\widehat{B}(t) - u(t, \omega)dt) \\
&= [\beta(t, \omega) - \theta(t, \omega)u(t, \omega)]dt + \theta(t, \omega)d\widehat{B}(t) \\
&= \alpha(t, \omega)dt + \theta(t, \omega)d\widehat{B}(t) .
\end{aligned}
$$

$\square$

Note that if $n = m$ and $\theta \in \mathbf{R}^{n \times n}$ is invertible, then the process $u(t, \omega)$ satisfying (8.6.17) is given uniquely by

$$u(t, \omega) = \theta^{-1}(t, \omega)[\beta(t, \omega) - \alpha(t, \omega)] . \tag{8.6.23}$$

Remark. In most applications, e.g. in finance (see Chapter 12) the process $\alpha(t, \omega)$ is chosen to be 0. Then the process $Y(t)$ gets the form (see (8.6.21))

$$dY(t) = \theta(t, \omega)d\widetilde{B}(t),$$

which implies that $Y(t)$ is a *local martingale* w.r.t. Q. In this case Q is called an *equivalent local martingale measure*. See Chapter 12.

Example 8.6.7 Suppose $Y(t) = \begin{bmatrix} Y_1(t) \\ Y_2(t) \end{bmatrix} \in \mathbf{R}^2$ is given by

$$
\begin{aligned}
dY_1(t) &= 2dt + dB_1(t) + dB_2(t) \\
dY_2(t) &= 4dt + dB_1(t) - dB_2(t)
\end{aligned}
$$

i.e.
$$dY(t) = \begin{bmatrix} 2 \\ 4 \end{bmatrix} dt + \begin{bmatrix} 1 & 1 \\ 1 & -1 \end{bmatrix} dB(t) ; \qquad B(t) = \begin{bmatrix} B_1(t) \\ B_2(t) \end{bmatrix} .$$

Choose $\alpha(t, \omega) = 0$. Then equation (8.6.17) gets the form

$$\begin{bmatrix} 1 & 1 \\ 1 & -1 \end{bmatrix} \begin{bmatrix} u_1 \\ u_2 \end{bmatrix} = \begin{bmatrix} 2 \\ 4 \end{bmatrix}$$

which has the unique solution

$$u_1 = 3, \quad u_2 = -1 .$$

Hence we put

$$dQ(\omega) = \exp(-3B_1(T) + B_2(T) - 5T)dP(\omega) \qquad \text{on } \mathcal{F}_T^{(2)}$$

and

$$d\widehat{B}(t) = \begin{bmatrix} 3 \\ -1 \end{bmatrix} dt + dB(t) \ .$$

Then trivially the Novikov condition holds (see Example 8.6.5) and we conclude that $\widehat{B}(t)$ is a Brownian motion w.r.t. the probability measure Q and

$$dY(t) = \begin{bmatrix} 1 & 1 \\ 1 & -1 \end{bmatrix} d\widehat{B}(t) \ .$$

Thus in this case $Y(t)$ is in fact a *martingale* w.r.t. Q, i.e. Q is an *equivalent martingale measure* for $Y(t)$.

Finally we formulate a diffusion version:

Theorem 8.6.8 (The Girsanov theorem III)
Let $X(t) = X^x(t) \in \mathbf{R}^n$ and $Y(t) = Y^x(t) \in \mathbf{R}^n$ be an Itô diffusion and an Itô process, respectively, of the forms

$$dX(t) = b(X(t))dt + \sigma(X(t))dB(t) \ ; \qquad t \leq T, \ \ X(0) = x \qquad (8.6.24)$$
$$dY(t) = [\gamma(t,\omega) + b(Y(t))]dt + \sigma(Y(t))dB(t) \ ; \quad t \leq T, \ \ Y(0) = x \quad (8.6.25)$$

where the functions $b \colon \mathbf{R}^n \to \mathbf{R}^n$ and $\sigma \colon \mathbf{R}^n \to \mathbf{R}^{n \times m}$ satisfy the conditions of Theorem 5.2.1 and $\gamma(t,\omega) \in \mathcal{W}_{\mathcal{H}}^n$, $x \in \mathbf{R}^n$. Suppose there exists a process $u(t,\omega) \in \mathcal{W}_{\mathcal{H}}^m$ such that

$$\sigma(Y(t))u(t,\omega) = \gamma(t,\omega). \qquad (8.6.26)$$

Define M_t, Q and $\widehat{B}(t)$ as in (8.6.18), (8.6.19) and (8.6.20). Assume that M_t is a martingale w.r.t. $\mathcal{F}_t^{(m)}$ and P. Then Q is a probability measure on $\mathcal{F}_T^{(m)}$ and

$$dY(t) = b(Y(t))dt + \sigma(Y(t))d\widehat{B}(t) \ . \qquad (8.6.27)$$

Therefore,
$$\begin{aligned} &\text{the } Q\text{-law of } Y^x(t) \text{ is the same as} \\ &\text{the } P\text{-law of } X^x(t); \ \ t \leq T \ . \end{aligned} \qquad (8.6.28)$$

Proof. The representation (8.6.27) follows by applying Theorem 8.6.6 to the case $\theta(t,\omega) = \sigma(Y(t))$, $\beta(t,\omega) = \gamma(t,\omega) + b(Y(t))$, $\alpha(t,\omega) = b(Y(t))$. Then the conclusion (8.6.28) follows from the weak uniqueness of solutions of stochastic differential equations (Lemma 5.3.1). □

The Girsanov theorem III can be used to produce weak solutions of stochastic differential equations. To illustrate this, suppose Y_t is a known weak or strong solution to the equation

$$dY_t = b(Y_t)dt + \sigma(Y_t)dB(t) \tag{8.6.29}$$

where $b \colon \mathbf{R}^n \to \mathbf{R}^n$, $\sigma \colon \mathbf{R}^n \to \mathbf{R}^{n \times m}$ and $B(t) \in \mathbf{R}^m$. We wish to find a weak solution $X(t)$ of a related equation

$$dX_t = a(X_t)dt + \sigma(X_t)dB(t) \tag{8.6.30}$$

where the drift function is changed to $a \colon \mathbf{R}^n \to \mathbf{R}^n$. Suppose we can find a function $u_0 \colon \mathbf{R}^n \to \mathbf{R}^m$ such that

$$\sigma(y)u_0(y) = b(y) - a(y) \ ; \qquad y \in \mathbf{R}^n \ .$$

(If $n = m$ and σ is invertible we choose

$$u_0 = \sigma^{-1} \cdot (b - a) \ .)$$

Then if $u(t, \omega) = u_0(Y_t(\omega))$ satisfies Novikov's conditions, we have, with Q and $\widehat{B}_t = \widehat{B}(t)$ as in (8.6.20) and (8.6.21), that

$$dY_t = a(Y_t)dt + \sigma(Y_t)d\widehat{B}_t \ . \tag{8.6.31}$$

Thus we have found a Brownian motion $(\widehat{B}_t, Q)$ such that Y_t satisfies (8.6.31). Therefore $(Y_t, \widehat{B}_t)$ is a weak solution of (8.6.30).

Example 8.6.9 Let $a \colon \mathbf{R}^n \to \mathbf{R}^n$ be a bounded, measurable function. Then we can construct a weak solution $X_t = X_t^x$ of the stochastic differential equation

$$dX_t = a(X_t)dt + dB_t \ ; \qquad X_0 = x \in \mathbf{R}^n \ . \tag{8.6.32}$$

We proceed according to the procedure above, with $\sigma = I$, $b = 0$ and

$$dY_t = dB_t \ ; \qquad Y_0 = x \ .$$

Choose

$$u_0 = \sigma^{-1} \cdot (b - a) = -a$$

and define

$$M_t = \exp\left\{ -\int\limits_0^t u_0(Y_s)dB_s - \tfrac{1}{2}\int\limits_0^t u_0^2(Y_s)ds \right\}$$

i.e.

$$M_t = \exp\left\{ \int\limits_0^t a(B_s)dB_s - \tfrac{1}{2}\int\limits_0^t a^2(B_s)ds \right\} \ .$$

Fix $T < \infty$ and put

$$dQ = M_T dP \qquad \text{on } \mathcal{F}_T^{(m)} \ .$$

Then

$$\widehat{B}_t := -\int_0^t a(B_s)ds + B_t$$

is a Brownian motion w.r.t. Q for $t \leq T$ and

$$dB_t = dY_t = a(Y_t)dt + d\widehat{B}_t \ .$$

Hence if we set $Y_0 = x$ the pair $(Y_t, \widehat{B}_t)$ is a weak solution of (8.6.32) for $t \leq T$. By weak uniqueness the Q-law of $Y_t = B_t$ coincides with the P-law of X_t^x, so that

$$\begin{aligned}
E[f_1(X_{t_1}^x)\ldots f_k(X_{t_k}^x)] &= E_Q[f_1(Y_{t_1})\ldots f_k(Y_{t_k})] \\
&= E[M_T f_1(B_{t_1})\ldots f_k(B_{t_k})] \qquad (8.6.33)
\end{aligned}$$

for all $f_1, \ldots, f_k \in C_0(\mathbf{R}^n)$; $t_1, \ldots, t_k \leq T$.

Exercises

8.1.* Let Δ denote the Laplace operator on $\mathbf{R}^n$.

    a)   Write down (in terms of Brownian motion) a bounded solution g of the Cauchy problem

$$\begin{cases} \dfrac{\partial g(t,x)}{\partial t} - \frac{1}{2}\Delta_x g(t,x) = 0 & \text{for } t > 0, \ x \in \mathbf{R}^n \\ g(0,x) = \phi(x) \end{cases}$$

        where $\phi \in C_0^2$ is given. (From general theory it is known that the solution is unique.)

    b)   Let $\psi \in C_b(\mathbf{R}^n)$ and $\alpha > 0$. Find a bounded solution u of the equation

$$(\alpha - \tfrac{1}{2}\Delta)u = \psi \qquad \text{in } \mathbf{R}^n \ .$$

    Prove that the solution is unique.

8.2.   Show that the solution $u(t,x)$ of the initial value problem

$$\frac{\partial u}{\partial t} = \tfrac{1}{2}\beta^2 x^2 \frac{\partial^2 u}{\partial x^2} + \alpha x \frac{\partial u}{\partial x} \ ; \qquad t > 0, \ x \in \mathbf{R}$$

$$u(0,x) = f(x) \qquad (f \in C_0^2(\mathbf{R}) \text{ given})$$

can be expressed as follows:

$$\begin{aligned}
u(t,x) &= E[f(x \cdot \exp\{\beta B_t + (\alpha - \tfrac{1}{2}\beta^2)t\})] \\
&= \frac{1}{\sqrt{2\pi t}} \int_{\mathbf{R}} f(x \cdot \exp\{\beta y + (\alpha - \tfrac{1}{2}\beta^2)t\}) \exp\left(-\frac{y^2}{2t}\right) dy \ ; \quad t > 0 \ .
\end{aligned}$$

8.3.   **(Kolmogorov's forward equation)**
Let X_t be an Itô diffusion in $\mathbf{R}^n$ with generator

$$Af(y) = \sum_{i,j} a_{ij}(y)\frac{\partial^2 f}{\partial y_i \partial y_j} + \sum_i b_i(y)\frac{\partial f}{\partial y_i} \; ; \qquad f \in C_0^2$$

and assume that the transition measure of X_t has a *density* $p_t(x, y)$, i.e. that

$$E^x[f(X_t)] = \int_{\mathbf{R}^n} f(y)p_t(x, y)dy \; ; \qquad f \in C_0^2 \; . \qquad (8.6.34)$$

Assume that $y \to p_t(x, y)$ is smooth for each t, x. Prove that $p_t(x, y)$ satisfies the *Kolmogorov forward equation*

$$\frac{d}{dt}p_t(x, y) = A_y^* p_t(x, y) \qquad \text{for all } x, y \; , \qquad (8.6.35)$$

where A_y^* operates on the variable y and is given by

$$A_y^* \phi(y) = \sum_{i,j} \frac{\partial^2}{\partial y_i \partial y_j}(a_{ij}\phi) - \sum_i \frac{\partial}{\partial y_i}(b_i\phi) \; ; \qquad \phi \in C^2 \quad (8.6.36)$$

(i.e. A_y^* is the *adjoint* of A_y.)
(Hint: By (8.6.34) and Dynkin's formula we have

$$\int_{\mathbf{R}^n} f(y)p_t(x, y)dy = f(x) + \int_0^t \int_{\mathbf{R}^n} A_y f(y)p_s(x, y)dy\, ds \; ; \qquad f \in C_0^2 \; .$$

Now differentiate w.r.t. t and use that

$$\langle A\phi, \psi \rangle = \langle \phi, A^*\psi \rangle \qquad \text{for } \phi \in C_0^2, \; \psi \in C^2 \; ,$$

where $\langle \cdot, \cdot \rangle$ denotes inner product in $L^2(dy)$.)

8.4.   Let B_t be n-dimensional Brownian motion ($n \geq 1$) and let F be a Borel set in $\mathbf{R}^n$. Prove that the expected total length of time t that B_t stays in F is zero if and only if the Lebesgue measure of F is zero. Hint: Consider the resolvent R_α for $\alpha > 0$ and then let $\alpha \to 0$.

8.5.   Show that the solution $u(t, x)$ of the initial value problem

$$\begin{cases} \frac{\partial u}{\partial t} = \rho\, u + \frac{1}{2}\Delta u \; t > 0 \; ; \quad x \in \mathbf{R}^n \\ u(0, x) = f(x) \quad (f \in C_0^2(\mathbf{R}^n) \text{ given}) \end{cases}$$

(where $\rho \in \mathbf{R}$ is a constant) can be expressed by

$$u(t, x) = (2\pi t)^{-n/2} \exp(\rho t) \int_{\mathbf{R}^n} f(y) \exp \left(-\frac{(x - y)^2}{2t} \right) dy \ .$$

8.6. In connection with the deduction of the Black-Scholes formula for the price of an option (see Chapter 12) the following partial differential equation appears:

$$\begin{cases} \frac{\partial u}{\partial t} = -\rho u + \alpha x \frac{\partial u}{\partial x} + \frac{1}{2} \beta^2 x^2 \frac{\partial^2 u}{\partial x^2} \ ; \ t > 0 \ , \ x \in \mathbf{R} \\ u(0, x) = (x - K)^+ \ ; \qquad\qquad\qquad x \in \mathbf{R} \ , \end{cases}$$

where $\rho > 0$, α, β and $K > 0$ are constants and

$$(x - K)^+ = \max(x - K, 0) \ .$$

Use the Feynman-Kac formula to prove that the solution u of this equation is given by

$$u(t, x) = \frac{e^{-\rho t}}{\sqrt{2\pi t}} \int_{\mathbf{R}} (x \cdot \exp\{(\alpha - \tfrac{1}{2}\beta^2)t + \beta y\} - K)^+ e^{-\frac{y^2}{2t}} \, dy \ ; \quad t > 0 \ .$$

(This expression can be simplified further. See Exercise 12.13.)

8.7. Let X_t be a sum of Itô integrals of the form

$$X_t = \sum_{k=1}^{n} \int_0^t v_k(s, \omega) dB_k(s) \ ,$$

where $(B_1, \ldots, B_n)$ is n-dimensional Brownian motion. Assume that

$$\beta_t := \int_0^t \sum_{k=1}^{n} v_k^2(s, \omega) ds \to \infty \qquad \text{as } t \to \infty, \text{ a.s.}$$

Prove that

$$\limsup_{t \to \infty} \frac{X_t}{\sqrt{2\beta_t \log \log \beta_t}} = 1 \quad \text{a.s.}$$

(Hint: Use the law of iterated logarithm.)

8.8. Let Z_t be a 1-dimensional Itô process of the form

$$dZ_t = u(t, \omega)dt + dB_t \ .$$

Let $\mathcal{G}_t$ be the σ-algebra generated by $\{Z_s(\cdot); s \leq t\}$ and define

$$dN_t = (u(t, \omega) - E[u|\mathcal{G}_t])dt + dB_t \ .$$

Use Corollary 8.4.5 to prove that N_t is a Brownian motion. (If we interpret Z_t as the *observation* process, then N_t is the *innovation* process. See Lemma 6.2.6.)

8.9. Define $\alpha(t) = \frac{1}{2}\ln(1 + \frac{2}{3}t^3)$. If B_t is a Brownian motion, prove that there exists another Brownian motion $\widetilde{B}_r$ such that

$$\int_0^{\alpha_t} e^s dB_s = \int_0^t r d\widetilde{B}_r \ .$$

8.10. Let B_t be a Brownian motion in $\mathbf{R}$. Show that

$$X_t := B_t^2$$

is a weak solution of the stochastic differential equation

$$dX_t = dt + 2\sqrt{|X_t|}d\widetilde{B}_t \ . \tag{8.6.37}$$

(Hint: Use Itô's formula to express X_t as a stochastic integral and compare with (8.6.37) by using Corollary 8.4.5.)

8.11.* a)   Let $Y(t) = t + B(t); \ t \geq 0$. For each $T > 0$ find a probability measure Q_T on $\mathcal{F}_T$ such that $Q_T \sim P$ and $\{Y(t)\}_{t \leq T}$ is Brownian motion w.r.t. Q_T. Use (8.6.9) to prove that there exists a probability measure Q on $\mathcal{F}_\infty$ such that

$$Q|\mathcal{F}_T = Q_T \qquad \text{for all } T > 0 \ .$$

b)   Show that

$$P\left(\lim_{t \to \infty} Y(t) = \infty\right) = 1$$

while

$$Q\left(\lim_{t \to \infty} Y(t) = \infty\right) = 0 \ .$$

Why does not this contradict the Girsanov theorem?

8.12.* Let

$$dY(t) = \begin{bmatrix} 0 \\ 1 \end{bmatrix} dt + \begin{bmatrix} 1 & 3 \\ -1 & -2 \end{bmatrix} \begin{bmatrix} dB_1(t) \\ dB_2(t) \end{bmatrix} \ ; \qquad t \leq T \ .$$

Find a probability measure Q on $\mathcal{F}_T^{(2)}$ such that $Q \sim P$ and such that

$$\widetilde{B}(t) := \begin{bmatrix} -3t \\ t \end{bmatrix} + \begin{bmatrix} B_1(t) \\ B_2(t) \end{bmatrix}$$

is a Brownian motion w.r.t. Q and

$$dY(t) = \begin{bmatrix} 1 & 3 \\ -1 & -2 \end{bmatrix} \begin{bmatrix} d\tilde{B}_1(t) \\ d\tilde{B}_2(t) \end{bmatrix} .$$

8.13. Let $b: \mathbf{R} \to \mathbf{R}$ be a Lipschitz-continuous function and define
$X_t = X_t^x \in \mathbf{R}$ by

$$dX_t = b(X_t)dt + dB_t, X_0 = x \in \mathbf{R} .$$

a) Use the Girsanov theorem to prove that for all $M < \infty$, $x \in \mathbf{R}$
and $t > 0$ we have

$$P[X_t^x \geq M] > 0 .$$

b) Choose $b(x) = -r$ where $r > 0$ is constant. Prove that for all x

$$X_t^x \to -\infty \qquad \text{as } t \to \infty \text{ a.s.}$$

Compare this with the result in a).

8.14. **(Polar sets for the graph of Brownian motion)**
Let B_t be 1-dimensional Brownian motion starting at $x \in \mathbf{R}$.

a) Prove that for every fixed time $t_0 > 0$ we have

$$P^x[B_{t_0} = 0] = 0 .$$

b) Prove that for every (non-trivial) closed interval $J \subset \mathbf{R}^+$ we
have

$$P^x[\exists t \in J \quad \text{such that } B_t = 0] > 0 .$$

(Hint: If $J = [t_1, t_2]$ consider $P^x[B_{t_1} < 0 \ \& \ B_{t_2} > 0]$ and then
use the intermediate value theorem.)

c) In view of a) and b) it is natural to ask what closed sets $F \subset \mathbf{R}^+$
have the property that

$$P^x[\exists t \in F \quad \text{such that } B_t = 0] = 0 . \qquad (8.6.38)$$

To investigate this question more closely we introduce the graph
X_t of Brownian motion, given by

$$dX_t = \begin{bmatrix} 1 \\ 0 \end{bmatrix} dt + \begin{bmatrix} 0 \\ 1 \end{bmatrix} dB_t ; \quad X_0 = \begin{bmatrix} t_0 \\ x_0 \end{bmatrix}$$

i.e.

$$X_t = X_t^{t_0, x_0} = \begin{bmatrix} t_0 + t \\ B_t^{x_0} \end{bmatrix} \quad \text{where } B_0^{x_0} = x_0 \text{ a.s.}$$

Then F satisfies (8.6.38) iff $K := F \times \{0\}$ is *polar* for X_t, in the
sense that

$$P^{t_0, x_0}[\exists t > 0 ; \ X_t \in K] = 0 \quad \text{for all } t_0, x_0 . \qquad (8.6.39)$$

The key to finding polar sets for a diffusion is to consider its *Green operator* R, which is simply the resolvent R_α with $\alpha = 0$:

$$Rf(t_0, x_0) = E^{t_0, x_0}\left[\int_{t_0}^{\infty} f(X_s)ds\right] \qquad \text{for } f \in C_0(\mathbf{R}^2) \ .$$

Show that

$$Rf(t_0, x_0) = \int_{\mathbf{R}^2} G(t_0, x_0; t, x)f(t, x)dt \, dx \ ,$$

where

$$G(t_0, x_0; t, x) = \mathcal{X}_{t > t_0} \cdot (2\pi(t - t_0))^{-\frac{1}{2}} \exp\left(-\frac{|x - x_0|^2}{2(t - t_0)}\right) \quad (8.6.40)$$

(G is the *Green function* of X_t.)

d) The *capacity* of K, $C(K) = C_G(K)$, is defined by

$$C(K) = \sup\{\mu(K); \mu \in M_G(K)\} \ ,$$

where $M_G(K) = \{\mu; \mu \text{ measure on } K \text{ s.t. } \int_K G(t_0, x_0; t, x)d\mu(t, x) \leq$ 1 for all $t_0, x_0\}$.

A general result from stochastic potential theory states that

$$P^{t_0, x_0}[X_t \text{ hits } K] = 0 \Leftrightarrow C(K) = 0 \ . \qquad (8.6.41)$$

See e.g. Blumenthal and Getoor (1968, Prop. VI.4.3). Use this to prove that

$$\Lambda_{\frac{1}{2}}(F) = 0 \Rightarrow P^{x_0}[\exists t \in F \quad \text{such that} \quad B_t = 0] = 0 \ ,$$

where $\Lambda_{\frac{1}{2}}$ denotes *1/2-dimensional Hausdorff measure* (Folland (1984, §10.2)).

8.15. Let $f \in C_0^2(\mathbf{R}^n)$ and $\alpha(x) = (\alpha_1(x), \ldots, \alpha_n(x))$ with $\alpha_i \in C_0^2(\mathbf{R}^n)$ be given functions and consider the partial differential equation

$$\begin{cases} \dfrac{\partial u}{\partial t} = \displaystyle\sum_{i=1}^n \alpha_i(x)\dfrac{\partial u}{\partial x_i} + \dfrac{1}{2}\sum_{i=1}^n \dfrac{\partial^2 u}{\partial x_i^2} \ ; \ t > 0, \ x \in \mathbf{R}^n \\ u(0, x) = f(x) \ ; \hspace{3.2cm} x \in \mathbf{R}^n \ . \end{cases}$$

a) Use the Girsanov theorem to show that the unique bounded solution $u(t, x)$ of this equation can be expressed by

$$u(t,x) = E^x\left[\exp\left(\int_0^t \alpha(B_s)dB_s - \frac{1}{2}\int_0^t \alpha^2(B_s)ds\right)f(B_t)\right],$$

where E^x is the expectation w.r.t. P^x.

b) Now assume that α is a gradient, i.e. that there exists $\gamma \in C^1(\mathbf{R}^n)$ such that

$$\nabla\gamma = \alpha .$$

Assume for simplicity that $\gamma \in C_0^2(\mathbf{R}^n)$. Use Itô's formula to prove that (see Exercise 4.8)

$$u(t,x) = \exp(-\gamma(x))E^x\left[\exp\left\{-\frac{1}{2}\int_0^t \left(\nabla\gamma^2(B_s)\right.\right.\right.$$

$$\left.\left.\left.+\Delta\gamma(B_s)\right)ds\right\}\exp(\gamma(B_t))f(B_t)\right].$$

c) Put $v(t,x) = \exp(\gamma(x))u(t,x)$. Use the Feynman-Kac formula to show that $v(t,x)$ satisfies the partial differential equation

$$\begin{cases} \frac{\partial v}{\partial t} = -\frac{1}{2}(\nabla\gamma^2 + \Delta\gamma)\cdot v + \frac{1}{2}\Delta v \; ; \; t > 0 \; ; \; x \in \mathbf{R}^n \\ v(0,x) = \exp(\gamma(x))f(x) \; ; \qquad\qquad x \in \mathbf{R}^n . \end{cases}$$

(See also Exercise 8.16.)

8.16. (A connection between B.m. with drift and killed B.m.)
Let B_t denote Brownian motion in $\mathbf{R}^n$ and consider the diffusion X_t in $\mathbf{R}^n$ defined by

$$dX_t = \nabla h(X_t)dt + dB_t \; ; \qquad X_0 = x \in \mathbf{R}^n . \qquad (8.6.42)$$

where $h \in C_0^1(\mathbf{R}^n)$.

a) There is an important connection between this process and the process Y_t obtained by *killing B_t at a certain rate V*. More precisely, first prove that for $f \in C_0(\mathbf{R}^n)$ we have

$$E^x[f(X_t)] = E^x\left[\exp\left(-\int_0^t V(B_s)ds\right)\cdot\exp(h(B_t)-h(x))\cdot f(B_t)\right],$$

$$(8.6.43)$$

where

$$V(x) = \frac{1}{2}|\nabla h(x)|^2 + \frac{1}{2}\Delta h(x) . \qquad (8.6.44)$$

(Hint: Use the Girsanov theorem to express the left hand side of
(8.6.43) in terms of B_t. Then use the Itô formula on $Z_t = h(B_t)$
to achieve (8.6.44).)

b)   Then use the Feynman-Kac formula to restate (8.6.43) as follows
(assuming $V \geq 0$):

$$T_t^X(f, x) = \exp(-h(x)) \cdot T_t^Y(f \cdot \exp h, x),$$

where T_t^X, T_t^Y denote the *transition operators* of the processes
X and Y, respectively, i.e.

$$T_t^X(f, x) = E^x[f(X_t)] \quad \text{and similarly for } Y.$$

8.17.   Suppose $Y(t) = \begin{bmatrix} Y_1(t) \\ Y_2(t) \end{bmatrix} \in \mathbf{R}^2$ is given by

$$dY_1(t) = \beta_1(t)dt + dB_1(t) + 2dB_2(t) + 3dB_3(t)$$
$$dY_2(t) = \beta_2(t)dt + dB_1(t) + 2dB_2(t) + 2dB_3(t)$$

where β_1, β_2 are bounded adapted processes.

Show that there are infinitely many equivalent martingale measures
Q for $Y(t)$. (See Example 8.6.7.)

8.18.   **(The Girsanov theorem in stochastic control)**
The following exercise is inspired by useful communications with
Jerome Stein. It is mainly based on an example given in Fleming
(1999), Chapter 2, Section 2.5. See also Platen and Rebolledo (1996).

Suppose we have a financial market with two investment possibilities:

(i)   a risk free asset, where the unit price $S_0(t)$ at time t is given by

$$dS_0(t) = \rho(t)S_0(t)dt; \qquad S_0(0) = 1$$

(ii)   a risky asset, where the unit price $S_1(t)$ at time t is given by

$$dS_1(t) = S_1(t)[\mu(t)dt + \sigma(t)dB(t)]; \qquad S_1(0) > 0.$$

We assume that the coefficients $\rho(t), \mu(t)$ and $\sigma(t) \neq 0$ are *deterministic* and satisfy

$$\int_0^T \{|\rho(t)| + |\mu(t)| + \sigma^2(t)\}dt < \infty.$$

A *portfolio* in this market can be represented by an $\mathcal{F}_t$-adapted
stochastic process $\pi(t)$ which gives the fraction of the total wealth
$X(t)$ invested in the risky asset at time t. If we assume that $\pi(t)$ is

self-financing (see Chapter 12) then the corresponding wealth process $X(t) = X_\pi(t)$ will have the dynamics

$$dX(t) = (1 - \pi(t))X(t)\rho(t)dt + \pi(t)X(t)[\mu(t)dt + \sigma(t)dB(t)]$$
$$X(0) = x > 0 \quad \text{(constant)}.$$

The problem is to find the portfolio π^* which maximizes the expected utility of the terminal wealth,

$$E[U(X_\pi(T)]$$

in the case when U is the *power utility function*, i.e.

$$U(x) = x^\gamma \quad \text{for some constant } \gamma \in (0, 1).$$

(See Chapter 11 and 12 for further discussions of this type of optimal portfolio problems.)

a)   Show that we can write

$$E[X_\pi^\gamma(T)] = K\, E\Big[Z_\pi \exp\Big(\gamma \int\limits_0^T \{(\mu(t) - \rho(t))\pi(t)$$

$$-\tfrac{1}{2}(1 - \gamma)\sigma^2(t)\pi^2(t)\}dt\Big)\Big]$$

where

$$Z_\pi = Z_\pi(\omega) = \exp\Big(\int\limits_0^T \gamma\sigma(t)\pi(t)dB(t) - \tfrac{1}{2}\gamma^2 \int\limits_0^T \sigma^2(t)\pi^2(t)dt\Big)$$

and

$$K = x^\gamma \exp\Big(\gamma \int\limits_0^T \rho(t)dt\Big) \quad \text{(does not depend on } \pi).$$

Define a new measure Q_π on $\mathcal{F}_T$ by

$$dQ_\pi(\omega) = Z_\pi(\omega)dP(\omega).$$

Then we can write

$$E[X_\pi^\gamma(T)] = K\, E_{Q_\pi}[F(\pi)],$$

where

$$F(\pi) = \exp\left(\gamma \int_0^T \{(\mu(t) - \rho(t))\pi(t) - \tfrac{1}{2}(1-\gamma)\sigma^2(t)\pi^2(t)\}dt\right)$$

b)   By maximizing the integral in the exponent of $F(\pi)$ pointwise for each t, ω, show that

$$F(\pi) \le F(\pi^*) = \exp\left(\frac{\gamma}{2} \int_0^T \frac{(\mu(t) - \rho(t))^2}{(1-\gamma)\sigma^2(t)} dt\right),$$

which is attained for

$$\pi(t) = \pi^*(t) := \frac{\mu(t) - \rho(t)}{(1-\gamma)\sigma^2(t)}$$

c)   Now assume, in addition to our previous assumptions, that the portfolios π we consider satisfy the Novikov condition

$$E\left[\exp\left(\frac{\gamma^2}{2} \int_0^T \sigma^2(t)\pi^2(t)dt\right)\right] < \infty.$$

Use this to conclude that

$$\sup_{\pi} E[X_{\pi}^{\gamma}(T)] = K \exp\left(\frac{\gamma}{2} \int_0^T \frac{(\mu(t) - \rho(t))^2}{(1-\gamma)\sigma^2(t)} dt\right),$$

and if this is finite, then the optimal portfolio is $\pi^*(t)$, given in b) above.

Chapter 9
Applications to Boundary Value Problems

9.1 The Combined Dirichlet-Poisson Problem. Uniqueness

We now use results from the preceding chapters to solve the following generalization of the Dirichlet problem stated in the introduction:

Let D be a domain (open connected set) in $\mathbf{R}^n$ and let L denote a *semi-elliptic* partial differential operator on $C^2(\mathbf{R}^n)$ of the form

$$L = \sum_{i=1}^{n} b_i(x) \frac{\partial}{\partial x_i} + \sum_{i,j=1}^{n} a_{ij}(x) \frac{\partial^2}{\partial x_i \partial x_j} \qquad (9.1.1)$$

where $b_i(x)$ and $a_{ij}(x) = a_{ji}(x)$ are continuous functions (see below). (By saying that L is *semi-elliptic* (resp. *elliptic*) we mean that all the eigenvalues of the symmetric matrix $a(x) = [a_{ij}(x)]_{i,j=1}^{n}$ are *non-negative* (resp. *positive*) for all x.)

The Combined Dirichlet-Poisson Problem

Let $\phi \in C(\partial D)$ and $g \in C(D)$ be given functions. Find $w \in C^2(D)$ such that

(i) $\qquad Lw = -g \qquad$ in D $\qquad\qquad\qquad\qquad\qquad$ (9.1.2)

and

(ii) $\qquad \lim_{\substack{x \to y \\ x \in D}} w(x) = \phi(y) \qquad$ for all $y \in \partial D$. $\qquad\qquad$ (9.1.3)

The idea of the solution is the following: First we find an Itô diffusion $\{X_t\}$ whose generator A coincides with L on $C_0^2(\mathbf{R}^n)$. To achieve this we simply choose a *square root* $\sigma(x) \in \mathbf{R}^{n \times n}$ of the matrix $2a(x)$, i.e.

$$\tfrac{1}{2}\sigma(x)\sigma^T(x) = [a_{ij}(x)] \ . \tag{9.1.4}$$

We assume that $\sigma(x)$ and $b(x) = [b_i(x)]$ satisfy conditions (5.2.1) and (5.2.2) of Theorem 5.2.1. (For example, if each $a_{ij} \in C^2(D)$ is bounded and has bounded first and second partial derivatives, then such a square root σ can be found. See Fleming and Rishel (1975).) Next we let X_t be the solution of

$$dX_t = b(X_t)dt + \sigma(X_t)dB_t \tag{9.1.5}$$

where B_t is n-dimensional Brownian motion. As usual we let E^x denote expectation with respect to the probability law Q^x of X_t starting at $x \in \mathbf{R}^n$. Then our candidate for the solution w of (9.1.2), (9.1.3) is

$$w(x) = E^x[\phi(X_{\tau_D}) \cdot \mathcal{X}_{\{\tau_D < \infty\}}] + E^x\left[\int_0^{\tau_D} g(X_t)dt\right] \tag{9.1.6}$$

provided that ϕ is bounded and

$$E^x\left[\int_0^{\tau_D} |g(X_t)|dt\right] < \infty \qquad \text{for all } x \ . \tag{9.1.7}$$

The Dirichlet-Poisson problem consists of two parts:

(i)   Existence of solution.
(ii)   Uniqueness of solution.

The uniqueness problem turns out to be simpler and therefore we handle this first. In this section we prove two easy and useful uniqueness results. Then in the next sections we discuss the existence of solution and other uniqueness questions.

Theorem 9.1.1 (Uniqueness theorem (1))
Suppose ϕ is bounded and g satisfies (9.1.7). Suppose $w \in C^2(D)$ is bounded and satisfies

(i) $Lw = -g \qquad in \ D$ $\hspace{4cm}$ (9.1.8)

and

(ii)' $\lim_{t\uparrow\tau_D} w(X_t) = \phi(X_{\tau_D}) \cdot \mathcal{X}_{\{\tau_D < \infty\}} \qquad a.s. \ Q^x \ for \ all \ x \ .$ (9.1.9)

Then

$$w(x) = E^x[\phi(X_{\tau_D}) \cdot \mathcal{X}_{\{\tau_D < \infty\}}] + E^x\left[\int_0^{\tau_D} g(X_t)dt\right] \ . \tag{9.1.10}$$

Proof. Let $\{D_k\}_{k=1}^{\infty}$ be an increasing sequence of open sets D_k such that $D_k \subset\subset D$ and $D = \bigcup_{k=1}^{\infty} D_k$. Define

$$\alpha_k = k \wedge \tau_{D_k} \, ; \qquad k = 1, 2, \ldots$$

Then by the Dynkin formula and (9.1.8)

$$w(x) = E^x[w(X_{\alpha_k})] - E^x\left[\int_0^{\alpha_k} Lw(X_t)dt\right]$$

$$= E^x[w(X_{\alpha_k})] + E^x\left[\int_0^{\alpha_k} g(X_t)dt\right] . \qquad (9.1.11)$$

By (9.1.9) $w(X_{\alpha_k}) \to \phi(X_{\tau_D}) \cdot \mathcal{X}_{\{\tau_D < \infty\}}$ pointwise boundedly a.s. Q^x. Hence

$$E^x[w(X_{\alpha_k})] \to E^x[\phi(X_{\tau_D}) \cdot \mathcal{X}_{\{\tau_D < \infty\}}] \qquad \text{as } k \to \infty . \qquad (9.1.12)$$

Moreover,

$$E^x\left[\int_0^{\alpha_k} g(X_t)dt\right] \to E^x\left[\int_0^{\tau_D} g(X_t)dt\right] \qquad \text{as } k \to \infty , \qquad (9.1.13)$$

since

$$\int_0^{\alpha_k} g(X_t)dt \to \int_0^{\tau_D} g(X_t)dt \quad \text{a.s.}$$

and

$$\left|\int_0^{\alpha_k} g(X_t)dt\right| \le \int_0^{\tau_D} |g(X_t)|dt , \quad \text{which is } Q^x\text{-integrable by (9.1.7)}.$$

Combining (9.1.12) and (9.1.13) with (9.1.11) we get (9.1.10). $\qquad\square$

An immediate consequence is:

Corollary 9.1.2 (Uniqueness theorem (2))
Suppose ϕ is bounded and g satisfies (9.1.7). Suppose

$$\tau_D < \infty \qquad \text{a.s. } Q^x \text{ for all } x . \qquad (9.1.14)$$

Then if $w \in C^2(D)$ is a bounded solution of the combined Dirichlet-Poisson problem (9.1.2), (9.1.3) we have

$$w(x) = E^x[\phi(X_{\tau_D})] + E^x\left[\int_0^{\tau_D} g(X_t)dt\right] . \qquad (9.1.15)$$

Example 9.1.3 (The classical Dirichlet problem)
Let D be a bounded open set in $\mathbf{R}^n$ and let ϕ be a bounded function on ∂D. Suppose there is a function $w \in C^2(D)$ such that

(i) $\qquad \Delta w = 0 \quad$ in D $\hspace{4cm}$ (9.1.16)
    and
(ii) $\qquad \lim_{\substack{x \to y \\ x \in D}} w(x) = \phi(y) \quad$ for all $y \in \partial D$ $\hspace{2.2cm}$ (9.1.17)

Then
$$w(x) = E^x[\phi(B_{\tau_D})].$$

This follows from Corollary 9.1.2, since

$$\tfrac{1}{2}\Delta = \tfrac{1}{2}\sum_{i=1}^n \frac{\partial^2}{\partial x_i^2} \qquad \text{is the generator of } B(t)$$

and we know from Example 7.4.2 that $\tau_D < \infty$ a.s.

Example 9.1.4 (The classical heat equation)
Consider the heat operator

$$L = \frac{\partial}{\partial s} + \frac{1}{2}\frac{\partial^2}{\partial x^2} ; \qquad (s, x) \in \mathbf{R} \times \mathbf{R}.$$

This is the generator of

$$X_t = X_t^{s,x} = (s + t, B_t^x) ; \qquad t \ge 0$$

where B_t^x is Brownian motion starting at $x \in \mathbf{R}$ (see Example 7.3.5). Therefore, if there exists a solution $w(s, x) \in C^2(\mathbf{R}^2)$ of the heat equation

(i) $\qquad \frac{\partial w}{\partial s} + \frac{1}{2}\frac{\partial^2 w}{\partial x^2} = 0 \qquad (s, x) \in (0, T) \times \mathbf{R} =: D$ $\hspace{1.5cm}$ (9.1.18)

(ii) $\qquad \lim_{t \to \tau_D} w(X_t) = \phi(X_{\tau_D}) \qquad$ a.s. $\hspace{3cm}$ (9.1.19)

where $\phi : \{T\} \times \mathbf{R} \to \mathbf{R}$ is a given bounded function, then it is given by

$$w(s, x) = E^{s,x}[\phi(X_{\tau_D})] = E^{s,x}[\phi(s + \tau_D, B_{\tau_D}^x)].$$

Here
$$\begin{aligned} \tau_D &= \inf\{t > 0; (s + t, B^x(t)) \notin [0, T] \times \mathbf{R}\} \\ &= \inf\{t > 0; s + t \notin [0, T]\} = T - s . \end{aligned} \qquad (9.1.20)$$

Therefore the solution of the heat equation is

$$w(s, x) = E^{s,x}[\phi(T, B^x_{T-s})]. \tag{9.1.21}$$

9.2 The Dirichlet Problem. Regular Points

We now consider the more complicated question of existence of solution. It is convenient to split the combined Dirichlet-Poisson problem in two parts: The Dirichlet problem and the Poisson problem:

The Dirichlet Problem

Let $\phi \in C(\partial D)$ be a given function. Find $u \in C^2(D)$ such that

(I) $Lu = 0$ in D (9.2.1)

and

(II) $\lim_{\substack{x \to y \\ x \in D}} u(x) = \phi(y)$ for all $y \in \partial D$.   (9.2.2)

The Poisson Problem

Let $g \in C(D)$ be a given function. Find $v \in C^2(D)$ such that

(a) $Lv = -g$ in D (9.2.3)

and

(b) $\lim_{\substack{x \to y \\ x \in D}} v(x) = 0$ for all $y \in \partial D$.   (9.2.4)

Note that if u and v solve the Dirichlet and the Poisson problem, respectively, then $w := u + v$ solves the combined Dirichlet-Poisson problem.

We first consider the Dirichlet problem and proceed to study the Poisson problem in the next section.

For simplicity we assume in this section that (9.1.14) holds.

In view of Corollary 9.1.2 the question of existence of a solution of the Dirichlet problem (9.2.1), (9.2.2) can be restated as follows: When is

$$u(x) := E^x[\phi(X_{\tau_D})] \tag{9.2.5}$$

a solution?

Unfortunately, in general this function u need not be in $C^2(D)$. In fact, it need not even be continuous. Moreover, it need not satisfy (9.2.2), either. Consider the following example:

Example 9.2.1 Let $X(t) = (X_1(t), X_2(t))$ be the solution of the equations

$$dX_1(t) = dt$$
$$dX_2(t) = 0$$

so that $X(t) = X(0) + t(1, 0) \in \mathbf{R}^2$; $t \geq 0$. Let

$$D = ((0, 1) \times (0, 1)) \cup ((0, 2) \times (0, \tfrac{1}{2}))$$

and let ϕ be a continuous function on ∂D such that

$$\begin{aligned} \phi &= 1 \quad \text{on } \{1\} \times [\tfrac{1}{2}, 1] \quad \text{and} \\ \phi &= 0 \quad \text{on } \{2\} \times [0, \tfrac{1}{2}] \\ \phi &= 0 \quad \text{on } \{0\} \times [0, 1] \, . \end{aligned}$$

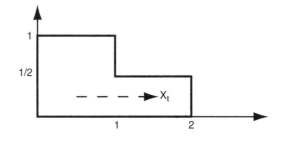

Then

$$u(t, x) = E^{t,x}[\phi(X_{\tau_D})] = \begin{cases} 1 & \text{if } x \in (\tfrac{1}{2}, 1) \\ 0 & \text{if } x \in (0, \tfrac{1}{2}) \, , \end{cases}$$

so u is not even continuous. Moreover,

$$\lim_{t \to 0^+} u(t, x) = 1 \neq \phi(0, x) \qquad \text{if } \tfrac{1}{2} < x < 1$$

so (9.2.2) does not hold.

However, the function $u(x)$ defined by (9.2.5) will solve the Dirichlet problem in a weaker, stochastic sense: The boundary condition (9.2.2) is replaced by the stochastic (pathwise) boundary condition (9.1.9) and the condition (9.2.1) ($Lu = 0$) is replaced by a condition related to the condition

$$\mathcal{A}u = 0$$

where $\mathcal{A}$ is the characteristic operator of X_t (Section 7.5).

We now explain this in more detail:

Definition 9.2.2 Let f be a locally bounded, measurable function on D. Then f is called X-harmonic in D if

$$f(x) = E^x[f(X_{\tau_U})]$$

for all $x \in D$ and all bounded open sets U with $\overline{U} \subset D$.

We make two important observations:

Lemma 9.2.3
a) Let f be X-harmonic in D. Then $\mathcal{A}f = 0$ in D.
b) Conversely, suppose $f \in C^2(D)$ and $\mathcal{A}f = 0$ in D. Then f is X-harmonic.

Proof.
a)   follows directly from the formula for $\mathcal{A}$.
b)   follows from the Dynkin formula: Choose U as in Definition 9.2.2. Then

$$E^x[f(X_{\tau_U})] = \lim_{k \to \infty} E^x[f(X_{\tau_U \wedge k})]$$

$$= f(x) + \lim_{k \to \infty} E^x\left[\int_0^{\tau_U \wedge k} (Lf)(X_s)ds \right] = f(x) ,$$

since $Lf = \mathcal{A}f = 0$ in U.   □

The most important examples of X-harmonic functions are given in the next result:

Lemma 9.2.4 Let ϕ be a bounded measurable function on ∂D and put

$$u(x) = E^x[\phi(X_{\tau_D})] ; \qquad x \in D .$$

Then u is X-harmonic. Thus, in particular, $\mathcal{A}u = 0$.

Proof. From the mean value property (7.2.9) we have, if $\overline{V} \subset D$

$$u(x) = \int_{\partial V} u(y)Q^x[X_{\tau_V} \in dy] = E^x[u(X_{\tau_V})] .$$

                                                                                □

We are now ready to formulate the weak, stochastic version:

The Stochastic Dirichlet Problem

Given a bounded measurable function ϕ on ∂D, find a function u on D such that

(i)$_s$ u is X-harmonic   (9.2.6)
(ii)$_s$ $\lim_{t \uparrow \tau_D} u(X_t) = \phi(X_{\tau_D})$ a.s. Q^x, $x \in D$.   (9.2.7)

We first solve the stochastic Dirichlet problem (9.2.6), (9.2.7) and then relate it to the original problem (9.2.1), (9.2.2).

Theorem 9.2.5 (Solution of the stochastic Dirichlet problem)
Let ϕ be a bounded measurable function on ∂D.

a) *(Existence) Define*

$$u(x) = E^x[\phi(X_{\tau_D})] \ . \tag{9.2.8}$$

Then u solves the stochastic Dirichlet problem (9.2.6), (9.2.7).
b) *(Uniqueness) Suppose g is a bounded function on D such that*

   (1)   *g is X-harmonic*
   (2) $\lim\limits_{t \uparrow \tau_D} g(X_t) = \phi(X_{\tau_D})$ *a.s. Q^x, $x \in D$.*

   Then $g(x) = E^x[\phi(X_{\tau_D})]$, $x \in D$.

Proof. a) It follows from Lemma 9.2.4 that $(i)_s$ holds. Fix $x \in D$. Let $\{D_k\}$ be an increasing sequence of open sets such that $D_k \subset\subset D$ and $D = \bigcup\limits_k D_k$.
Put $\tau_k = \tau_{D_k}$, $\tau = \tau_D$. Then by the strong Markov property

$$\begin{aligned} u(X_{\tau_k}) &= E^{X_{\tau_k}}[\phi(X_\tau)] = E^x[\theta_{\tau_k}(\phi(X_\tau))|\mathcal{F}_{\tau_k}] \\ &= E^x[\phi(X_\tau)|\mathcal{F}_{\tau_k}] \ . \end{aligned} \tag{9.2.9}$$

Now $M_k = E^x[\phi(X_\tau)|\mathcal{F}_{\tau_k}]$ is a bounded (discrete time) martingale so by the martingale convergence theorem Corollary C.9 (Appendix C) we get that

$$\lim_{k \to \infty} u(X_{\tau_k}) = \lim_{k \to \infty} E^x[\phi(X_\tau)|\mathcal{F}_{\tau_k}] = \phi(X_\tau) \tag{9.2.10}$$

both pointwise for a.a. ω and in $L^p(Q^x)$, for all $p < \infty$. Moreover, by (9.2.9) it follows that for each k the process

$$N_t = u(X_{\tau_k \vee (t \wedge \tau_{k+1})}) - u(X_{\tau_k}) \ ; \qquad t \ge 0$$

is a martingale w.r.t. $\mathcal{G}_t = \mathcal{F}_{\tau_k \vee (t \wedge \tau_{k+1})}$.
   So by the martingale inequality

$$Q^x\left[\sup_{\tau_k \le r \le \tau_{k+1}} |u(X_r) - u(X_{\tau_k})| > \epsilon \right] \le \frac{1}{\epsilon^2} E^x[|u(X_{\tau_{k+1}}) - u(X_{\tau_k})|^2]$$

$$\to 0 \quad \text{as } k \to \infty, \text{ for all } \epsilon > 0 \ . \tag{9.2.11}$$

From (9.2.10) and (9.2.11) we conclude that $(ii)_s$ holds.

b)   Let D_k, τ_k be as in a). Then since g is X-harmonic we have

$$g(x) = E^x[g(X_{\tau_k})]$$

for all k. So by (2) and bounded convergence

$$g(x) = \lim_{k \to \infty} E^x[g(X_{\tau_k})] = E^x[\phi(X_{\tau_D})] , \quad \text{as asserted} .$$

□

Finally we return to the original Dirichlet problem (9.2.1), (9.2.2). We have already seen that a solution need not exist. However, it turns out that for a large class of processes X_t we do get a solution (for all D) if we reduce the requirement in (9.2.2) to hold only for a subset of the boundary points $y \in \partial D$ called the **regular** boundary points. Before we define regular points and state the result precisely, we need the following auxiliary lemmas:

(As before we let $\mathcal{M}_t$ and $\mathcal{M}_\infty$ denote the σ-algebras generated by X_s; $s \le t$ and by X_s; $s \ge 0$ respectively).

Lemma 9.2.6 (The 0–1 law) *Let* $H \in \bigcap_{t>0} \mathcal{M}_t$. *Then either* $Q^x(H) = 0$ *or* $Q^x(H) = 1$.

Proof. From the Markov property we have

$$E^x[\theta_t \eta | \mathcal{M}_t] = E^{X_t}[\eta]$$

for all bounded, $\mathcal{M}_\infty$-measurable $\eta \colon \Omega \to \mathbf{R}$. This implies that

$$\int_H \theta_t \eta \cdot dQ^x = \int_H E^{X_t}[\eta]dQ^x , \qquad \text{for all } t .$$

First assume that $\eta = \eta_k = g_1(X_{t_1}) \cdots g_k(X_{t_k})$, where each g_i is bounded and continuous. Then letting $t \to 0$ we obtain

$$\int_H \eta dQ^x = \lim_{t \to 0} \int_H \theta_t \eta dQ^x = \lim_{t \to 0} \int_H E^{X_t}[\eta]dQ^x = Q^x(H)E^x[\eta]$$

by Feller continuity (Lemma 8.1.4) and bounded convergence. Approximating the general η by functions η_k as above we conclude that

$$\int_H \eta dQ^x = Q^x(H)E^x[\eta]$$

for all bounded $\mathcal{M}_\infty$-measurable η. If we put $\eta = \mathcal{X}_H$ we obtain $Q^x(H) = (Q^x(H))^2$, which completes the proof.

□

Corollary 9.2.7 *Let* $y \in \mathbf{R}^n$. *Then*

$$either \quad Q^y[\tau_D = 0] = 0 \quad or \quad Q^y[\tau_D = 0] = 1 .$$

Proof. $H = \{\omega; \tau_D = 0\} \in \bigcap_{t>0} \mathcal{M}_t .$

□

In other words, either a.a. paths X_t starting from y stay within D for a positive period of time or a.a. paths X_t starting from y leave D immediately. In the last case we call the point y *regular*, i.e.

Definition 9.2.8 *A point $y \in \partial D$ is called* regular *for D (w.r.t. X_t) if*

$$Q^y[\tau_D = 0] = 1 .$$

Otherwise the point y is called irregular.

Example 9.2.9 Corollary 9.2.7 may seem hard to believe at first glance. For example, if X_t is a 2-dimensional Brownian motion B_t and $\overline{D}$ is the square $[0, 1] \times [0, 1]$ one might think that, starting from $(\frac{1}{2}, 0)$, say, half of the paths will stay in the upper half plane and half in the lower, for a positive period of time. However, Corollary 9.2.7 says that this is not the case: Either they all stay in D initially or they all leave D immediately. Symmetry considerations imply that the first alternative is impossible. Thus $(\frac{1}{2}, 0)$, and similarly all the other points of ∂D, are regular for D w.r.t. B_t.

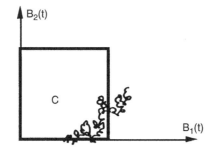

Example 9.2.10 Let $D = [0, 1] \times [0, 1]$ and let L be the parabolic differential operator

$$Lf(t, x) = \frac{\partial f}{\partial t} + \frac{1}{2} \cdot \frac{\partial^2 f}{\partial x^2} ; \qquad (t, x) \in \mathbf{R}^2 .$$

(See Example 7.3.5)

Here

$$b = \begin{pmatrix} 1 \\ 0 \end{pmatrix} \quad \text{and} \quad a = [a_{ij}] = \frac{1}{2} \begin{pmatrix} 0 & 0 \\ 0 & 1 \end{pmatrix} .$$

So, for example, if we choose $\sigma = \begin{pmatrix} 0 & 0 \\ 1 & 0 \end{pmatrix}$, we have $\frac{1}{2}\sigma\sigma^T = a$. This gives the following stochastic differential equation for the Itô diffusion X_t associated with L:

$$dX_t = \begin{pmatrix} 1 \\ 0 \end{pmatrix} dt + \begin{pmatrix} 0 & 0 \\ 1 & 0 \end{pmatrix} \begin{pmatrix} dB_t^{(1)} \\ dB_t^{(2)} \end{pmatrix} .$$

In other words,

$$X_t = \begin{pmatrix} t + t_0 \\ B_t \end{pmatrix}, \quad X_0 = \begin{pmatrix} t_0 \\ x \end{pmatrix}$$

where B_t is 1-dimensional Brownian motion. So we end up with the graph of Brownian motion, which we started with in Example 7.3.5. In the case it is not hard to see that the irregular points of ∂D consist of the open line $\{0\} \times (0,1)$, the rest of the boundary points being regular.

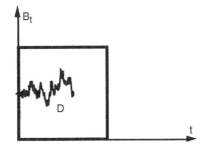

Example 9.2.11 Let $\Delta = \{(x,y);\ x^2 + y^2 < 1\} \subset \mathbf{R}^2$ and let $\{\Delta_n\}$ be a sequence of disjoint open discs in Δ centered at $(2^{-n}, 0)$, respectively, $n = 1, 2, \ldots$. Put

$$D = \Delta \setminus \overline{\left(\bigcup_{n=1}^{\infty} \Delta_n \right)}.$$

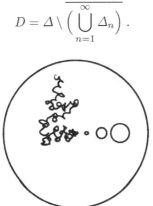

Then it is easy to see that all the points of $\partial \Delta \cup \bigcup_{n=1}^{\infty} \partial \Delta_n$ are regular for D w.r.t. 2-dimensional Brownian motion B_t, using a similar argument as in Example 9.2.9. But what about the point 0? The answer depends on the sizes of the discs Δ_n. More precisely, if r_n is the radius of Δ_n then 0 is a regular point for D if and only if

$$\sum_{n=1}^{\infty} \frac{n}{\log \frac{1}{r_n}} = \infty. \tag{9.2.12}$$

This is a consequence of the famous Wiener criterion. See Port and Stone (1979), p. 225.

Having defined regular points we now formulate the announced generalized version of the Dirichlet problem:

The Generalized Dirichlet Problem

Given a domain $D \subset \mathbf{R}^n$ and L and ϕ as before, find a function $u \in C^2(D)$ such that

(i) $\qquad Lu = 0 \qquad$ in D $\hfill$ (9.2.13)

and

(ii) $\qquad \lim_{\substack{x \to y \\ x \in D}} u(x) = \phi(y) \qquad$ for all *regular* $y \in \partial D$. $\hfill$ (9.2.14)

First we establish that if a solution of this problem exists, it must be the solution of the stochastic Dirichlet problem found in Theorem 9.2.5, provided that X_t satisfies *Hunt's condition* (H):

$\qquad$ (H): Every semipolar set for X_t is polar for X_t . $\hfill$ (9.2.15)

A *semipolar* set is a countable union of *thin* sets and a measurable set $G \subset \mathbf{R}^n$ is called *thin* (for X_t) if $Q^x[T_G = 0] = 0$ for all x, where $T_G = \inf\{t > 0; X_t \in G\}$ is the first hitting time of G. (Intuitively: For all starting points the process does not hit G immediately, a.s). A measurable set $F \subset \mathbf{R}^n$ is called *polar* (for X_t) if $Q^x[T_F < \infty] = 0$ for all x. (Intuitively: For all starting points the process *never* hits F, a.s.). Clearly every polar set is semipolar, but the converse need not be true (consider the process in Example 9.2.1). However, condition (H) does hold for Brownian motion (See Blumenthal and Getoor (1968)). It follows from the Girsanov theorem that condition (H) holds for all Itô diffusions whose diffusion coefficient matrix has a bounded inverse and whose drift coefficient satisfies the Novikov condition for all $T < \infty$.

We also need the following result, the proof of which can be found in Blumenthal and Getoor (1968, Prop. II.3.3):

Lemma 9.2.12 *Let $U \subset D$ be open and let I denote the set of irregular points of U. Then I is a semipolar set.*

Theorem 9.2.13 *Suppose X_t satisfies Hunt's condition (H). Let ϕ be a bounded continuous function on ∂D. Suppose there exists a bounded $u \in C^2(D)$ such that*

(i) $\quad Lu = 0$ *in* D
(ii)$_s$ $\quad \lim_{\substack{x \to y \\ x \in D}} u(x) = \phi(y)$ *for all regular* $y \in \partial D$

Then $u(x) = E^x[\phi(X_{\tau_D})]$.

Proof. Let $\{D_k\}$ be as in the proof Theorem 9.1.1. By Lemma 9.2.3 b) u is X-harmonic and therefore

$$u(x) = E^x[u(X_{\tau_k})] \qquad \text{for all } x \in D_k \text{ and all } k \ .$$

If $k \to \infty$ then $X_{\tau_k} \to X_{\tau_D}$ and so $u(X_{\tau_k}) \to \phi(X_{\tau_D})$ if X_{τ_D} is regular. From the Lemma 9.2.12 we know that the set I of irregular points of ∂D is semipolar. So by condition (H) the set I is polar and therefore $X_{\tau_D} \notin I$ a.s. Q^x. Hence

$$u(x) = \lim E^x[u(X_{\tau_k})] = E^x[\phi(X_{\tau_D})] \ , \quad \text{as claimed .}$$

$\square$

Under what conditions is the solution u of the stochastic Dirichlet problem (9.2.6), (9.2.7) also a solution of the generalized Dirichlet problem (9.2.13), (9.2.14)? This is a difficult question and we will content ourselves with the following partial answer:

Theorem 9.2.14 *Suppose L is uniformly elliptic in D, i.e. the eigenvalues of $[a_{ij}]$ are bounded away from 0 in D. Let ϕ be a bounded continuous function on ∂D. Put*

$$u(x) = E^x[\phi(X_{\tau_D})] \ .$$

Then $u \in C^{2+\alpha}(D)$ for all $\alpha < 1$ and u solves the Dirichlet problem (9.2.13), (9.2.14), i.e.

(i) $Lu = 0$ *in* D.
(ii)$_r$ $\lim_{\substack{x \to y \\ x \in D}} u(x) = \phi(y)$ *for all regular* $y \in \partial D$.

Remark. If k is an integer, $\alpha > 0$ and G is an open set $C^{k+\alpha}(G)$ denotes the set of functions on G whose partial derivatives up to k'th order is Lipschitz (Hölder) continuous with exponent α.

Proof. Choose an open ball Δ with $\overline{\Delta} \subset D$ and let $f \in C(\partial\Delta)$. Then, from the general theory of partial differential equations, for all $\alpha < 1$ there exists a continuous function u on $\overline{\Delta}$ such that $u|\Delta \in C^{2+\alpha}(\Delta)$ and

$$Lu = 0 \qquad \text{in } \Delta \tag{9.2.16}$$

$$u = f \qquad \text{on } \partial\Delta \tag{9.2.17}$$

(see e.g. Dynkin (1965 II, p. 226)). Since $u|\Delta \in C^{2+\alpha}(\Delta)$ we have: If K is any compact subset of Δ there exists a constant C only depending on K and the C^α-norms of the coefficients of L such that

$$\|u\|_{C^{2+\alpha}(K)} \leq C(\|Lu\|_{C^\alpha(\Delta)} + \|u\|_{C(\Delta)}) \ . \tag{9.2.18}$$

(See Bers, John and Schechter (1964, Theorem 3, p. 232).) Combining (9.2.16), (9.2.17) and (9.2.18) we obtain

$$\|u\|_{C^{2+\alpha}(K)} \leq C\|f\|_{C(\partial\Delta)} \ . \tag{9.2.19}$$

By uniqueness (Theorem 9.2.13) we know that

$$u(x) = \int u(y)d\mu_x(y) \ , \tag{9.2.20}$$

where $d\mu_x = Q^x[X_{\tau_\Delta} \in dy]$ is the first exit distribution of X_t from Δ. From (9.2.19) it follows that

$$\left| \int f d\mu_{x_1} - \int f d\mu_{x_2} \right| \leq C\|f\|_{C(\partial\Delta)}|x_1 - x_2|^\alpha \ ; \quad x_1, x_2 \in K \ . \tag{9.2.21}$$

By approximating a given continuous function on $\partial\Delta$ uniformly by functions in $C^\infty(\partial\Delta)$ we see that (9.2.21) holds for all functions $f \in C(\partial\Delta)$. Therefore

$$\|\mu_{x_1} - \mu_{x_2}\| \leq C|x_1 - x_2|^\alpha \ ; \quad x_1, x_2 \in K \tag{9.2.22}$$

where $\| \ \|$ denotes the operator norm on measures on $\partial\Delta$. So if g is any bounded measurable function on $\partial\Delta$ we know that the function

$$\widehat{g}(x) = \int g(y)d\mu_x(y) = E^x[g(X_{\tau_\Delta})]$$

belongs to the class $C^\alpha(K)$. Since $u(x) = E^x[u(X_{\tau_U})]$ for all open sets U with $\overline{U} \subset D$ and $x \in U$ (Lemma 9.2.4) this applies to $g = u$ and we conclude that $u \in C^\alpha(M)$ for any compact subset M of D.

We may therefore apply the solution to the problem (9.2.16), (9.2.17) once more, this time with $f = u$ and this way we obtain that

$$u(x) = E^x[u(X_{\tau_D})] \qquad \text{belongs to } C^{2+\alpha}(M)$$

for any compact $M \subset D$. Therefore (i) holds by Lemma 9.2.3 a).

To obtain (ii)$_r$ we apply a theorem from the theory of parabolic differential equations: The Kolmogorov backward equation

$$Lv = \frac{\partial v}{\partial t}$$

has a fundamental solution $v = p(t, x, y)$ jointly continuous in t, x, y for $t > 0$ and bounded in x, y for each fixed $t > 0$ (See Dynkin (1965 II), Theorem 0.4 p. 227). It follows (by bounded convergence) that the process X_t is a *strong Feller process*, in the sense that the function

$$x \to E^x[f(X_t)] = \int\limits_{\mathbf{R}^n} f(y)p(t,x,y)dy$$

is continuous, for all $t > 0$ and all *bounded, measurable functions* f. In general we have:

> If X_t is a strong Feller Itô diffusion and $D \subset \mathbf{R}^n$ is open then
> $$\lim_{\substack{x \to y \\ x \in D}} E^x[\phi(X_{\tau_D})] = \phi(y)$$
> for all regular $y \in \partial D$ and bounded $\phi \in C(\partial D)$. $\hspace{2cm}$ (9.2.23)

(See Theorem 13.3 p. 32–33 in Dynkin (1965 II)).

Therefore u satisfies property (ii)$_r$ and the proof is complete. $\hspace{1cm}$ $\square$

Example 9.2.15 We have already seen (Example 9.2.1) that condition (9.1.3) does not hold in general. This example shows that it need not hold even when L is elliptic: Consider Example 9.2.11 again, in the case when the point 0 is not regular. Choose $\phi \in C(\partial D)$ such that

$$\phi(0) = 1, 0 \le \phi(y) < 1 \qquad \text{for } y \in \partial D \setminus \{0\} .$$

Since $\{0\}$ is polar for B_t (see Exercise 9.7 a) we have $B^0_{\tau_D} \ne 0$ a.s and therefore

$$u(0) = E^0[\phi(B_{\tau_D})] < 1 .$$

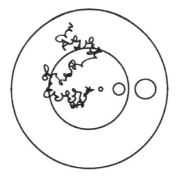

By a slight extension of the mean value property (7.2.9) (see Exercise 9.4) we get

$$E^0[u(X_{\sigma_k})] = E^0[\phi(X_{\tau_D})] = u(0) < 1 \hspace{2cm} (9.2.24)$$

where

$$\sigma_k = \inf\left\{t > 0; B_t \notin D \cap \left\{|x| < \frac{1}{k}\right\}\right\}, \qquad k = 1, 2, \ldots$$

This implies that it is impossible that $u(x) \to 1$ as $x \to 0$. Therefore (9.1.3) does not hold in this case.

In general one can show that the regular points for Brownian motion are exactly the regular points in the classical potential theoretic sense, i.e. the points y on ∂D where the limit of the generalized Perron-Wiener-Brelot solution coincide with $\phi(y)$, for all $\phi \in C(\partial D)$. See Doob (1984), Port and Stone (1979) or Rao (1977).

Example 9.2.16 Let D denote the infinite strip

$$D = \{(t, x) \in \mathbf{R}^2; x < R\} , \qquad \text{where } R \in \mathbf{R}$$

and let L be the differential operator

$$Lf(t, x) = \frac{\partial f}{\partial t} + \frac{1}{2} \frac{\partial^2 f}{\partial x^2} ; \qquad f \in C^2(D) .$$

An Itô diffusion whose generator coincides with L on $C_0^2(\mathbf{R}^2)$ is (see Example 9.2.10)

$$X_t = (s + t, B_t) ; \qquad t \geq 0 ,$$

and all the points of ∂D are regular for this process. It is not hard to see that in this case (9.1.14) holds, i.e.

$$\tau_D < \infty \quad \text{a.s.}$$

(see Exercise 7.4).

Assume that ϕ is a bounded continuous function on $\partial D = \{(t, R); t \in \mathbf{R}\}$. Then by Theorem 9.2.5 the function

$$u(s, x) = E^{s,x}[\phi(X_{\tau_D})]$$

is the solution of the stochastic Dirichlet problem (9.2.6), (9.2.7), where $E^{s,x}$ denotes expectation w.r.t. the probability law $Q^{s,x}$ for X starting at (s, x). Does u also solve the problem (9.2.13), (9.2.14)? Using the Laplace transform it is possible to find the distribution of the first exit point on ∂D for X, i.e. to find the distribution of the first time $t = \hat{\tau}$ that B_t reaches the value R. (See Karlin and Taylor (1975), p. 363. See also Exercise 7.19.) The result is

$$P^x[\hat{\tau} \in dt] = g(x, t)dt ,$$

where

$$g(x, t) = \begin{cases} (R - x)(2\pi t^3)^{-1} \exp(-\frac{(R-x)^2}{2t}) ; & t > 0 \\ \qquad\qquad 0 ; & t \leq 0 . \end{cases} \qquad (9.2.25)$$

Thus the solution u may be written

$$u(s,x) = \int_0^\infty \phi(s+t,R)g(x,t)dt = \int_s^\infty \phi(r,R)g(x,r-s)dr \ .$$

From the explicit formula for u it is clear that $\frac{\partial u}{\partial s}$ and $\frac{\partial^2 u}{\partial x^2}$ are continuous and we conclude that $Lu = 0$ in D by Lemma 9.2.3. So u satisfies (9.2.13). What about property (9.2.14)? It is not hard to see that for $t > 0$

$$E^{t_0,x}[f(X_t)] = (2\pi t)^{-\frac{1}{2}} \int_R f(t_0+t,y) \exp\left(-\frac{|x-y|^2}{2t}\right) dy$$

for all bounded, (t,x)-measurable functions f. (See (2.2.2)). Therefore X_t is *not* a strong Feller process, so we cannot appeal to (9.2.23) to obtain (9.2.14). However, it is easy to verify directly that if $|y| = R$, $t_1 > 0$ then for all $\epsilon > 0$ there exists $\delta > 0$ such that $|x-y| < \delta$, $|t-t_1| < \delta \Rightarrow Q^{t,x}[X_{\tau_D} \in N] \geq 1-\epsilon$, where $N = [t_1 - \epsilon, t_1 + \epsilon] \times \{y\}$. And this is easily seen to imply (9.2.14).

Remark. As the above example (and Example 9.2.1) shows, an Itô diffusion need not be a *strong* Feller process. However, we have seen that it is always a Feller process (Lemma 8.1.4).

9.3 The Poisson Problem

Let $L = \sum a_{ij} \frac{\partial^2}{\partial x_i \partial x_j} + \sum b_i \frac{\partial}{\partial x_i}$ be a semi-elliptic partial differential operator on a domain $D \subset \mathbf{R}^n$ as before and let X_t be an associated Itô diffusion, described by (9.1.4) and (9.1.5). In this section we study the Poisson problem (9.2.3), (9.2.4). For the same reasons as in Section 9.2 we generalize the problem to the following:

The Generalized Poisson Problem

Given a continuous function g on D find a C^2 function v in D such that

a) $Lv = -g$ in D (9.3.1)

b) $\lim_{\substack{x \to y \\ x \in D}} v(x) = 0$ for all *regular* $y \in \partial D$ (9.3.2)

Again we will first study a stochastic version of the problem and then investigate the relation between the corresponding stochastic solution and the solution (if it exists) of (9.3.1), (9.3.2):

Theorem 9.3.1 (Solution of the stochastic Poisson problem)
Assume that

$$E^x\left[\int_0^{\tau_D}|g(X_s)|ds\right]<\infty \qquad \text{for all } x\in D .\qquad (9.3.3)$$

(This occurs, for example, if g is bounded and $E^x[\tau_D]<\infty$ for all $x\in D$).
Define

$$v(x)=E^x\left[\int_0^{\tau_D}g(X_s)ds\right] .\qquad (9.3.4)$$

Then

$$\mathcal{A}v=-g \qquad \text{in } D ,\qquad (9.3.5)$$

and

$$\lim_{t\uparrow\tau_D}v(X_t)=0 \quad a.s. \ Q^x, \text{ for all } x\in D .\qquad (9.3.6)$$

Proof. Choose U open, $x\in U\subset\subset D$. Put $\eta=\int_0^{\tau_D}g(X_s)ds$, $\tau=\tau_U$.
    Then by the strong Markov property (7.2.5)

$$\frac{E^x[v(X_\tau)]-v(x)}{E^x[\tau]}=\frac{1}{E^x[\tau]}(E^x[E^{X_\tau}[\eta]]-E^x[\eta])$$

$$=\frac{1}{E^x[\tau]}(E^x[E^x[\theta_\tau\eta|\mathcal{F}_\tau]]-E^x[\eta])=\frac{1}{E^x[\tau]}(E^x[\theta_\tau\eta-\eta]) .$$

Approximate η by sums of the form

$$\eta^{(k)}=\sum g(X_{t_i})\mathcal{X}_{\{t_i<\tau_D\}}\Delta t_i .$$

Since

$$\theta_t\eta^{(k)}=\sum g(X_{t_i+t})\mathcal{X}_{\{t_i+t<\tau_D^t\}}\Delta t_i \quad \text{for all } k$$

(see the argument for (7.2.6)) we see that

$$\theta_\tau\eta=\int_\tau^{\tau_D}g(X_s)ds .\qquad (9.3.7)$$

Therefore

$$\frac{E^x[v(X_\tau)]-v(x)}{E^x[\tau]}=\frac{-1}{E^x[\tau]}E^x\left[\int_0^\tau g(X_s)ds\right]\to -g(x) \quad \text{as } U\downarrow x ,$$

since g is continuous. This proves (9.3.5).
    Put $H(x)=E^x[\int_0^{\tau_D}|g(X_s)|ds]$. Let D_k,τ_k be as in the proof of Theorem 9.2.5. Then by the same argument as above we get

$$E^x[H(X_{\tau_k \wedge t})] = E^x[E^x[\int_{\tau_k \wedge t}^{\tau_D} |g(X_s)|ds|\mathcal{F}_{\tau_k \wedge t}]]$$

$$= E^x\left[\int_{\tau_k \wedge t}^{\tau_D} |g(X_s)|ds\right] \to 0 \quad \text{as } t \to \tau_D, \; k \to \infty$$

by dominated convergence. This implies (9.3.6). $\qquad\qquad\qquad\qquad\qquad$ $\square$

Remark. For functions g satisfying (9.3.3) define the operator $\mathcal{R}$ by

$$(\mathcal{R}g)(x) = \check{g}(x) = E^x\left[\int_0^{\tau_D} g(X_s)ds\right].$$

Then (9.3.5) can be written

$$\mathcal{A}(\mathcal{R}g) = -g \qquad\qquad\qquad\qquad\qquad (9.3.8)$$

i.e. the operator $-\mathcal{R}$ is a right inverse of the operator $\mathcal{A}$. Similarly, if we define

$$\mathcal{R}_\alpha g(x) = E^x\left[\int_0^{\tau_D} g(X_s)e^{-\alpha s}ds\right] \qquad \text{for } \alpha \geq 0 \qquad (9.3.9)$$

then the same method of proof as in Theorem 8.1.5 gives that

$$(\mathcal{A} - \alpha)\mathcal{R}_\alpha g = -g \, ; \qquad \alpha \geq 0 \, . \qquad\qquad\qquad (9.3.10)$$

(If $\alpha > 0$ then the assumption (9.3.3) can be replaced by the assumption that g is bounded (and continuous as before)).

Thus we may regard the operator $\mathcal{R}_\alpha$ as a generalization of the resolvent operator R_α discussed in Chapter 8, and formula (9.3.10) as the analogue of Theorem 8.1.5 b).

Next we establish that if a solution v of the generalized problem (9.3.1), (9.3.2) exists, then v is the solution (9.3.4) of the stochastic problem (9.3.5), (9.3.6):

Theorem 9.3.2 (Uniqueness theorem for the Poisson equation)
Assume that X_t satisfies Hunt's condition (H) ((9.2.15)). Assume that (9.3.3) holds and that there exists a function $v \in C^2(D)$ and a constant C such that

$$|v(x)| \leq C\left(1 + E^x\left[\int_0^{\tau_D} |g(X_s)|ds\right]\right) \qquad \text{for all } x \in D \qquad (9.3.11)$$

and with the properties

$$Lv = -g \qquad in \ D \ , \tag{9.3.12}$$

$$\lim_{\substack{x \to y \\ x \in D}} v(x) = 0 \quad for \ all \ regular \ points \ y \in \partial D \ . \tag{9.3.13}$$

Then $v(x) = E^x [\int_0^{\tau_D} g(X_s) ds]$.

Proof. Let D_k, τ_k be as in the proof of Theorem 9.2.5. Then by Dynkin's formula

$$E^x[v(X_{\tau_k})] - v(x) = E^x \left[\int_0^{\tau_k} (Lv)(X_s) ds \right] = -E^x \left[\int_0^{\tau_k} g(X_s) ds \right] \ .$$

By dominated convergence we obtain

$$v(x) = \lim_{k \to \infty} \left(E^x[v(X_{\tau_k})] + E^x \left[\int_0^{\tau_k} g(X_s) ds \right] \right) = E^x \left[\int_0^{\tau_D} g(X_s) ds \right] \ ,$$

since X_{τ_D} is a regular point a.s. by condition (H) and Lemma 9.2.12.   □

Finally we combine the Dirichlet and Poisson problem and obtain the following result:

Theorem 9.3.3 (Solution of the combined stochastic Dirichlet and Poisson problem).
Assume that (9.1.14) holds. Let $\phi \in C(\partial D)$ be bounded and let $g \in C(D)$ satisfy

$$E^x \left[\int_0^{\tau_D} |g(X_s)| ds \right] < \infty \qquad for \ all \ x \in D \ . \tag{9.3.14}$$

Define

$$w(x) = E^x[\phi(X_{\tau_D})] + E^x \left[\int_0^{\tau_D} g(X_s) ds \right] \ , \qquad x \in D \ . \tag{9.3.15}$$

a) *Then*

$$\mathcal{A}w = -g \qquad in \ D \tag{9.3.16}$$

and

$$\lim_{t \uparrow \tau_D} w(X_t) = \phi(X_{\tau_D}) \quad a.s. \ Q^x, \ for \ all \ x \in D \ . \tag{9.3.17}$$

b) *Moreover, if there exists a function $w_1 \in C^2(D)$ and a constant C such that*

$$|w_1(x)| < C \left(1 + E^x \left[\int_0^{\tau_D} |g(X_s)| ds \right] \right) \ , \qquad x \in D \ , \tag{9.3.18}$$

and w_1 satisfies (9.3.16) and (9.3.17), then $w_1 = w$.

Remark. With an approach similar to the one used in Theorem 9.2.14 one can prove that if L is uniformly elliptic in D and $g \in C^\alpha(D)$ (for some $\alpha > 0$) is bounded, then the function w given by (9.3.15) solves the Dirichlet-Poisson problem, i.e.

$$Lw = -g \qquad \text{in } D \tag{9.3.19}$$

and

$$\lim_{\substack{x \to y \\ x \in D}} w(x) = \phi(y) \qquad \text{for all regular } y \in \partial D \ . \tag{9.3.20}$$

The Green Measure

The solution v given by the formula (9.3.4) may be rewritten as follows:

Definition 9.3.4 *The* Green measure *(of X_t w.r.t. D at x), $G(x, \cdot)$ is defined by*

$$G(x, H) = E^x \left[\int_0^{\tau_D} \mathcal{X}_H(X_s) ds \right] , \qquad H \subset \mathbf{R}^n \quad \textit{Borel set} \tag{9.3.21}$$

or

$$\int f(y) G(x, dy) = E^x \left[\int_0^{\tau_D} f(X_s) ds \right] , \qquad \textit{f bounded, continuous .} \tag{9.3.22}$$

In other words, $G(x, H)$ is the expected length of time the process stays in H before it exits from D. If X_t is Brownian motion, then

$$G(x, H) = \int_H G(x, y) dy \ ,$$

where $G(x, y)$ is the classical Green function w.r.t. D and dy denotes Lebesque measure. See Doob (1984), Port and Stone (1979) or Rao (1977). See also Example 9.3.6 below.

Note that using the Fubini theorem we obtain the following relation between the Green measure G and the *transition measure* for X_t in D, $Q_t^D(x, H) = Q^x[X_t \in H, t < \tau_D]$:

$$G(x, H) = E^x \left[\int_0^\infty \mathcal{X}_H(X_s) \cdot \mathcal{X}_{[0, \tau_D)}(s) ds \right] = \int_0^\infty Q_t^D(x, H) dt \ . \tag{9.3.23}$$

From (9.3.22) we get

$$v(x) = E^x \left[\int_0^{\tau_D} g(X_s)ds \right] = \int_D g(y)G(x, dy) , \qquad (9.3.24)$$

which is the familiar formula for the solution of the Poisson equation in the classical case.

Also note that by using the Green function, we may regard the Dynkin formula as a generalization of the classical Green formula:

Corollary 9.3.5 (The Green formula) *Let $E^x[\tau_D] < \infty$ for all $x \in D$ and assume that $f \in C_0^2(\mathbf{R}^n)$. Then*

$$f(x) = E^x[f(X_{\tau_D})] - \int_D (L_X f)(y)G(x, dy) . \qquad (9.3.25)$$

In particular, if $f \in C_0^2(D)$ we have

$$f(x) = - \int_D (L_X f)(y)G(x, dy) . \qquad (9.3.26)$$

(As before $L_X = \sum b_i \frac{\partial}{\partial x_i} + \frac{1}{2} \sum (\sigma\sigma^T)_{ij} \frac{\partial^2}{\partial x_i \partial x_j}$ when

$$dX_t = b(X_t)dt + \sigma(X_t)dB_t .)$$

Proof. By Dynkin's formula and (9.3.24) we have

$$E^x[f(X_{\tau_D})] = f(x) + E^x \left[\int_0^{\tau_D} (L_X f)(X_s)ds \right] = f(x) + \int_D (L_X f)(y)G(x, dy) .$$

Remark. Combining (9.3.8) with (9.3.26) we see that if $E^x[\tau_K] < \infty$ for all compacts $K \subset D$ and all $x \in D$, then $-\mathcal{R}$ is the inverse of the operator $\mathcal{A}$ on $C_0^2(D)$:

$$\mathcal{A}(\mathcal{R}f) = \mathcal{R}(\mathcal{A}f) = -f , \qquad \text{for all } f \in C_0^2(D) . \qquad (9.3.27)$$

More generally, for all $\alpha \geq 0$ we get the following analogue of Theorem 8.1.5:

$$(\mathcal{A} - \alpha)(\mathcal{R}_\alpha f) = \mathcal{R}_\alpha(\mathcal{A} - \alpha)f = -f \qquad \text{for all } f \in C_0^2(D) . \qquad (9.3.28)$$

The first part of this is already established in (9.3.10) and the second part follows from the following useful extension of the Dynkin formula

$$E^x[e^{-\alpha\tau}f(X_\tau)] = f(x) + E^x\left[\int_0^\tau e^{-\alpha s}(A-\alpha)f(X_s)ds\right] . \qquad (9.3.29)$$

If $\alpha > 0$ this is valid for all stopping times $\tau \leq \infty$ and all $f \in C_0^2(\mathbf{R}^n)$. (See Exercise 9.6.)

Example 9.3.6 If $X_t = B_t$ is 1-dimensional Brownian motion in a bounded interval $(a, b) \subset \mathbf{R}$ then we can compute the Green function $G(x, y)$ explicitly. To this end, choose a bounded continuous function $g: (a, b) \to \mathbf{R}$ and let us compute

$$v(x) := E^x\left[\int_0^{\tau_D} g(B_t)dt\right] .$$

By Corollary 9.1.2 we know that v is the solution of the differential equation

$$\tfrac{1}{2}v''(x) = -g(x) ; \qquad x \in (a, b)$$
$$v(a) = v(b) = 0 .$$

Integrating twice and using the boundary conditions we get

$$v(x) = \frac{2(x-a)}{b-a}\int_a^b\left(\int_a^y g(z)dz\right)dy - 2\int_a^x\left(\int_a^y g(z)dz\right)dy .$$

Changing the order of integration we can rewrite this as

$$v(x) = \int_a^b g(y)G(x, y)dy$$

where

$$G(x, y) = \frac{2(x-a)(b-y)}{b-a} - 2(x-y)\cdot\mathcal{X}_{(-\infty,x)}(y) . \qquad (9.3.30)$$

We conclude that *the Green function of Brownian motion in the interval (a, b) is given by (9.3.30)*.

In higher dimension n the Green function $y \to G(x, y)$ of Brownian motion starting at x will not be continuous at x. It will have a logarithmic singularity (i.e. a singularity of order $\ln\frac{1}{|x-y|}$) for $n = 2$ and a singularity of the order $|x - y|^{2-n}$ for $n > 2$.

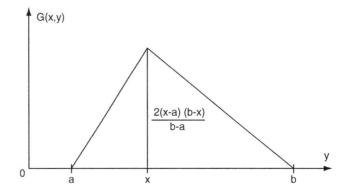

Exercises

9.1.* In each of the cases below find an Itô diffusion whose generator coincides with L on C_0^2:

a) $Lf(t, x) = \alpha \frac{\partial f}{\partial t} + \frac{1}{2}\beta^2 \frac{\partial^2 f}{\partial x^2}$; α, β constants

b) $Lf(x_1, x_2) = a\frac{\partial f}{\partial x_1} + b\frac{\partial f}{\partial x_2} + \frac{1}{2}(\frac{\partial^2 f}{\partial x_1^2} + \frac{\partial^2 f}{\partial x_2^2})$; a, b constants

c) $Lf(x) = \alpha x f'(x) + \frac{1}{2}\beta^2 f''(x)$; α, β constants

d) $Lf(x) = \alpha f'(x) + \frac{1}{2}\beta^2 x^2 f''(x)$; α, β constants

e) $Lf(x_1, x_2) = \ln(1 + x_1^2)\frac{\partial f}{\partial x_1} + x_2\frac{\partial f}{\partial x_2} + x_2^2\frac{\partial^2 f}{\partial x_1^2} + 2x_1x_2\frac{\partial^2 f}{\partial x_1 \partial x_2} + 2x_1^2\frac{\partial^2 f}{\partial x_2^2}$.

9.2.* Use Theorem 9.3.3 to find the bounded solutions of the following boundary value problems:

(i)
$$\begin{cases} \frac{\partial u}{\partial t} + \frac{1}{2} \cdot \frac{\partial^2 u}{\partial x^2} = \phi(t, x) ; & 0 < t < T, \ x \in \mathbf{R} \\ u(T, x) = \psi(x) ; & x \in \mathbf{R} \end{cases}$$

where ϕ, ψ are given bounded, continuous functions.

(ii)
$$\begin{cases} \alpha x u'(x) + \frac{1}{2}\beta^2 x^2 u''(x) = 0 ; & 0 < x < x_0 \\ u(x_0) = x_0^2 \end{cases}$$

where α, β are given constants, $\alpha \geq \frac{1}{2}\beta^2$.

(iii) If $\alpha < \frac{1}{2}\beta^2$ there are infinitely many bounded solutions of (ii), and an additional boundary condition e.g. at $x = 0$ is needed to provide uniqueness. Explain this in view of Theorem 9.3.3.

9.3.* Write down (using Brownian motion) and compare the solutions $u(t,x)$ of the following two boundary value problems:

a) $\begin{cases} \frac{\partial u}{\partial t} + \frac{1}{2}\Delta u = 0 & \text{for } 0 < t < T, \ x \in \mathbf{R}^n \\ u(T,x) = \phi(x) & \text{for } x \in \mathbf{R}^n \ . \end{cases}$

b) $\begin{cases} \frac{\partial u}{\partial t} - \frac{1}{2}\Delta u = 0 & \text{for } 0 < t < T, \ x \in \mathbf{R}^n \\ u(0,x) = \psi(x) & \text{for } x \in \mathbf{R}^n \ . \end{cases}$

9.4. Let G and H be bounded open subsets of $\mathbf{R}^n, G \subset H$, and let B_t be n-dimensional Brownian motion. Use the property (H) for B_t to prove that

$$\inf\{t > 0; B_t \notin H\} = \inf\{t > \tau_G; B_t \notin H\} \quad \text{a.s.}$$

i.e., with the terminology of (7.2.6),

$$\tau_H = \tau_H^\alpha \qquad \text{where } \alpha = \tau_G \ .$$

Use this to prove that if $X_t = B_t$ then the mean value property (7.2.9) holds for all bounded open $G \subset H$, i.e. it is not necessary to require $G \subset\subset H$ in this case. This verifies the statement (9.2.24).

9.5. **(The eigenvalues of the Laplacian)**
Let $D \subset \mathbf{R}^n$ be open, bounded and let $\lambda \in \mathbf{R}$.

a) Suppose there exists a solution $u \in C^2(D) \cap C(\overline{D})$, u not identically zero, such that

$$\begin{cases} -\frac{1}{2}\Delta u = \lambda u & \text{in} \quad D \\ u = 0 & \text{on} \quad \partial D \ . \end{cases} \qquad (9.3.31)$$

Show that we must have $\lambda > 0$. (Hint: If $\frac{1}{2}\Delta u = -\lambda u$ in D then

$$\langle \tfrac{1}{2}\Delta u, u \rangle = \langle -\lambda u, u \rangle$$

where

$$\langle u, v \rangle = \int_D u(x)v(x)dx \ .$$

Now use integration by parts.)

b) It can be shown that if D is smooth then there exist $0 < \lambda_0 < \lambda_1 < \cdots < \lambda_n < \cdots$ where $\lambda_n \to \infty$ such that (9.3.31) holds for $\lambda = \lambda_n$, $n = 0, 1, 2, \ldots$, and for no other values of λ. The numbers $\{\lambda_n\}$ are called the *eigenvalues* of the operator $-\frac{1}{2}\Delta$ in the domain D and the corresponding (nontrivial) solutions u_n of (9.3.31) are called the *eigenfunctions*. There is an interesting probabilistic interpretation of the lowest eigenvalue λ_0. The following result indicates this:

Put $\tau = \tau_D = \inf\{t > 0; B_t \notin D\}$, choose $\rho > 0$ and define

$$w_\rho(x) = E^x[\exp(\rho\tau)] \; ; \qquad x \in D \; .$$

Prove that if $w_\rho(x) < \infty$ for all $x \in D$ then ρ is not an eigenvalue for $-\frac{1}{2}\Delta$. (Hint: Let u be a solution of (9.3.31) with $\lambda = \rho$. Apply Dynkin's formula to the process $dY_t = (dt, dB_t)$ and the function $f(t, x) = e^{\rho t}u(x)$ to deduce that $u(x) = 0$ for $x \in D$).

c)   Conclude that

$$\lambda_0 \geq \sup\{\rho; E^x[\exp(\rho\tau)] < \infty \text{ for all } x \in D\} \; .$$

(We have in fact equality here. See for example Durrett (1984), Chap. 8B).

9.6.   Prove formula (9.3.29), for example by applying the Dynkin formula to the process

$$dY_t = \begin{bmatrix} dt \\ dX_t \end{bmatrix}$$

and the function $g(y) = g(t, x) = e^{-\alpha t}f(x)$.

9.7.   a)   Let B_t be Brownian motion in $\mathbf{R}^2$. Prove that

$$P^x[\exists t > 0; B_t = y] = 0 \qquad \text{for all } x, y \in \mathbf{R}^2 \; .$$

(Hint: First assume $x \neq y$. We may choose $y = 0$. One possible approach would be to apply Dynkin's formula with $f(u) = \ln |u|$ and $\tau = \inf\{t > 0; |B_t| \leq \rho \text{ or } |B_t| \geq R\}$, where $0 < \rho < R$. Let $\rho \to 0$ and then $R \to \infty$. If $x = y$ consider $P^x[\exists t > \epsilon; B_t = x]$ and use the Markov property.)

b)   Let $B_t = (B_t^{(1)}, B_t^{(2)})$ be Brownian motion in $\mathbf{R}^2$. Prove that $\tilde{B}_t = (-B_t^{(1)}, B_t^{(2)})$ is also a Brownian motion.

c)   Prove that $0 \in \mathbf{R}^2$ is a *regular* boundary point (for Brownian motion) of the plane region

$$D = \{(x_1, x_2) \in \mathbf{R}^2; x_1^2 + x_2^2 < 1\} \setminus \{(x_1, 0); x_1 \geq 0\} \; .$$

d)   Prove that $0 \in \mathbf{R}^3$ is an *irregular* boundary point (for Brownian motion) of the 3-dimensional region

$$U = \{(x_1, x_2, x_3) \in \mathbf{R}^3, x_1^2 + x_2^2 + x_3^2 < 1\} \setminus \{(x_1, 0, 0); x_1 \geq 0\} \; .$$

9.8.*   a)   Find an Itô diffusion X_t and a measurable set G which is semipolar but not polar for X_t.

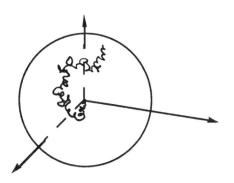

b) Find an Itô diffusion X_t and a countable family of thin sets H_k;
    $k = 1, 2, \ldots$ such that $\bigcup\limits_{k=1}^{\infty} H_k$ is not thin.

9.9.   a) Let X_t be an Itô diffusion in $\mathbf{R}^n$ and assume that g is a non-
                constant locally bounded real X_t-harmonic function on a con-
                nected open set $G \subset \mathbf{R}^n$. Prove that g satisfies the following
                weak form of *the maximum principle*: g does not have a (local
                or global) maximum at any point of G. (Similarly g satisfies *the
                minimum principle*).

           b) Give an example to show that a non-constant bounded X_t-
                harmonic function g can have a (non-strict) global maximum.
                (Hint: Consider uniform motion to the right.)

9.10.*   Find the (stochastic) solution $f(t, x)$ of the boundary value problem

$$\begin{cases} K(x)e^{-\rho t} + \dfrac{\partial f}{\partial t} + \alpha x \dfrac{\partial f}{\partial x} + \tfrac{1}{2}\beta^2 x^2 \dfrac{\partial^2 f}{\partial x^2} = 0 & \text{for } x > 0,\ 0 < t < T \\ f(T, x) = e^{-\rho T}\phi(x) & \text{for } x > 0 \end{cases}$$

where K, ϕ are given functions and T, ρ, α, β are constants, $\rho > 0$,
$T > 0$. (Hint: Consider $dY_t = (dt, dX_t)$ where X_t is a geometric
Brownian motion).

9.11.   a) The *Poisson kernel* is defined by

$$P_r(\theta) = \frac{1 - r^2}{1 - 2r\cos\theta + r^2} = \frac{1 - |z|^2}{|1 - z|^2}$$

           where $r \geq 0$, $\theta \in [0, 2\pi]$, $z = re^{i\theta} \in \mathbf{C}$ $(i = \sqrt{-1}\,)$.
           The *Poisson formula* states that if D denotes the open unit disk
           in the plane $\mathbf{R}^2 = \mathbf{C}$ and $h \in C(\overline{D})$ satisfies $\Delta h = 0$ in D then

$$h(re^{i\theta}) = \frac{1}{2\pi} \int_0^{2\pi} P_r(t - \theta)h(e^{it})dt \ .$$

Prove that the probability that Brownian motion, starting from $z \in D$, first exits from D at a set $F \subset \partial D$ is given by

$$\frac{1}{2\pi} \int_F P_r(t - \theta)dt \ , \quad \text{where } z = re^{i\theta} \ .$$

b) The function

$$w = \phi(z) = i\frac{1 + z}{1 - z}$$

maps the disc $D = \{|z| < 1\}$ conformally onto the half plane $H = \{w = u + iv; v > 0\}$, $\phi(\partial D) = \mathbf{R}$ and $\phi(0) = i$. Use Example 8.5.9 to prove that if μ denotes the harmonic measure for Brownian motion at the point $i = (0, 1)$ for the half plane H then

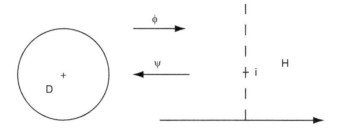

$$\int_{\mathbf{R}} f(\xi)d\mu(\xi) = \frac{1}{2\pi} \int_0^{2\pi} f(\phi(e^{it}))dt = \frac{1}{2\pi i} \int_{\partial D} \frac{f(\phi(z))}{z}dz \ .$$

c) Substitute $w = \phi(z)$ (i.e. $z = \psi(w) := \phi^{-1}(w) = \frac{w-i}{w+i}$) in the integral above to show that

$$\int_{\mathbf{R}} f(\xi)d\mu(\xi) = \frac{1}{\pi} \int_{\partial H} f(w)\frac{dw}{|w - i|^2} = \frac{1}{\pi} \int_{-\infty}^{\infty} f(x)\frac{dx}{x^2 + 1} \ .$$

d) Show that the harmonic measure μ_H^w for Brownian motion in H at the point $w = u + iv \in H$ is given by

$$d\mu_H^w(x) = \frac{1}{\pi} \cdot \frac{v}{(x-u)^2 + v^2} dx \ .$$

9.12. **(A Feynman-Kac formula for boundary value problems)**
Let X_t be an Itô diffusion on $\mathbf{R}^n$ whose generator coincides with a given partial differential operator L on $C_0^2(\mathbf{R}^n)$. Let D, ϕ and g be as in Theorem 9.3.3 and let $q(x) \geq 0$ be a continuous function on $\mathbf{R}^n$. Consider the boundary value problem: Find $h \in C^2(D) \cap C(\overline{D})$ such that

$$\begin{cases} Lh(x) - q(x)h(x) = -g(x) & \text{on } D \\ \lim_{x \to y} h(x) = \phi(y) \ ; & y \in \partial D \ . \end{cases}$$

Show that if a bounded solution h exists, then

$$h(x) = E^x \left[\int_0^{\tau_D} e^{-\int_0^t q(X_s)ds} g(X_t)dt + e^{-\int_0^{\tau_D} q(X_s)ds} \phi(X_{\tau_D}) \right] \ .$$

(Compare with the Feynman-Kac formula.)
Hint: Proceed as in the proof of Theorem 8.2.1 b).

For more information on stochastic solutions of boundary value problems see Freidlin (1985).

9.13. Let $D = (a, b)$ be a bounded interval.

a) For $x \in \mathbf{R}$ define

$$X_t = X_t^x = x + \mu t + \sigma B_t \ ; \qquad t \geq 0$$

where μ, σ are constants, $\sigma \neq 0$. Use Corollary 9.1.2 to compute

$$w(x) := E^x[\phi(X_{\tau_D})] + E^x \left[\int_0^{\tau_D} g(X_t)dt \right]$$

when $\phi : \{a, b\} \to \mathbf{R}$ and $g : (a, b) \to \mathbf{R}$ are given functions, g bounded and continuous.

b) Use the results in a) to compute the Green function $G(x, y)$ of the process X_t.
(Hint: Choose $\phi = 0$ and proceed as in Example 9.3.6.)

9.14. Let $D = (a, b) \subset (0, \infty)$ be a bounded interval and let

$$dX_t = rX_tdt + \alpha X_tdB_t \ ; \qquad X_0 = x \in (a, b)$$

be a geometric Brownian motion.

a) Use Corollary 9.1.2 to find

$$Q^x[X_{\tau_D} = b] .$$

(Hint: Choose $g = 0$ and $\phi(a) = 0$, $\phi(b) = 1$.)

b) Use Corollary 9.1.2 to compute

$$w(x) = E^x[\phi(X_{\tau_D})] + E^x\left[\int_0^{\tau_D} g(X_t)dt\right]$$

for given functions $\phi\colon \{a, b\} \to \mathbf{R}$ and $g\colon (a, b) \to \mathbf{R}$, g bounded
and continuous.
(Hint: The substitution $t = \ln x$, $w(x) = h(\ln x)$ transforms the
differential equation

$$\tfrac{1}{2}\alpha^2 x^2 w''(x) + r x w'(x) = -g(x) ; \qquad x > 0$$

into the differential equation

$$\tfrac{1}{2}\alpha^2 h''(t) + (r - \tfrac{1}{2}\alpha^2)h'(t) = -g(e^t) ; \qquad t \in \mathbf{R} .)$$

9.15.* a) Let $D = (a, b) \subset \mathbf{R}$ be a bounded interval and let $X_t = B_t$ be
1-dimensional Brownian motion. Use Corollary 9.1.2 to compute

$$h(x) = E^x[e^{-\rho\tau_D}\psi(B_{\tau_D})] + E^x\left[\theta\int_0^{\tau_D} e^{-\rho t}B_t^2 dt\right]$$

for a given function $\psi\colon \{a, b\} \to \mathbf{R}$, when $\rho > 0$ and $\theta \in \mathbf{R}$ are
constants.
(Hint: Consider the Itô diffusion

$$dY_t = \begin{bmatrix} dY_t^{(1)} \\ dY_t^{(2)} \end{bmatrix} = \begin{bmatrix} dt \\ dB_t \end{bmatrix} = \begin{bmatrix} 1 \\ 0 \end{bmatrix} dt + \begin{bmatrix} 0 \\ 1 \end{bmatrix} dB_t; \quad Y_0 = y = (s, x) .$$

Then

$$h(x) = w(0, x)$$

where

$$w(s, x) = w(y) = E^y[\phi(Y_{\tau_D})] + E^y\left[\int_0^{\tau_D} g(Y_t)dt\right]$$

with $\phi(y) = \phi(s, x) = e^{-\rho s}\psi(x)$
and $g(y) = g(s, x) = \theta e^{-\rho s}x^2 .$

Note that

$$\tau_D = \inf\{t > 0; B_t \notin (a,b)\} = \inf\{t > 0; Y_t^{(2)} \notin (a,b)\}$$
$$= \inf\{t > 0; Y_t \notin \mathbf{R} \times (a,b)\} .$$

To find $w(s,x)$ solve the boundary value problem

$$\begin{cases} \frac{1}{2}\frac{\partial^2 w}{\partial x^2} + \frac{\partial w}{\partial s} = -\theta\, e^{-\rho s} x^2 \,; & a < x < b \\ w(s,a) = e^{-\rho s}\psi(a)\,, & w(s,b) = e^{-\rho s}\psi(b)\,. \end{cases}$$

To this end, try $w(s,x) = e^{-\rho s}h(x)$.)

b)   Use the method in a) to find $E^x[e^{-\rho \tau_D}]$.
(Compare with Exercise 7.19.)

9.16.   (The Black-Scholes equation)

a)   Let $D = (0,T) \times (0,\infty) \subset \mathbf{R}^2$, where $T > 0$ is a constant. Show that the unique solution $w \in C^{1,2}(D) \cap C(\bar{D})$ of the boundary value problem

$$\begin{cases} -rw(s,x) + \frac{\partial w}{\partial s}(s,x) + rx\frac{\partial w}{\partial x}(s,x) + \frac{1}{2}\sigma^2 x^2 \frac{\partial^2 w}{\partial x^2}(s,x) = 0\,; & (s,x) \in D \\ w(T,x) = (x-K)^+\,; & x \geq 0 \end{cases}$$

(where $r > 0$, $\sigma \neq 0$ are constants) is given by

$$w(s,x) = E^{s,x}[e^{-r(T-s)}(X_{T-s} - K)^+] \qquad\qquad (9.3.32)$$

where $X_t = X_t^x$ is the geometric Brownian motion

$$dX_t = rX_t dt + \sigma X_t dB_t\,; \qquad X_0 = x > 0\,.$$

[Hint: Put $u(s,x) = e^{-rs}w(s,x)$. Then apply Theorem 9.2.13 and Theorem 9.1.1 to the boundary value problem

$$\begin{cases} \frac{\partial u}{\partial s}(s,x) + rx\frac{\partial u}{\partial x}(s,x) + \frac{1}{2}\sigma^2 x^2 \frac{\partial^2 u}{\partial x^2}(s,x) = 0\,; & (s,x) \in D \\ \lim_{t \to \tau_D} u(Y_t) = e^{-r(s+\tau_D)}(X_{\tau_D} - K)^+\,; & x \geq 0 \end{cases}$$

where $Y_t = Y_t^{s,x} = (s+t, X_t^x)$ and $\tau_D = \inf\{t > 0; Y_t \notin D\} = T - s$.]

b)   Use (9.3.32) to show that

$$w(0,x) = x\Phi\big(\eta + \tfrac{1}{2}\sigma\sqrt{T}\,\big) - Ke^{-rT}\Phi\big(\eta - \tfrac{1}{2}\sigma\sqrt{T}\,\big)$$

where

$$\eta = \sigma^{-1} T^{-1/2} \left(\ln \frac{x}{K} + rT \right)$$

and

$$\Phi(t) = \frac{1}{\sqrt{2\pi}} \int_{-\infty}^{t} e^{-\frac{s^2}{2}} ds$$

is the normal distribution function.

This is the celebrated *Black-Scholes option pricing formula*. See Corollary 12.3.8.

Chapter 10
Application to Optimal Stopping

10.1 The Time-Homogeneous Case

Problem 5 in the introduction is a special case of a problem of the following type:

Problem 10.1.1 (The optimal stopping problem)
Let X_t be an Itô diffusion on $\mathbf{R}^n$ and let g (*the reward function*) be a given function on $\mathbf{R}^n$, satisfying

a) $g(\xi) \geq 0$ for all $\xi \in \mathbf{R}^n$ $\qquad\qquad\qquad\qquad\qquad\qquad$ (10.1.1)
b) g is continuous.

Find a stopping time $\tau^* = \tau^*(x, \omega)$ (called an *optimal stopping time*) for $\{X_t\}$ such that

$$E^x[g(X_{\tau^*})] = \sup_\tau E^x[g(X_\tau)] \qquad \text{for all } x \in \mathbf{R}^n , \qquad (10.1.2)$$

the sup being taken over all stopping times τ for $\{X_t\}$. We also want to find the corresponding optimal expected reward

$$g^*(x) = E^x[g(X_{\tau^*})] . \qquad\qquad\qquad\qquad (10.1.3)$$

Here $g(X_\tau)$ is to be interpreted as 0 at the points $\omega \in \Omega$ where $\tau(\omega) = \infty$ and as usual E^x denotes the expectation with respect to the probability law Q^x of the process X_t; $t \geq 0$ starting at $X_0 = x \in \mathbf{R}^n$.

We may regard X_t as the state of a game at time t, each ω corresponds to one sample of the game. For each time t we have the option of stopping the game, thereby obtaining the reward $g(X_t)$, or continue the game in the hope that stopping it at a later time will give a bigger reward. The problem is of course that we do not know what state the game is in at future times, only the probability distribution of the "future". Mathematically, this means

that the possible "stopping" times we consider really are stopping times in
the sense of Definition 7.2.1: The decision whether $\tau \leq t$ or not should only
depend on the behaviour of the Brownian motion B_r (driving the process X)
up to time t, or perhaps only on the behaviour of X_r up to time t. So, among
all possible stopping times τ we are asking for the optimal one, τ^*, which
gives the best result "in the long run", i.e. the biggest expected reward in
the sense of (10.1.2).

In the following we will outline how a solution to this problem can be
obtained using the material from the preceding chapter. Later in this chapter
we shall see that our discussion of problem (10.1.2)–(10.1.3) also covers the
apparently more general problems

$$g^*(s, x) = \sup_{\tau} E^{(s,x)}[g(s + \tau, X_\tau)] = E^{(s,x)}[g(s + \tau^*, X_{\tau^*})] \qquad (10.1.4)$$

and

$$G^*(s, x) = \sup_{\tau} E^{(s,x)}\left[\int_0^\tau f(s + t, X_t)dt + g(s + \tau, X_\tau) \right]$$

$$= E^{(s,x)}\left[\int_0^{\tau^*} f(s + t, X_t)dt + g(s + \tau^*, X_{\tau^*}) \right] \qquad (10.1.5)$$

where f is a given *profit rate (or reward rate) function* (satisfying certain
conditions).

We shall also discuss possible extensions of problem (10.1.2)–(10.1.3) to
cases where g is not necessarily continuous or where g may assume negative
values.

A basic concept in the solution of (10.1.2)–(10.1.3) is the following:

Definition 10.1.2 *A measurable function* $f : \mathbf{R}^n \to [0, \infty]$ *is called* supermeanvalued *(w.r.t.* X_t*) if*

$$f(x) \geq E^x[f(X_\tau)] \qquad (10.1.6)$$

for all stopping times τ *and all* $x \in \mathbf{R}^n$.

If, in addition, f *is also* lower semicontinuous, *then* f *is called* l.s.c. superharmonic *or just* superharmonic *(w.r.t.* X_t*).*

Note that if $f : \mathbf{R}^n \to [0, \infty]$ is lower semicontinuous then by the Fatou
lemma

$$f(x) \leq E^x[\varliminf_{k \to \infty} f(X_{\tau_k})] \leq \varliminf_{k \to \infty} E^x[f(X_{\tau_k})], \qquad (10.1.7)$$

for any sequence $\{\tau_k\}$ of stopping times such that $\tau_k \to 0$ a.s. P. Combining
this with (10.1.6) we see that if f is (l.s.c.) superharmonic, then

$$f(x) = \lim_{k \to \infty} E^x[f(X_{\tau_k})] \qquad \text{for all } x, \qquad (10.1.8)$$

for all such sequences τ_k.

Remarks. 1) In the literature (see e.g. Dynkin (1965 II)) one often finds a weaker concept of X_t-superharmonicity, defined by the supermeanvalued property (10.1.6) plus the stochastic continuity requirement (10.1.8). This weaker concept corresponds to the X_t-harmonicity defined in Chapter 9.

   2) If $f \in C^2(\mathbf{R}^n)$ it follows from Dynkin's formula that f is superharmonic w.r.t. X_t if and only if

$$\mathcal{A}f \leq 0$$

where $\mathcal{A}$ is the characteristic operator of X_t. This is often a useful criterion (See e.g. Example 10.2.1).

   3) If $X_t = B_t$ is Brownian motion in $\mathbf{R}^n$ then the superharmonic functions for X_t coincide with the (nonnegative) superharmonic functions in classical potential theory. See Doob (1984) or Port and Stone (1979).

   We state some useful properties of superharmonic and supermeanvalued functions.

Lemma 10.1.3 a)   *If f is superharmonic (supermeanvalued) and $\alpha > 0$, then αf is superharmonic (supermeanvalued).*

b)   *If f_1, f_2 are superharmonic (supermeanvalued), then $f_1 + f_2$ is superharmonic (supermeanvalued).*

c)   *If $\{f_j\}_{j \in J}$ is a family of supermeanvalued functions, then $f(x) := \inf_{j \in J}\{f_j(x)\}$ is supermeanvalued if it is measurable (J is any set).*

d)   *If $f_1, f_2, \cdots$ are superharmonic (supermeanvalued) functions and $f_k \uparrow f$ pointwise, then f is superharmonic (supermeanvalued).*

e)   *If f is supermeanvalued and $\sigma \leq \tau$ are stopping times, then $E^x[f(X_\sigma)] \geq E^x[f(X_\tau)]$.*

f)   *If f is supermeanvalued and H is a Borel set, then $\widetilde{f}(x) := E^x[f(X_{\tau_H})]$ is supermeanvalued.*

Proof of Lemma 10.1.3.

a)   and b) are straightforward.

c)   Suppose f_j is supermeanvalued for all $j \in J$. Then

$$f_j(x) \geq E^x[f_j(X_\tau)] \geq E^x[f(X_\tau)] \qquad \text{for all } j .$$

So $f(x) = \inf f_j(x) \geq E^x[f(X_\tau)]$, as required.

d)   Suppose f_j is supermeanvalued, $f_j \uparrow f$. Then

$$f(x) \geq f_j(x) \geq E^x[f_j(X_\tau)] \qquad \text{for all } j, \text{ so}$$
$$f(x) \geq \lim_{j \to \infty} E^x[f_j(X_\tau)] = E^x[f(X_\tau)] ,$$

by monotone convergence. Hence f is supermeanvalued. If each f_j is also lower semicontinuous then if $y_k \to x$ as $k \to \infty$ we have

$$f_j(x) \leq \varliminf_{k \to \infty} f_j(y_k) \leq \varliminf_{k \to \infty} f(y_k) \qquad \text{for each } j .$$

Hence, by letting $j \to \infty$,

$$f(x) \leq \varliminf_{k \to \infty} f(y_k) \;.$$

e)   If f is supermeanvalued we have by the Markov property when $t > s$

$$E^x[f(X_t)|\mathcal{F}_s] = E^{X_s}[f(X_{t-s})] \leq f(X_s) \;, \qquad (10.1.9)$$

i.e. the process

$$\zeta_t = f(X_t)$$

is a *supermartingale* w.r.t. the σ-algebras $\mathcal{F}_t$ generated by $\{B_r; r \leq t\}$. (Appendix C). Therefore, by Doob's optional sampling theorem (see Gihman and Skorohod (1975, Theorem 6 p. 11)) we have

$$E^x[f(X_\sigma)] \geq E^x[f(X_\tau)]$$

for all stopping times σ, τ with $\sigma \leq \tau$ a.s. Q^x.

f)   Suppose f is supermeanvalued. By the strong Markov property (7.2.2) and formula (7.2.6) we have, for any stopping time α,

$$
\begin{aligned}
E^x[\widetilde{f}(X_\alpha)] &= E^x[E^{X_\alpha}[f(X_{\tau_H})]] = E^x[E^x[\theta_\alpha f(X_{\tau_H})|\mathcal{F}_\alpha]] \\
&= E^x[\theta_\alpha f(X_{\tau_H})] = E^x[f(X_{\tau_H^\alpha})]
\end{aligned}
\qquad (10.1.10)
$$

where $\tau_H^\alpha = \inf\{t > \alpha; X_t \notin H\}$. Since $\tau_H^\alpha \geq \tau_H$ we have by e)

$$E^x[\widetilde{f}(X_\alpha)] \leq E^x[f(X_{\tau_H})] = \widetilde{f}(x) \;,$$

so $\widetilde{f}$ is supermeanvalued.   □

The following concepts are fundamental:

Definition 10.1.4 *Let h be a real measurable function on $\mathbf{R}^n$. If f is a superharmonic (supermeanvalued) function and $f \geq h$ we say that f is a superharmonic (supermeanvalued) majorant of h (w.r.t. X_t). The function*

$$\overline{h}(x) = \inf_f f(x); \quad x \in \mathbf{R}^n \;, \qquad (10.1.11)$$

the inf being taken over all supermeanvalued majorants f of h, is called the least supermeanvalued majorant of h.

Similarly, suppose there exists a function $\widehat{h}$ such that

(i) $\widehat{h}$ is a superharmonic majorant of h and

(ii)   if f is any other superharmonic majorant of h then $\widehat{h} \leq f$.

Then $\widehat{h}$ is called *the least superharmonic majorant* of h.

Note that by Lemma 10.1.3 c) the function $\overline{h}$ is supermeanvalued if it is measurable. Moreover, if $\overline{h}$ is lower semicontinuous, then $\widehat{h}$ exists and $\widehat{h} = \overline{h}$. Later we will prove that if g is nonnegative (or lower bounded) and lower semicontinuous, then $\widehat{g}$ exists and $\widehat{g} = \overline{g}$ (Theorem 10.1.7).

Let $g \geq 0$ and let f be a supermeanvalued majorant of g. Then if τ is a stopping time

$$f(x) \geq E^x[f(X_\tau)] \geq E^x[g(X_\tau)] .$$

So

$$f(x) \geq \sup_\tau E^x[g(X_\tau)] = g^*(x) .$$

Therefore we always have, if $\widehat{g}$ exists,

$$\widehat{g}(x) \geq g^*(x) \qquad \text{for all } x \in \mathbf{R}^n . \tag{10.1.12}$$

What is not so easy to see is that the converse inequality also holds, i.e. that in fact

$$\widehat{g} = g^* . \tag{10.1.13}$$

We will prove this after we have established a useful iterative procedure for calculating $\widehat{g}$. Before we give such a procedure let us introduce a concept which is related to superharmonic functions:

Definition 10.1.5 *A lower semicontinuous function $f\colon \mathbf{R}^n \to [0, \infty]$ is called excessive (w.r.t. X_t) if*

$$f(x) \geq E^x[f(X_s)] \qquad \text{for all } s \geq 0, \ x \in \mathbf{R}^n . \tag{10.1.14}$$

It is clear that a superharmonic function must be excessive. What is not so obvious, is that the converse also holds:

Theorem 10.1.6 *Let $f\colon \mathbf{R}^n \to [0, \infty]$. Then f is excessive w.r.t. X_t if and only if f is superharmonic w.r.t. X_t.*

Proof in a special case. Let L be the differential operator associated to X (given by the right hand side of (7.3.3)), so that L coincides with the generator A of X on C_0^2. We only prove the theorem in the special case when $f \in C^2(\mathbf{R}^n)$ and Lf is bounded: Then by Dynkin's formula we have

$$E^x[f(X_t)] = f(x) + E^x\left[\int_0^t Lf(X_r)dr\right] \qquad \text{for all } t \geq 0 ,$$

so if f is excessive then $Lf \leq 0$. Therefore, if τ is a stopping time we get

$$E^x[f(X_{t \wedge \tau})] \leq f(x) \qquad \text{for all } t \geq 0 .$$

Letting $t \to \infty$ we see that f is superharmonic. □

A proof in the general case can be found in Dynkin (1965 II, p. 5).

The first iterative procedure for the least superharmonic majorant $\widehat{g}$ of g is the following:

Theorem 10.1.7 (Construction   of   the   least   superharmonic majorant)

Let $g = g_0$ be a nonnegative, lower semicontinuous function on $\mathbf{R}^n$ and define inductively

$$g_n(x) = \sup_{t \in S_n} E^x[g_{n-1}(X_t)] \,, \tag{10.1.15}$$

where $S_n = \{k \cdot 2^{-n}; 0 \leq k \leq 4^n\}$, $n = 1, 2, \ldots$. Then $g_n \uparrow \widehat{g}$ and $\widehat{g}$ is the least *superharmonic* majorant of g. Moreover, $\widehat{g} = \overline{g}$.

Proof. Note that $\{g_n\}$ is increasing. Define $\breve{g}(x) = \lim\limits_{n \to \infty} g_n(x)$. Then

$$\breve{g}(x) \geq g_n(x) \geq E^x[g_{n-1}(X_t)] \qquad \text{for all } n \text{ and all } t \in S_n \,.$$

Hence

$$\breve{g}(x) \geq \lim_{n \to \infty} E^x[g_{n-1}(X_t)] = E^x[\breve{g}(X_t)] \tag{10.1.16}$$

for all $t \in S = \bigcup\limits_{n=1}^{\infty} S_n$.

Since $\breve{g}$ is an increasing limit of lower semicontinuous functions (Lemma 8.1.4) $\breve{g}$ is lower semicontinuous. Fix $t \in \mathbf{R}$ and choose $t_k \in S$ such that $t_k \to t$. Then by (10.1.16), the Fatou lemma and lower semicontinuity

$$\breve{g}(x) \geq \varliminf_{k \to \infty} E^x[\breve{g}(X_{t_k})] \geq E^x[\varliminf_{k \to \infty} \breve{g}(X_{t_k})] \geq E^x[\breve{g}(X_t)] \,.$$

So $\breve{g}$ is an excessive function. Therefore $\breve{g}$ is superharmonic by Theorem 10.1.6 and hence $\breve{g}$ is a superharmonic majorant of g. On the other hand, if f is any supermeanvalued majorant of g, then clearly by induction

$$f(x) \geq g_n(x) \qquad \text{for all } n$$

and so $f(x) \geq \breve{g}(x)$. This proves that $\breve{g}$ is the least supermeanvalued majorant $\overline{g}$ of g. So $\breve{g} = \widehat{g}$.   □

It is a consequence of Theorem 10.1.7 that we may replace the finite sets S_n by the whole interval $[0, \infty]$:

Corollary 10.1.8 *Define $h_0 = g$ and inductively*

$$h_n(x) = \sup_{t \geq 0} E^x[h_{n-1}(X_t)] \,; \qquad n = 1, 2, \ldots$$

Then $h_n \uparrow \widehat{g}$.

Proof. Let $h = \lim h_n$. Then clearly $h \geq \check{g} = \widehat{g}$. On the other hand, since $\widehat{g}$ is excessive we have

$$\widehat{g}(x) \geq \sup_{t \geq 0} E^x[\widehat{g}(X_t)].$$

So by induction

$$\widehat{g} \geq h_n \qquad \text{for all } n .$$

Thus $\widehat{g} = h$ and the proof is complete.

We are now ready for our first main result on the optimal stopping problem. The following result is basically due to Dynkin (1963) (and, in a martingale context, Snell (1952)):

Theorem 10.1.9 (Existence theorem for optimal stopping)
Let g^* denote the optimal reward and $\widehat{g}$ the least superharmonic majorant of a continuous reward function $g \geq 0$.

a)   *Then*

$$g^*(x) = \widehat{g}(x) . \tag{10.1.17}$$

b)   *For $\epsilon > 0$ let*

$$D_\epsilon = \{x; g(x) < \widehat{g}(x) - \epsilon\} . \tag{10.1.18}$$

Suppose g is bounded. *Then stopping at the first time τ_ϵ of exit from D_ϵ is close to being optimal, in the sense that*

$$|g^*(x) - E^x[g(X_{\tau_\epsilon})]| \leq \epsilon \tag{10.1.19}$$

for all x.

c)   *For arbitrary continuous $g \geq 0$ let*

$$D = \{x; g(x) < g^*(x)\} \quad \text{(the continuation region)} . \tag{10.1.20}$$

For $N = 1, 2, \ldots$ define $g_N = g \wedge N$, $D_N = \{x; g_N(x) < \widehat{g_N}(x)\}$ and $\sigma_N = \tau_{D_N}$. Then $D_N \subset D_{N+1}$, $D_N \subset D \cap g^{-1}([0, N))$, $D = \bigcup_N D_N$. If $\sigma_N < \infty$ a.s. Q^x for all N then

$$g^*(x) = \lim_{N \to \infty} E^x[g(X_{\sigma_N})] . \tag{10.1.21}$$

d)   *In particular, if $\tau_D < \infty$ a.s. Q^x and the family $\{g(X_{\sigma_N})\}_N$ is uniformly integrable w.r.t. Q^x (Appendix C), then*

$$g^*(x) = E^x[g(X_{\tau_D})]$$

and $\tau^* = \tau_D$ is an optimal stopping time.

Proof. First assume that g is bounded and define

$$\widetilde{g}_\epsilon(x) = E^x[\widehat{g}(X_{\tau_\epsilon})] \qquad \text{for } \epsilon > 0 . \tag{10.1.22}$$

Then $\widetilde{g}_\epsilon$ is supermeanvalued by Lemma 10.1.3 f). We claim that

$$g(x) \leq \widetilde{g}_\epsilon(x) \qquad \text{for all } x \,. \tag{10.1.23}$$

To see this define

$$\beta := \sup_x \{g(x) - \widetilde{g}_\epsilon(x)\} \,. \tag{10.1.24}$$

Then $\widetilde{g}_\epsilon(x) + \beta$ is a supermeanvalued majorant of g. Hence

$$\widehat{g}(x) < \widetilde{g}_\epsilon(x) + \beta \quad \text{for all } x \,. \tag{10.1.25}$$

Choose α such that $0 < \alpha < \epsilon$. Then there exists x_0 such that

$$\beta - \alpha < g(x_0) - \widetilde{g}_\epsilon(x_0) \,.$$

Clearly

$$0 \leq \widehat{g}(x_0) - g(x_0) \leq \widetilde{g}_\epsilon(x) + \beta - g(x_0) < \alpha \,.$$

Hence

$$x_0 \in D_\alpha^C \subseteq D_\epsilon^C \tag{10.1.26}$$

and therefore

$$\widetilde{g}_\epsilon(x) = E^{x_0}[\widehat{g}(X_{\tau_\epsilon})] = \widehat{g}(x_0) \,.$$

This gives

$$\beta - \alpha < g(x_0) - \widetilde{g}_\epsilon(x_0) \leq \widehat{g}(x_0) - \widetilde{g}_\epsilon(x_0) = 0 \,. \tag{10.1.27}$$

Letting $\alpha \downarrow 0$, we conclude that

$$\beta \leq 0 \,,$$

which proves the claim (10.1.23).

We conclude that $\widetilde{g}_\epsilon$ is a supermeanvalued majorant of g. Therefore

$$\widehat{g} \leq \widetilde{g}_\epsilon = E[\widehat{g}(X_{\tau_\epsilon})] \leq E[(g + \epsilon)(X_{\tau_\epsilon})] \leq g^* + \epsilon \tag{10.1.28}$$

and since ϵ was arbitrary we have by (10.1.12)

$$\widehat{g} = g^* \,.$$

If g is not bounded, let

$$g_N = \min(N, g) \,, \qquad N = 1, 2, \ldots$$

and as before let $\widehat{g_N}$ be the least superharmonic majorant of g_N. Then

$$g^* \geq g_N^* = \widehat{g_N} \uparrow h \qquad \text{as } N \to \infty \,, \text{ where } h \geq \widehat{g}$$

since h is a superharmonic majorant of g. Thus $h = \widehat{g} = g^*$ and this proves
(10.1.17) for general g. From (10.1.28) and (10.1.17) we obtain (10.1.19).

Finally, to obtain c) and d) let us again first assume that g is bounded.
Then, since

$$\tau_\epsilon \uparrow \tau_D \qquad \text{as } \epsilon \downarrow 0$$

and $\tau_D < \infty$ a.s we have

$$E^x[g(X_{\tau_\epsilon})] \rightarrow E^x[g(X_{\tau_D})] \qquad \text{as } \epsilon \downarrow 0 , \qquad (10.1.29)$$

and hence by (10.1.28) and (10.1.17)

$$g^*(x) = E^x[g(X_{\tau_D})] \qquad \text{if } g \text{ is bounded .} \qquad (10.1.30)$$

Finally, if g is not bounded define

$$h = \lim_{N \to \infty} \widehat{g_N} .$$

Then h is superharmonic by Lemma 10.1.3 d) and since $\widehat{g_N} \leq \widehat{g}$ for all N we
have $h \leq \widehat{g}$. On the other hand $g_N \leq \widehat{g_N} \leq h$ for all N and therefore $g \leq h$.
Since $\widehat{g}$ is the least superharmonic majorant of g we conclude that

$$h = \widehat{g} . \qquad (10.1.31)$$

Hence by (10.1.30), (10.1.31) we obtain (10.1.21):

$$g^*(x) = \lim_{N \to \infty} \widehat{g_N}(x) = \lim_{N \to \infty} E^x[g_N(X_{\sigma_N})] \leq \lim_{N \to \infty} E^x[g(X_{\sigma_N})] \leq g^*(x) .$$

Note that $\widehat{g_N} \leq N$ everywhere, so if $g_N(x) < \widehat{g_N}(x)$ then $g_N(x) < N$ and therefore $g(x) = g_N(x) < \widehat{g_N}(x) \leq \widehat{g}(x)$ and $g_{N+1}(x) = g_N(x) < \widehat{g_N}(x) \leq \widehat{g_{N+1}}(x)$.
Hence $D_N \subset D \cap \{x; g(x) < N\}$ and $D_N \subset D_{N+1}$ for all N. So by (10.1.31)
we conclude that D is the increasing union of the sets $D_N; N = 1, 2, \ldots$
Therefore

$$\tau_D = \lim_{N \to \infty} \sigma_N .$$

So by (10.1.21) and uniform integrability we have

$$\widehat{g}(x) = \lim_{N \to \infty} \widehat{g_N}(x) = \lim_{N \to \infty} E^x[g_N(X_{\sigma_N})]$$
$$= E^x[\lim_{N \to \infty} g_N(X_{\sigma_N})] = E^x[g(X_{\tau_D})] ,$$

and the proof of Theorem 10.1.9 is complete.   □

Remarks.

1) Note that the sets D, D_ϵ and D_N are open, since $\widehat{g} = g^*$ is lower semi-
    continuous and g is continuous.

2)   By inspecting the proof of a) we see that (10.1.17) holds under the
weaker assumption that $g \geq 0$ is lower semicontinuous.

The following consequence of Theorem 10.1.9 is often useful:

Corollary 10.1.10 *Suppose there exists a Borel set H such that*

$$\tilde{g}_H(x):= E^x[g(X_{\tau_H})]$$

is a supermeanvalued majorant of g. Then

$$g^*(x) = \tilde{g}_H(x) , \quad so \ \tau^* = \tau_H \ is \ optimal .$$

Proof. If $\tilde{g}_H$ is a supermeanvalued majorant of g then clearly

$$\bar{g}(x) \leq \tilde{g}_H(x) .$$

On the other hand we of course have

$$\tilde{g}_H(x) \leq \sup_{\tau} E^x[g(X_\tau)] = g^*(x) ,$$

so $g^* = \tilde{g}_H$ by Theorem 10.1.7 and Theorem 10.1.9 a).   □

Corollary 10.1.11 *Let*

$$D = \{x; g(x) < \hat{g}(x)\}$$

and put

$$\tilde{g}(x) = \tilde{g}_D(x) = E^x[g(X_{\tau_D})] .$$

If $\tilde{g} \geq g$ then $\tilde{g} = g^*$.

Proof. Since $X_{\tau_D} \notin D$ we have $g(X_{\tau_D}) \geq \hat{g}(X_{\tau_D})$ and therefore $g(X_{\tau_D}) = \hat{g}(X_{\tau_D})$, a.s. Q^x. So $\tilde{g}(x) = E^x[\hat{g}(X_{\tau_D})]$ is supermeanvalued since $\hat{g}$ is, and the result follows from Corollary 10.1.10.   □

Theorem 10.1.9 gives a sufficient condition for the existence of an optimal stopping time τ^*. Unfortunately, τ^* need not exist in general. For example, if

$$X_t = t \quad for \ t \geq 0 \quad (\text{deterministic})$$

and

$$g(\xi) = \frac{\xi^2}{1 + \xi^2} ; \quad \xi \in \mathbf{R}$$

then $g^*(x) = 1$, but there is no stopping time τ such that

$$E^x[g(X_\tau)] = 1 .$$

However, we can prove that if an optimal stopping time τ^* exists, then the stopping time given in Theorem 10.1.9 is optimal:

Theorem 10.1.12 (Uniqueness theorem for optimal stopping)
Define as before

$$D = \{x; g(x) < g^*(x)\} \subset \mathbf{R}^n .$$

Suppose there exists an optimal stopping time $\tau^* = \tau^*(x, \omega)$ for the problem (10.1.2) for all x. Then

$$\tau^* \geq \tau_D \qquad \text{for all } x \in D \tag{10.1.32}$$

and

$$g^*(x) = E^x[g(X_{\tau_D})] \qquad \text{for all } x \in \mathbf{R}^n . \tag{10.1.33}$$

Hence τ_D is an optimal stopping time for the problem (10.1.2).

Proof. Choose $x \in D$. Let τ be an $\mathcal{F}_t$-stopping time and assume $Q^x[\tau < \tau_D] > 0$. Since $g(X_\tau) < g^*(X_\tau)$ if $\tau < \tau_D$ and $g \leq g^*$ always, we have

$$E^x[g(X_\tau)] = \int_{\tau < \tau_D} g(X_\tau)dQ^x + \int_{\tau \geq \tau_D} g(X_\tau)dQ^x$$

$$< \int_{\tau < \tau_D} g^*(X_\tau)dQ^x + \int_{\tau \geq \tau_D} g^*(X_\tau)dQ^x = E^x[g^*(X_\tau)] \leq g^*(x) ,$$

since g^* is superharmonic. This proves (10.1.32).

To obtain (10.1.33) we first choose $x \in D$. Since $\hat{g}$ is superharmonic we have by (10.1.32) and Lemma 10.1.3 e)

$$g^*(x) = E^x[g(X_{\tau^*})] \leq E^x[\hat{g}(X_{\tau^*})] \leq E^x[\hat{g}(X_{\tau_D})]$$
$$= E^x[g(X_{\tau_D})] \leq g^*(x) , \qquad \text{which proves (10.1.33) for } x \in D .$$

Next, choose $x \in \partial D$ to be an *irregular* boundary point of D. Then $\tau_D > 0$ a.s. Q^x. Let $\{\alpha_k\}$ be a sequence of stopping times such that $0 < \alpha_k < \tau_D$ and $\alpha_k \to 0$ a.s. Q^x, as $k \to \infty$. Then $X_{\alpha_k} \in D$ so by (10.1.32), (7.2.6) and the strong Markov property (7.2.2)

$$E^x[g(X_{\tau_D})] = E^x[\theta_{\alpha_k}g(X_{\tau_D})] = E^x[E^{X_{\alpha_k}}[g(X_{\tau_D})]] = E^x[g^*(X_{\alpha_k})] \quad \text{for all } k .$$

Hence by lower semicontinuity and the Fatou lemma

$$g^*(x) \leq E^x[\varliminf_{k \to \infty} g^*(X_{\alpha_k})] \leq \varliminf_{k \to \infty} E^x[g^*(X_{\alpha_k})] = E^x[g(X_{\tau_D})] .$$

Finally, if $x \in \partial D$ is a *regular* boundary point of D or if $x \notin \overline{D}$ we have $\tau_D = 0$ a.s. Q^x and hence $g^*(x) = E^x[g(X_{\tau_D})]$. $\square$

Remark. The following observation is sometimes useful:
Let $\mathcal{A}$ be the characteristic operator of X. Assume $g \in C^2(\mathbf{R}^n)$. Define

$$U = \{x; \mathcal{A}g(x) > 0\} . \tag{10.1.34}$$

Then, with D as before, (10.1.20),

$$U \subset D. \tag{10.1.35}$$

Consequently, from (10.1.32) we conclude that it is *never optimal to stop the process before it exits from U*. But there may be cases when $U \neq D$, so that it is optimal to proceed beyond U before stopping. (This is in fact the typical situation.) See e.g. Example 10.2.2.

To prove (10.1.35) choose $x \in U$ and let τ_0 be the first exit time from a bounded open set $W \ni x$, $W \subset U$. Then by Dynkin's formula, for $u > 0$

$$E^x[g(X_{\tau_0 \wedge u})] = g(x) + E^x\left[\int_0^{\tau_0 \wedge u} Ag(X_s)ds\right] > g(x)$$

so $g(x) < g^*(x)$ and therefore $x \in D$.

Example 10.1.13 Let $X_t = B_t$ be a Brownian motion in $\mathbf{R}^2$. Using that B_t is recurrent in $\mathbf{R}^2$ (Example 7.4.2) one can show that the only (nonnegative) superharmonic functions in $\mathbf{R}^2$ are the constants (Exercise 10.2).

Therefore

$$g^*(x) = \|g\|_\infty := \sup\{g(y); y \in \mathbf{R}^2\} \qquad \text{for all } x.$$

So if g is unbounded then $g^* = \infty$ and no optimal stopping time exists. Assume therefore that g is bounded. The continuation region is

$$D = \{x; g(x) < \|g\|_\infty\},$$

so if ∂D is a *polar set* i.e. cap $(\partial D) = 0$, where cap denotes the *logarithmic capacity* (see Port and Stone (1979)), then $\tau_D = \infty$ a.s. and no optimal stopping exists. On the other hand, if cap$(\partial D) > 0$ then $\tau_D < \infty$ a.s. and

$$E^x[g(B_{\tau_D})] = \|g\|_\infty = g^*(x),$$

so $\tau^* = \tau_D$ is optimal.

Example 10.1.14 The situation is different in $\mathbf{R}^n$ for $n \geq 3$.

a)   To illustrate this let $X_t = B_t$ be Brownian motion in $\mathbf{R}^3$ and let the reward function be

$$g(\xi) = \begin{cases} |\xi|^{-1} & \text{for } |\xi| \geq 1 \\ 1 & \text{for } |\xi| < 1 \end{cases}; \qquad \xi \in \mathbf{R}^3.$$

Then g is superharmonic (in the classical sense) in $\mathbf{R}^3$, so $g^* = g$ everywhere and the best policy is to stop immediately, no matter where the starting point is.

b)   Let us change g to

$$h(x) = \begin{cases} |x|^{-\alpha} & \text{for } |x| \geq 1 \\ 1 & \text{for } |x| < 1 \end{cases}$$

for some $\alpha > 1$. Let $H = \{x; |x| > 1\}$ and define

$$\widetilde{h}(x) = E^x[h(B_{\tau_H})] = P^x[\tau_H < \infty] .$$

Then by Example 7.4.2

$$\widetilde{h}(x) = \begin{cases} 1 & \text{if } |x| \leq 1 \\ |x|^{-1} & \text{if } |x| > 1 , \end{cases}$$

i.e. $\widetilde{h} = g$ (defined in a)), which is a superharmonic majorant of h. Therefore by Corollary 10.1.10

$$h^* = \widetilde{h} = g ,$$

$H = D$ and $\tau^* = \tau_H$ is an optimal stopping time.

Reward Functions Assuming Negative Values

The results we have obtained so far regarding the problem (10.1.2)–(10.1.3) are based on the assumptions (10.1.1). To some extent these assumptions can be relaxed, although neither can be removed completely. For example, we have noted that Theorem 10.1.9 a) still holds if $g \geq 0$ is only assumed to be lower semicontinuous.

The nonnegativity assumption on g can also be relaxed. First of all, note that if g is bounded below, say $g \geq -M$ where $M > 0$ is a constant, then we can put

$$g_1 = g + M \geq 0$$

and apply the theory to g_1. Since

$$E^x[g(X_\tau)] = E^x[g_1(X_\tau)] - M \qquad \text{if } \tau < \infty \text{ a.s. },$$

we have $g^*(x) = g_1^*(x) - M$, so the problem can be reduced to the optimal stopping problem for the nonnegative function g_1. (See Exercise 10.4.)

If g is not bounded below, then problem (10.1.2)–(10.1.3) is not well-defined unless

$$E^x[g^-(X_\tau)] < \infty \qquad \text{for all } \tau \tag{10.1.36}$$

where

$$g^-(x) = -\min(g(x), 0) .$$

If we assume that g satisfies the stronger condition that

the family $\{g^-(X_\tau);\ \tau$ stopping time$\}$ is uniformly integrable   (10.1.37)

then basically all the theory from the nonnegative case carries over. We refer to the reader to Shiryaev (1978) for more information. See also Theorem 10.4.1.

10.2 The Time-Inhomogeneous Case

Let us now consider the case when the reward function g depends on both time and space, i.e.

$$g = g(t,x)\colon \mathbf{R} \times \mathbf{R}^n \to [0,\infty)\ ,\qquad g \text{ is continuous}\ . \tag{10.2.1}$$

Then the problem is to find $g_0(x)$ and τ^* such that

$$g_0(x) = \sup_\tau E^x[g(\tau, X_\tau)] = E^x[g(\tau^*, X_{\tau^*})]\ . \tag{10.2.2}$$

To reduce this case to the original case (10.1.2)–(10.1.3) we proceed as follows:
   Suppose the Itô diffusion $X_t = X_t^x$ has the form

$$dX_t = b(X_t)dt + \sigma(X_t)dB_t\ ;\qquad t \geq 0\ ,\ \ X_0 = x$$

where $b\colon \mathbf{R}^n \to \mathbf{R}^n$ and $\sigma\colon \mathbf{R}^n \to \mathbf{R}^{n\times m}$ are given functions satisfying the conditions of Theorem 5.2.1 and B_t is m-dimensional Brownian motion. Define the Itô diffusion $Y_t = Y_t^{(s,x)}$ in $\mathbf{R}^{n+1}$ by

$$Y_t = \begin{bmatrix} s+t \\ X_t^x \end{bmatrix}\ ;\qquad t \geq 0\ . \tag{10.2.3}$$

Then

$$dY_t = \begin{bmatrix} 1 \\ b(X_t) \end{bmatrix} dt + \begin{bmatrix} 0 \\ \sigma(X_t) \end{bmatrix} dB_t = \widehat{b}(Y_t)dt + \widehat{\sigma}(Y_t)dB_t \tag{10.2.4}$$

where

$$\widehat{b}(\eta) = \widehat{b}(t,\xi) = \begin{bmatrix} 1 \\ b(\xi) \end{bmatrix} \in \mathbf{R}^{n+1}\ ,\quad \widehat{\sigma}(\eta) = \widehat{\sigma}(t,\xi) = \begin{bmatrix} 0\cdots 0 \\ \text{----} \\ \sigma(\xi) \end{bmatrix} \in \mathbf{R}^{(n+1)\times m}\ ,$$

with $\eta = (t,\xi) \in \mathbf{R} \times \mathbf{R}^n$.
   So Y_t is an Itô diffusion starting at $y = (s,x)$. Let $R^y = R^{(s,x)}$ denote the probability law of $\{Y_t\}$ and let $E^y = E^{(s,x)}$ denote the expectation w.r.t. R^y.

In terms of Y_t the problem (10.2.2) can be written

$$g_0(x) = g^*(0, x) = \sup_\tau E^{(0,x)}[g(Y_\tau)] = E^{(0,x)}[g(Y_{\tau^*})] \qquad (10.2.5)$$

which is a special case of the problem

$$g^*(s, x) = \sup_\tau E^{(s,x)}[g(Y_\tau)] = E^{(s,x)}[g(Y_{\tau^*})] , \qquad (10.2.6)$$

which is of the form (10.1.2)–(10.1.3) with X_t replaced by Y_t.

Note that the characteristic operator $\widehat{\mathcal{A}}$ of Y_t is given by

$$\widehat{\mathcal{A}}\phi(s, x) = \frac{\partial \phi}{\partial s}(s, x) + \mathcal{A}\phi(s, x) ; \qquad \phi \in C^2(\mathbf{R} \times \mathbf{R}^n) \qquad (10.2.7)$$

where $\mathcal{A}$ is the characteristic operator of X_t (working on the x-variables).

Example 10.2.1 Let $X_t = B_t$ be 1-dimensional Brownian motion and let the reward function be

$$g(t, \xi) = e^{-\alpha t + \beta \xi} ; \qquad \xi \in \mathbf{R}$$

where $\alpha, \beta \geq 0$ are constants. The characteristic operator $\widehat{\mathcal{A}}$ of $Y_t^{s,x} = \begin{bmatrix} s+t \\ B_t^x \end{bmatrix}$ is given by

$$\widehat{\mathcal{A}}f(s, x) = \frac{\partial f}{\partial s} + \frac{1}{2} \cdot \frac{\partial^2 f}{\partial x^2} ; \qquad f \in C^2 .$$

Thus

$$\mathcal{A}g = (-\alpha + \tfrac{1}{2}\beta^2)g ,$$

so if $\beta^2 \leq 2\alpha$ then $g^* = g$ and the best policy is to stop immediately. If $\beta^2 > 2\alpha$ we have

$$U := \{(s, x); \widehat{\mathcal{A}}g(s, x) > 0\} = \mathbf{R}^2$$

and therefore by (10.1.35) $D = \mathbf{R}^2$ and hence τ^* does not exist. If $\beta^2 > 2\alpha$ we can use Theorem 10.1.7 to prove that $g^* = \infty$:

$$\sup_{t \in S_n} E^{(s,x)}[g(Y_t)] = \sup_{t \in S_n} E[e^{-\alpha(s+t) + \beta B_t^x}]$$

$$= \sup_{t \in S_n} [e^{-\alpha(s+t)} \cdot e^{\beta x + \frac{1}{2}\beta^2 t}] \qquad \text{(see the remark following (5.1.6))}$$

$$= \sup_{t \in S_n} g(s, x) \cdot e^{(-\alpha + \frac{1}{2}\beta^2)t} = g(s, x) \cdot \exp((-\alpha + \tfrac{1}{2}\beta^2)2^n) ,$$

so $g_n(s, x) \to \infty$ as $n \to \infty$.

Hence no optimal stopping exists in this case.

Example 10.2.2 (When is the right time to sell the stocks? (Part 1))
We now return to a specific version of Problem 5 in the introduction:

Suppose the price X_t at time t of a person's assets (e.g. a house, stocks, oil ...) varies according to a stochastic differential equation of the form

$$dX_t = rX_t dt + \alpha X_t dB_t, X_0 = x > 0 ,$$

where B_t is 1-dimensional Brownian motion and r, α are known constants. (The problem of estimating α and r from a series of observations can be approached using the quadratic variation $\langle X, X \rangle_t$ of the process $\{X_t\}$ (Exercise 4.7) and filtering theory (Example 6.2.11), respectively.) Suppose that connected to the sale of the assets there is a fixed fee/tax or transaction cost $a > 0$. Then if the person decides to sell at time t the discounted net of the sale is

$$e^{-\rho t}(X_t - a) ,$$

where $\rho > 0$ is given discounting factor. The problem is to find a stopping time τ that maximizes

$$E^{(s,x)}[e^{-\rho \tau}(X_\tau - a)] = E^{(s,x)}[g(\tau, X_\tau)] ,$$

where

$$g(t, \xi) = e^{-\rho t}(\xi - a) .$$

The characteristic operator $\widehat{\mathcal{A}}$ of the process $Y_t = (s + t, X_t)$ is given by

$$\widehat{\mathcal{A}}f(s, x) = \frac{\partial f}{\partial s} + rx\frac{\partial f}{\partial x} + \tfrac{1}{2}\alpha^2 x^2 \frac{\partial^2 f}{\partial x^2} ; \qquad f \in C^2(\mathbf{R}^2) .$$

Hence $\widehat{\mathcal{A}}g(s, x) = -\rho e^{-\rho s}(x - a) + rxe^{-\rho s} = e^{-\rho s}((r - \rho)x + \rho a)$. So

$$U := \{(s, x); \widehat{\mathcal{A}}g(s, x) > 0\} = \begin{cases} \mathbf{R} \times \mathbf{R}_+ & \text{if } r \geq \rho \\ \{(s, x); x < \frac{a\rho}{\rho - r}\} & \text{if } r < \rho . \end{cases}$$

So if $r \geq \rho$ we have $U = D = \mathbf{R} \times \mathbf{R}_+$ so τ^* does not exist. If $r > \rho$ then $g^* = \infty$ while if $r = \rho$ then

$$g^*(s, x) = xe^{-\rho s} .$$

(The proofs of these statements are left as Exercise 10.5.)

It remains to examine the case $r < \rho$. (If we regard ρ as the sum of interest rate, inflation and tax etc., this is not an unreasonable assumption in applications.) First we establish that the region D must be invariant w.r.t. t, in the sense that

$$D + (t_0, 0) = D \qquad \text{for all } t_0 . \tag{10.2.8}$$

To prove (10.2.8) consider

$$D + (t_0, 0) = \{(t + t_0, x); (t, x) \in D\} = \{(s, x); (s - t_0, x) \in D\}$$
$$= \{(s, x); g(s - t_0, x) < g^*(s - t_0, x)\} = \{(s, x); e^{\rho t_0}g(s, x) < e^{\rho t_0}g^*(s, x)\}$$
$$= \{(s, x); g(s, x) < g^*(s, x)\} = D ,$$

where we have used that

$$g^*(s - t_0, x) = \sup_\tau E^{(s-t_0,x)}[e^{-\rho\tau}(X_\tau - a)] = \sup_\tau E[e^{-\rho(\tau+(s-t_0))}(X_\tau^x - a)]$$

$$= e^{\rho t_0} \sup_\tau E[e^{-\rho(\tau+s)}(X_\tau^x - a)] = e^{\rho t_0} g^*(s, x) .$$

Therefore the connected component of D that contains U must have the form

$$D(x_0) = \{(t, x); 0 < x < x_0\} \qquad \text{for some } x_0 \geq \tfrac{a\rho}{\rho-r} .$$

Note that D cannot have any other components, for if V is a component of D disjoint from U then $\widehat{A}g < 0$ in V and so, if $y \in V$,

$$E^y[g(Y_\tau)] = g(y) + E^y\left[\int_0^\tau \widehat{A}g(Y_t)dt \right] < g(y)$$

for all exit times τ bounded by the exit time from an x-bounded strip in V. From this we conclude by Theorem 10.1.9 c) that $g^*(y) = g(y)$, which implies $V = \emptyset$.

Put $\tau(x_0) = \tau_{D(x_0)}$ and let us compute

$$\widetilde{g}(s, x) = \widetilde{g}_{x_0}(s, x) = E^{(s,x)}[g(Y_{\tau(x_0)})] . \tag{10.2.9}$$

From Chapter 9 we know that $f = \widetilde{g}$ is the (bounded) solution of the boundary value problem

$$\left. \begin{array}{r} \dfrac{\partial f}{\partial s} + rx\dfrac{\partial f}{\partial x} + \tfrac{1}{2}\alpha^2 x^2 \dfrac{\partial^2 f}{\partial x^2} = 0 \qquad \text{for } 0 < x < x_0 \\[2mm] f(s, x_0) = e^{-\rho s}(x_0 - a) . \end{array} \right\} \tag{10.2.10}$$

(Note that $\mathbf{R} \times \{0\}$ does not contain any regular boundary points of D w.r.t. $Y_t = (s + t, X_t)$.)

If we try a solution of (10.2.10) of the form

$$f(s, x) = e^{-\rho s}\phi(x)$$

we get the following 1-dimensional problem

$$\left. \begin{array}{r} -\rho\phi + rx\phi'(x) + \tfrac{1}{2}\alpha^2 x^2 \phi''(x) = 0 \quad \text{for } 0 < x < x_0 \\[2mm] \phi(x_0) = x_0 - a . \end{array} \right\} \tag{10.2.11}$$

The general solution ϕ of (10.2.11) is

$$\phi(x) = C_1 x^{\gamma_1} + C_2 x^{\gamma_2} ,$$

where C_1, C_2 are arbitrary constants and

$$\gamma_i = \alpha^{-2}\left[\tfrac{1}{2}\alpha^2 - r \pm \sqrt{(r - \tfrac{1}{2}\alpha^2)^2 + 2\rho\alpha^2} \right] \qquad (i = 1, 2), \; \gamma_2 < 0 < \gamma_1 .$$

Since $\phi(x)$ is bounded as $x \to 0$ we must have $C_2 = 0$ and the boundary requirement $\phi(x_0) = x_0 - a$ gives $C_1 = x_0^{-\gamma_1}(x_0 - a)$. We conclude that the bounded solution f of (10.2.10) is

$$\widetilde{g}_{x_0}(s, x) = f(s, x) = e^{-\rho s}(x_0 - a)\left(\frac{x}{x_0}\right)^{\gamma_1}. \tag{10.2.12}$$

If we fix (s, x) then the value of x_0 which maximizes $\widetilde{g}_{x_0}(s, x)$ is easily seen to be given by

$$x_0 = x_{\max} = \frac{a\gamma_1}{\gamma_1 - 1} \tag{10.2.13}$$

(note that $\gamma_1 > 1$ if and only if $r < \rho$).

Thus we have arrived at the candidate $\widetilde{g}_{x_{\max}}(s, x)$ for $g^*(s, x) = \sup_\tau E^{(s,x)}[e^{-\rho\tau}(X_\tau - a)]$. To verify that we indeed have $\widetilde{g}_{x_{\max}} = g^*$ it would suffice to prove that $\widetilde{g}_{x_{\max}}$ is a supermeanvalued majorant of g (see Corollary 10.1.10). This can be done, but we do not give the details here, since this problem can be solved more easily by Theorem 10.4.1 (see Example 10.4.2).

The conclusion is therefore that one should sell the assets the first time the price of them reaches the value $x_{\max} = \frac{a\gamma_1}{\gamma_1 - 1}$. The expected discounted profit obtained from this strategy is

$$g^*(s, x) = \widetilde{g}_{x_{\max}}(s, x) = e^{-\rho s}\left(\frac{\gamma_1 - 1}{a}\right)^{\gamma_1 - 1}\left(\frac{x}{\gamma_1}\right)^{\gamma_1}.$$

Remark. The reader is invited to check that the value $x_0 = x_{\max}$ is the only value of x_0 which makes the function

$$x \to \widetilde{g}_{x_0}(s, x) \qquad \text{(given by (10.2.9))}$$

continuously differentiable at x_0. This is not a coincidence. In fact, it illustrates a general phenomenon which is known as the *high contact* (or smooth fit) *principle*. See Samuelson (1965), McKean (1965), Bather (1970) and Shiryaev (1978). This principle is the basis of the fundamental connection between optimal stopping and *variational inequalities*. Later in this chapter we will discuss some aspects of this connection. More information can be found in Bensoussan and Lions (1978) and Friedman (1976). See also Brekke and Øksendal (1991).

10.3 Optimal Stopping Problems Involving an Integral

Let

$$dY_t = b(Y_t)dt + \sigma(Y_t)dB_t, \qquad Y_0 = y \tag{10.3.1}$$

be an Itô diffusion in $\mathbf{R}^k$. Let $g: \mathbf{R}^k \to [0, \infty)$ be continuous and let $f: \mathbf{R}^k \to [0, \infty)$ be Lipschitz continuous with at most linear growth. (These conditions can be relaxed. See (10.1.37) and Theorem 10.4.1.) Consider the optimal stopping problem: Find $\Phi(y)$ and τ^* such that

$$\Phi(y) = \sup_\tau E^y \left[\int_0^\tau f(Y_t)dt + g(Y_\tau) \right] = E^y \left[\int_0^{\tau^*} f(Y_t)dt + g(Y_{\tau^*}) \right] . \quad (10.3.2)$$

This problem can be reduced to our original problem (10.1.2)–(10.1.3) by proceeding as follows: Define the Itô diffusion Z_t in $\mathbf{R}^k \times \mathbf{R} = \mathbf{R}^{k+1}$ by

$$dZ_t = \begin{bmatrix} dY_t \\ dW_t \end{bmatrix} := \begin{bmatrix} b(Y_t) \\ f(Y_t) \end{bmatrix} dt + \begin{bmatrix} \sigma(Y_t) \\ 0 \end{bmatrix} dB_t ; \quad Z_0 = z = (y, w) . \quad (10.3.3)$$

Then we see that

$$\Phi(y) = \sup_\tau E^{(y,0)}[W_\tau + g(Y_\tau)] = \sup_\tau E^{(y,0)}[\tilde{g}(Z_\tau)]$$

with

$$\tilde{g}(z) := \tilde{g}(y, w) := g(y) + w ; \quad z = (y, w) \in \mathbf{R}^k \times \mathbf{R} . \quad (10.3.4)$$

This is again a problem of the type (10.1.2)–(10.1.3) with X_t replaced by Z_t and g replaced by $\tilde{g}$. Note that the connection between the characteristic operators $\mathcal{A}_Y$ of Y_t and $\mathcal{A}_Z$ of Z_t is given by

$$\mathcal{A}_Z \phi(z) = \mathcal{A}_Z \phi(y, w) = \mathcal{A}_Y \phi(y, w) + f(y)\frac{\partial \phi}{\partial w} , \quad \phi \in C^2(\mathbf{R}^{k+1}) . \quad (10.3.5)$$

In particular, if $\tilde{g}(y, w) = g(y) + w \in C^2(\mathbf{R}^{k+1})$ then

$$\mathcal{A}_Z \tilde{g}(y, w) = \mathcal{A}_Y g(y) + f(y) . \quad (10.3.6)$$

Hence, in this general case the domain U of (10.1.34) gets the form

$$U = \{y; \mathcal{A}_Y g(y) + f(y) > 0\} . \quad (10.3.7)$$

Example 10.3.1 Consider the optimal stopping problem

$$\Phi(x) = \sup_\tau E^x \left[\int_0^\tau \theta e^{-\rho t} X_t dt + e^{-\rho \tau} X_\tau \right] ,$$

where

$$dX_t = \alpha X_t dt + \beta X_t dB_t ; \quad X_0 = x > 0$$

is geometric Brownian motion (α, β, θ constants, $\theta > 0$). We put

$$dY_t = \begin{bmatrix} dt \\ dX_t \end{bmatrix} = \begin{bmatrix} 1 \\ \alpha X_t \end{bmatrix} dt + \begin{bmatrix} 0 \\ \beta X_t \end{bmatrix} dB_t ; \qquad Y_0 = (s, x)$$

and

$$dZ_t = \begin{bmatrix} dY_t \\ dW_t \end{bmatrix} = \begin{bmatrix} 1 \\ \alpha X_t \\ \theta e^{-\rho t} X_t \end{bmatrix} dt + \begin{bmatrix} 0 \\ \beta X_t \\ 0 \end{bmatrix} dB_t ; \qquad Z_0 = (s, x, w) .$$

Then with

$$f(y) = f(s, x) = \theta e^{-\rho s} x , \qquad g(y) = e^{-\rho s} x$$

and

$$\widetilde{g}(s, x, w) = g(s, x) + w = e^{-\rho s} x + w$$

we have

$$\mathcal{A}_Z \, \widetilde{g} = \frac{\partial \widetilde{g}}{\partial s} + \alpha x \frac{\partial \widetilde{g}}{\partial x} + \tfrac{1}{2} \beta^2 x^2 \frac{\partial^2 \widetilde{g}}{\partial x^2} + \theta e^{-\rho s} x \frac{\partial \widetilde{g}}{\partial w} = (-\rho + \alpha + \theta) e^{-\rho s} x .$$

Hence

$$U = \{(s, x, w); \mathcal{A}_Z \, \widetilde{g}(s, x, w) > 0\} = \begin{cases} \mathbf{R}^3 & \text{if } \rho < \alpha + \theta \\ \emptyset & \text{if } \rho \geq \alpha + \theta . \end{cases}$$

From this we conclude (see Exercise 10.6):

If $\rho \geq \alpha + \theta$ then $\tau^* = 0$

$$\text{and } \Phi(s, x, w) = \widetilde{g}(s, x, w) = e^{-\rho s} x + w . \tag{10.3.8}$$

If $\alpha < \rho < \alpha + \theta$ then τ^* does not exist

$$\text{and } \Phi(s, x, w) = \frac{\theta x}{\rho - \alpha} e^{-\rho s} + w . \tag{10.3.9}$$

If $\rho \leq \alpha$ then τ^* does not exist and $\Phi = \infty$. $\tag{10.3.10}$

10.4 Connection with Variational Inequalities

The 'high contact principle' says, roughly, that – under certain conditions –
the solution g^* of (10.1.2)–(10.1.3) is a C^1 function on $\mathbf{R}^n$ if $g \in C^2(\mathbf{R}^n)$.
This is a useful information which can help us to determine g^*. Indeed, this
principle is so useful that it is frequently applied in the literature also in cases
where its validity has not been rigorously proved.

Fortunately it turns out to be easy to prove a *sufficiency* condition of
high contact type, i.e. a kind of verification theorem for optimal stopping,
which makes it easy to verify that a given candidate for g^* (that we may have
found by guessing or intuition) is actually equal to g^*. The result below is a
simplified variant of a result in Brekke and Øksendal (1991):

In the following we fix a domain G in $\mathbf{R}^k$ and we let

$$dY_t = b(Y_t)dt + \sigma(Y_t)dB_t \ ; \qquad Y_0 = y \qquad (10.4.1)$$

be an Itô diffusion in $\mathbf{R}^k$. Define

$$\tau_G = \tau_G(y,\omega) = \inf\{t > 0; Y_t(\omega) \notin G\} \ . \qquad (10.4.2)$$

Let $f: \mathbf{R}^k \to \mathbf{R}$, $g: \mathbf{R}^k \to \mathbf{R}$ be continuous functions satisfying

(a) $E^y[\int_0^{\tau_G} f^-(Y_t)dt] < \infty \qquad$ for all $y \in \mathbf{R}^k \qquad (10.4.3)$

and

(b) the family $\{g^-(Y_\tau); \tau \text{ stopping time}, \tau \leq \tau_G\}$ is uniformly integrable
     w.r.t. R^y (the probability law of Y_t), for all $y \in \mathbf{R}^k$. $\qquad (10.4.4)$

Let $\mathcal{T}$ denote the set of all stopping times $\tau \leq \tau_G$. Consider the following problem: Find $\Phi(y)$ and $\tau^* \in \mathcal{T}$ such that

$$\Phi(y) = \sup_{\tau \in \mathcal{T}} J^\tau(y) = J^{\tau^*}(y) \ , \qquad (10.4.5)$$

where

$$J^\tau(y) = E^y\left[\int_0^\tau f(Y_t)dt + g(Y_\tau)\right] \qquad \text{for } \tau \in \mathcal{T} \ .$$

Note that since $J^0(y) = g(y)$ we have

$$\Phi(y) \geq g(y) \qquad \text{for all } y \in G \ . \qquad (10.4.6)$$

We can now formulate the variational inequalities. As usual we let

$$L = L_Y = \sum_{i=1}^k b_i(y)\frac{\partial}{\partial y_i} + \tfrac{1}{2}\sum_{i,j=1}^k (\sigma\sigma^T)_{ij}(y)\frac{\partial^2}{\partial y_i \partial y_j}$$

be the partial differential operator which coincides with the generator A_Y of Y_t on $C_0^2(\mathbf{R}^k)$.

Theorem 10.4.1 (Variational inequalities for optimal stopping)
a) *Suppose we can find a function* $\phi: \overline{G} \to \mathbf{R}$ *such that*

(i) $\phi \in C^1(G) \cap C(\overline{G})$
(ii) $\phi \geq g$ *on* G *and* $\lim_{t \to \tau_G^-} \phi(Y_t) = g(Y_{\tau_G})\mathcal{X}_{\{\tau_G < \infty\}}$ *a.s.*

   Define
$$D = \{x \in G; \phi(x) > g(x)\} \ .$$

Suppose Y_t spends 0 time on ∂D a.s., i.e.

(iii) $E^y\left[\int\limits_0^{\tau_G} \chi_{\partial D}(Y_t)dt\right] = 0$ *for all $y \in G$*

and suppose that

(iv)   *∂D is a Lipschitz surface, i.e. ∂D is locally the graph of a function $h: \mathbf{R}^{k-1} \to \mathbf{R}$ such that there exists $K < \infty$ with*

$$|h(x) - h(y)| \leq K|x - y| \qquad \text{for all } x, y .$$

Moreover, suppose the following:

(v)   *$\phi \in C^2(G \setminus \partial D)$ and the second order derivatives of ϕ are locally bounded near ∂D*

(vi)   *$L\phi + f \leq 0$ on $G \setminus D$.*
Then

$$\phi(y) \geq \Phi(y) \qquad \text{for all } y \in G.$$

b) *Suppose, in addition to the above, that*

(vii)   *$L\phi + f = 0$ on D*

(viii)   *$\tau_D := \inf\{t > 0; Y_t \notin D\} < \infty$ a.s. R^y for all $y \in G$*
and

(ix)   *the family $\{\phi(Y_\tau); \tau \leq \tau_D, \tau \in \mathcal{T}\}$ is uniformly integrable w.r.t. R^y, for all $y \in G$.*

Then

$$\phi(y) = \Phi(y) = \sup_{\tau \in \mathcal{T}} E^y\left[\int\limits_0^\tau f(Y_t)dt + g(Y_\tau)\right] ; \qquad y \in G \qquad (10.4.7)$$

and

$$\tau^* = \tau_D \qquad\qquad\qquad\qquad (10.4.8)$$

is an optimal stopping time for this problem.

Proof. By (i), (iv) and (v) we can find a sequence of functions $\phi_j \in C^2(G) \cap C(\overline{G})$, $j = 1, 2, \ldots$, such that

(a) $\phi_j \to \phi$ uniformly on compact subsets of $\overline{G}$, as $j \to \infty$
(b) $L\phi_j \to L\phi$ uniformly on compact subsets of $G \setminus \partial D$, as $j \to \infty$
(c) $\{L\phi_j\}_{j=1}^\infty$ is locally bounded on G.

(See Appendix D).

Let $\{G_R\}_{R=1}^\infty$ be a sequence of bounded open sets such that $G = \bigcup\limits_{R=1}^\infty G_R$. Put $T_R = \min(R, \inf\{t > 0; Y_t \notin G_R\})$ and let $\tau \leq \tau_G$ be a stopping time. Let $y \in G$. Then by Dynkin's formula

$$E^y[\phi_j(Y_{\tau \wedge T_R})] = \phi_j(y) + E^y\left[\int\limits_0^{\tau \wedge T_R} L\phi_j(Y_t)dt\right] \qquad (10.4.9)$$

Hence by (a), (b), (c) and (iii) and the bounded a.e. convergence

$$\phi(y) = \lim_{j \to \infty} E^y \left[\int_0^{\tau \wedge T_R} -L\phi_j(Y_t)dt + \phi_j(Y_{\tau \wedge T_R}) \right]$$

$$= E^y \left[\int_0^{\tau \wedge T_R} -L\phi(Y_t)dt + \phi(Y_{\tau \wedge T_R}) \right] . \qquad (10.4.10)$$

Therefore, by (ii), (iii) and (vi),

$$\phi(y) \geq E^y \left[\int_0^{\tau \wedge T_R} f(Y_t)dt + g(Y_{\tau \wedge T_R}) \right] .$$

Hence by the Fatou lemma and (10.4.3), (10.4.4)

$$\phi(y) \geq \lim_{R \to \infty} E^y \left[\int_0^{\tau \wedge T_R} f(Y_t)dt + g(Y_{\tau \wedge T_R}) \right] \geq E^y \left[\int_0^{\tau} f(Y_t)dt + g(Y_{\tau}) \right] .$$

Since $\tau \leq \tau_G$ was arbitrary, we conclude that

$$\phi(y) \geq \Phi(y) \qquad \text{for all } y \in G , \qquad (10.4.11)$$

which proves a).

We proceed to prove b): If $y \notin D$ then $\phi(y) = g(y) \leq \Phi(y)$ so by (10.4.11) we have

$$\phi(y) = \Phi(y) \quad \text{and} \quad \hat{\tau} = \hat{\tau}(y, \omega) := 0 \qquad \text{is optimal for } y \notin D . \qquad (10.4.12)$$

Next, suppose $y \in D$. Let $\{D_k\}_{k=1}^{\infty}$ be an increasing sequence of open sets D_k such that $\overline{D}_k \subset D$, $\overline{D}_k$ is compact and $D = \bigcup_{k=1}^{\infty} D_k$. Put $\tau_k = \inf\{t > 0; Y_t \notin D_k\}$, $k = 1, 2, \ldots$ By Dynkin's formula we have for $y \in D_k$,

$$\phi(y) = \lim_{j \to \infty} \phi_j(y) = \lim_{j \to \infty} E^y \left[\int_0^{\tau_k \wedge T_R} -L\phi_j(Y_t)dt + \phi_j(Y_{\tau_k \wedge T_R}) \right]$$

$$= E^y \left[\int_0^{\tau_k \wedge T_R} -L\phi(Y_t)dt + \phi(Y_{\tau_k \wedge T_R}) \right] = E^y \left[\int_0^{\tau_k \wedge T_R} f(Y_t)dt + \phi(Y_{\tau_k \wedge T_R}) \right]$$

So by uniform integrability and (ii), (vii), (viii) we get

$$\phi(y) = \lim_{R,k \to \infty} E^y \left[\int_0^{\tau_k \wedge \tau_R} f(Y_t)dt + \phi(Y_{\tau_k \wedge \tau_R}) \right]$$

$$= E^y \left[\int_0^{\tau_D} f(Y_t)dt + g(Y_{\tau_D}) \right] = J^{\tau_D}(y) \le \Phi(y) . \qquad (10.4.13)$$

Combining (10.4.11) and (10.4.13) we get

$$\phi(y) \ge \Phi(y) \ge J^{\tau_D}(y) = \phi(y)$$

so

$$\phi(y) = \Phi(y) \quad \text{and} \quad \widehat{\tau}(y,\omega) := \tau_D \quad \text{is optimal when } y \in D . \qquad (10.4.14)$$

From (10.4.12) and (10.4.14) we conclude that

$$\phi(y) = \Phi(y) \qquad \text{for all } y \in G .$$

Moreover, the stopping time $\widehat{\tau}$ defined by

$$\widehat{\tau}(y,\omega) = \begin{cases} 0 & \text{for } y \notin D \\ \tau_D & \text{for } y \in D \end{cases}$$

is optimal. By Theorem 10.1.12 we conclude that τ_D is optimal also.   □

Example 10.4.2 (When is the right time to sell the stocks? (Part 2))
To illustrate Theorem 10.4.1 let us apply it to reconsider Example 10.2.2:
    Rather than *proving* (10.2.8) and the following properties of D, we now simply *guess/assume* that D has the form

$$D = \{(s,x); 0 < x < x_0\}$$

for some $x_0 > 0$, which is intuitively reasonable. Then we solve (10.2.11) for arbitrary x_0 and we arrive at the following candidate ϕ for g^*:

$$\phi(s,x) = \begin{cases} e^{-\rho s}(x_0 - a)(\frac{x}{x_0})^{\gamma_1} & \text{for } 0 < x < x_0 \\ e^{-\rho s}(x - a) & \text{for } x \ge x_0 . \end{cases}$$

The requirement that $\phi \in C^1$ (Theorem 10.4.1 (i)) gives the value (10.2.13) for x_0. It is clear that $\phi \in C^2$ outside ∂D and by construction $L\phi = 0$ on D. Moreover, conditions (iii), (iv), (viii) and (ix) clearly hold. It remains to verify that

(ii) $\phi(s,x) > g(s,x)$ for $0 < x < x_0$, i.e. $\phi(s,x) > e^{-\rho s}(x - a)$ for $0 < x < x_0$
    and
(v) $L\phi(s,x) \le 0$ for $x > x_0$, i.e. $Lg(s,x) \le 0$ for $x > x_0$.

    This is easily done by direct calculation (assuming $r < \rho$).

We conclude that $\phi = g^*$ and $\tau^* = \tau_D$ is optimal (with the value (10.2.13) for x_0).

Exercises

10.1.* In each of the optimal stopping problems below find the supremum g^* and – if it exists – an optimal stopping time τ^*. (Here B_t denotes 1-dimensional Brownian motion)

    a) $g^*(x) = \sup\limits_{\tau} E^x[B_\tau^2]$

    b) $g^*(x) = \sup\limits_{\tau} E^x[|B_\tau|^p]$,

       where $p > 0$.

    c) $g^*(x) = \sup\limits_{\tau} E^x[e^{-B_\tau^2}]$

    d) $g^*(s,x) = \sup\limits_{\tau} E^{(s,x)}[e^{-\rho(s+\tau)}\cosh B_\tau]$

       where $\rho > 0$ and $\cosh \mathrm{x} = \frac{1}{2}(e^x + e^{-x})$.

10.2.* a) Prove that the only nonnegative $(B_t\text{-})$ superharmonic functions in $\mathbf{R}^2$ are the constants.

       (Hint: Suppose u is a nonnegative superharmonic function and that there exist $x, y \in \mathbf{R}^2$ such that

$$u(x) < u(y).$$

       Consider

$$E^x[u(B_\tau)],$$

       where τ is the first hitting time for B_t of a small disc centered at y).

    b) Prove that the only nonnegative superharmonic functions in $\mathbf{R}$ are the constants and use this to find $g^*(x)$ when

$$g(x) = \begin{cases} xe^{-x} & \text{for } x > 0 \\ 0 & \text{for } x \leq 0. \end{cases}$$

    c) Let $\gamma \in \mathbf{R}$, $n \geq 3$ and define, for $x \in \mathbf{R}^n$,

$$f_\gamma(x) = \begin{cases} |x|^\gamma & \text{for } |x| \geq 1 \\ 1 & \text{for } |x| < 1. \end{cases}$$

       For what values of γ is $f_\gamma(\cdot)$ $((B_t)\text{-})$ harmonic for $|x| > 1$? Prove that f_γ is superharmonic in $\mathbf{R}^n$ iff $\gamma \in [2-n, 0]$.

10.3.* Find g^*, τ^* such that

$$g^*(s,x) = \sup_\tau E^{(s,x)}[e^{-\rho(s+\tau)}B_\tau^2] = E^{(s,x)}[e^{-\rho(s+\tau^*)}B_{\tau^*}^2] \, ,$$

where B_t is 1-dimensional Brownian motion, $\rho > 0$ is constant.
Hint: First assume that the continuation region has the form

$$D = \{(s,x); -x_0 < x < x_0\}$$

for some x_0 and then try to determine x_0. Then apply Theorem 10.4.1.

10.4. Let X_t be an Itô diffusion on $\mathbf{R}^n$ and $g: \mathbf{R}^n \to \mathbf{R}^+$ a continuous reward function. Define

$$g^\diamond(x) = \sup\{E^x[g(X_\tau)]; \ \tau \text{ stopping time}, \ E^x[\tau] < \infty\} \, .$$

Show that $g^\diamond = g^*$.
[Hint: If τ is a stopping time put $\tau_k = \tau \wedge k$ for $k = 1, 2, \ldots$ and consider

$$E^x[g(X_\tau) \cdot \mathcal{X}_{\tau < \infty}] \le E^x[\lim_{k \to \infty} g(X_{\tau_k})]] \, .$$

10.5. With g, r, ρ as in Example 10.2.2 prove that

a) if $r > \rho$ then $g^* = \infty$,
b) if $r = \rho$ then $g^*(s,x) = xe^{-\rho s}$.

10.6. Prove statements (10.3.8), (10.3.9), (10.3.10) in Example 10.3.1.

10.7. As a supplement to Exercise 10.4 it is worth noting that if g is not bounded below then the two problems

$$g^*(x) = \sup\{E^x[g(X_\tau)]; \ \tau \text{ stopping time}\}$$

and

$$g^\diamond(x) = \sup\{E^x[g(X_\tau)]; \tau \text{ stopping time}, \ E^x[\tau] < \infty\}$$

need not have the same solution. For example, if $g(x) = x$, $X_t = B_t \in \mathbf{R}$ prove that

$$g^*(x) = \infty \qquad \text{for all } x \in \mathbf{R}$$

while

$$g^\diamond(x) = x \qquad \text{for all } x \in \mathbf{R} \, .$$

(See Exercise 7.4.)

10.8. Give an example with g not bounded below where Theorem 10.1.9 a) fails. (Hint: See Exercise 10.7.)

10.9.* Solve the optimal stopping problem

$$\Phi(x) = \sup_{\tau} E^x \left[\int_0^\tau e^{-\rho t} B_t^2 \, dt + e^{-\rho \tau} B_\tau^2 \right] .$$

10.10.   Prove the following simple, but useful, observation, which can be regarded as an extension of (10.1.35):
Let $W = \{(s,x); \exists \tau \text{ with } g(s,x) < E^{(s,x)}[g(s+\tau, X_\tau)]\}$.
Then $W \subset D$.

10.11.   Consider the optimal stopping problem

$$g^*(s,x) = \sup_{\tau} E^{(s,x)} \left[e^{-\rho(s+\tau)} B_\tau^+ \right] ,$$

where $B_t \in \mathbf{R}$ and $x^+ = \max\{x, 0\}$.

a)   Use the argument for (10.2.8) and Exercise 10.10 to prove that the continuation region D has the form

$$D = \{(s,x); x < x_0\}$$

for some $x_0 > 0$.

b)   Determine x_0 and find g^*.

c)   Verify the *high contact principle*:

$$\frac{\partial g^*}{\partial x} = \frac{\partial g}{\partial x} \qquad \text{when } (s,x) = (s, x_0) ,$$

where $g(t,x) = e^{-\rho t} x^+$.

10.12.*   The first time the high contact principle was formulated seems to be in a paper by Samuelson (1965), who studied the optimal time for selling an asset, if the reward obtained by selling at the time t and when price is ξ is given by

$$g(t, \xi) = e^{-\rho t} (\xi - 1)^+ .$$

The price process is assumed to be a geometric Brownian motion X_t given by

$$dX_t = rX_t dt + \alpha X_t dB_t , \qquad X_0 = x > 0 ,$$

where $r < \rho$.
In other words, the problem is to find g^*, τ^* such that

$$g^*(s,x) = \sup_{\tau} E^{(s,x)} \left[e^{-\rho(s+\tau)} (X_\tau - 1)^+ \right] = E^{(s,x)} \left[e^{-\rho(s+\tau^*)} (X_{\tau^*} - 1)^+ \right].$$

a)   Use the argument for (10.2.8) and Exercise 10.10 to prove that the continuation region D has the form

$$D = \{(s,x); 0 < x < x_0\}$$

for some $x_0 > \frac{\rho}{\rho-r}$.

b) For a given $x_0 > \frac{\rho}{\rho-r}$ solve the boundary value problem

$$\begin{cases} \frac{\partial f}{\partial s} + rx\frac{\partial f}{\partial x} + \frac{1}{2}\alpha^2 x^2 \frac{\partial^2 f}{\partial x^2} = 0 & \text{for } 0 < x < x_0 \\ f(s,0) = 0 \\ f(s,x_0) = e^{-\rho s}(x_0 - 1)^+ \end{cases}$$

by trying $f(s,x) = e^{-\rho s}\phi(x)$.

c) Determine x_0 by using the high contact principle, i.e. by using that

$$\frac{\partial f}{\partial x} = \frac{\partial g}{\partial x} \qquad \text{when } x = x_0 .$$

d) With f, x_0 as in b), c) define

$$\gamma(s,x) = \begin{cases} f(s,x) \ ; & x < x_0 \\ e^{-\rho s}(x-1)^+ \ ; & x \geq x_0 . \end{cases}$$

Use Theorem 10.4.1 to verify that $\gamma = g^*$ and that $\tau^* = \tau_D$ is optimal.

10.13.*   (A resource extraction problem)

Suppose the price P_t of one unit of a resource (e.g. gas, oil) at time t is varying like a geometric Brownian motion

$$dP_t = \alpha P_t dt + \beta P_t dB_t \ ; \qquad P_0 = p$$

where B_t is 1-dimensional Brownian motion and α, β are constants. Let Q_t denote the amount of remaining resources at time t. Assume that the rate of extraction is proportional to the remaining amount, so that

$$dQ_t = -\lambda Q_t dt \ ; \qquad Q_0 = q$$

where $\lambda > 0$ is a constant.

If the running cost rate is $K > 0$ and we stop the extraction at the time $\tau = \tau(\omega)$ then the expected total discounted profit is given by

$$J^\tau(s,p,q) = E^{(s,p,q)}\left[\int_0^\tau (\lambda P_t Q_t - K)e^{-\rho(s+t)}dt + e^{-\rho(s+\tau)}g(P_\tau, Q_\tau)\right],$$

where $\rho > 0$ is the discounting exponent and $g(p,q)$ is a given bequest function giving the value of the remaining resource amount q when the price is p.

a) Write down the characteristic operator $\mathcal{A}$ of the diffusion process

$$dX_t = \begin{bmatrix} dt \\ dP_t \\ dQ_t \end{bmatrix} ; \qquad X_0 = (s, p, q)$$

and formulate the variational inequalities of Theorem 10.4.1 corresponding to the optimal stopping problem

$$\Phi(s, p, q) = \sup_\tau J^\tau(s, p, q) = J^{\tau^*}(s, p, q) .$$

b) Assume that $g(p, q) = pq$ and find the domain U corresponding to (10.1.34), (10.3.7), i.e.

$$U = \{(s, p, q); \mathcal{A}(e^{-\rho s} g(p, q)) + f(s, p, q) > 0\} ,$$

where

$$f(s, p, q) = e^{-\rho s}(\lambda pq - K) .$$

Conclude that
(i)   if $\rho \geq \alpha$ then $\tau^* = 0$ and $\Phi(s, p, q) = pqe^{-\rho s}$
(ii) if $\rho < \alpha$ then $D \supset \{(s, p, q); pq > \frac{K}{\alpha - \rho}\}$.

c) As a candidate for Φ when $\rho < \alpha$ we try a function of the form

$$\phi(s, p, q) = \begin{cases} e^{-\rho s} pq ; & 0 < pq \leq y_0 \\ e^{-\rho s} \psi(pq) ; & pq > y_0 \end{cases}$$

for a suitable $\psi \colon \mathbf{R} \to \mathbf{R}$, and a suitable y_0. Use Theorem 10.4.1 to determine ψ, y_0 and to verify that with this choice of ψ, y_0 we have $\phi = \Phi$ and $\tau^* = \inf\{t > 0; P_t Q_t \leq y_0\}$, if $\rho < \alpha < \rho + \lambda$.

d) What happens if $\rho + \lambda \leq \alpha$?

10.14.* (Finding the optimal investment time (I))
Solve the optimal stopping problem

$$\Psi(s, p) = \sup_\tau E^{(s,p)} \left[\int_\tau^\infty e^{-\rho(s+t)} P_t dt - C e^{-\rho(s+\tau)} \right] ,$$

where

$$dP_t = \alpha P_t dt + \beta P_t dB_t ; \qquad P_0 = p ,$$

B_t is 1-dimensional Brownian motion and α, β, ρ, C are constants, $0 < \alpha < \rho$ and $C > 0$. (We may interpret this as the problem of finding the optimal time τ for investment in a project. The profit rate after investment is P_t and the cost of the investment is C. Thus Ψ gives the maximal expected discounted net profit.)

Hint: Write $\int\limits_{\tau}^{\infty} e^{-\rho(s+t)} P_t dt = e^{-\rho s}[\int\limits_{0}^{\infty} e^{-\rho t} P_t dt - \int\limits_{0}^{\tau} e^{-\rho t} P_t dt]$. Com-

pute $E[\int\limits_{0}^{\infty} e^{-\rho t} P_t dt]$ by using the solution formula for P_t (see Chap-

ter 5) and then apply Theorem 10.4.1 to the problem

$$\Phi(s,p) = \sup_{\tau} E^{(s,p)}\left[-\int\limits_{0}^{\tau} e^{-\rho(s+t)} P_t dt - Ce^{-\rho(s+\tau)}\right].$$

10.15. Let B_t be 1-dimensional Brownian motion and let $\rho > 0$ be constant.

a)   Show that the family

$$\{e^{-\rho\tau} B_\tau; \tau \text{ stopping time}\}$$

is uniformly integrable w.r.t. P^x.

b)   Solve the optimal stopping problem

$$\Phi(s,x) = \sup_{\tau} E^{(s,x)}[e^{-\rho(s+\tau)}(B_\tau - a)]$$

when $a > 0$ is constant. This may be regarded as a variation of
Example 10.2.2/10.4.2 with the price process represented by B_t
rather than X_t.

10.16. **(Finding the optimal investment time (II))**
Solve the optimal stopping problem

$$\Psi(s,p) = \sup_{\tau} E^{(s,p)}\left[\int\limits_{\tau}^{\infty} e^{-\rho(s+t)} P_t dt - Ce^{-\rho(s+\tau)}\right]$$

where

$$dP_t = \mu\,dt + \sigma\,dB_t ; \qquad P_0 = p$$

with $\mu, \sigma \neq 0$ constants. (Compare with Exercise 10.14.)

10.17. a)   Let

$$dX_t = \mu\,dt + \sigma\,dB_t ; \qquad X_0 = x \in \mathbf{R}$$

where μ and σ are constants. Prove that if $\rho > 0$ is constant then

$$E^x\left[\int\limits_{0}^{\infty} e^{-\rho t}|X_t|dt\right] < \infty \qquad \text{for all } x .$$

b)   Solve the optimal stopping problem

$$\Phi(s,x) = \sup_{\tau \geq 0} E^{s,x}\left[\int\limits_{0}^{\tau} e^{-\rho(s+t)}(X_t - a)dt\right],$$

where $a \geq 0$ is a constant.

Chapter 11
Application to Stochastic Control

11.1 Statement of the Problem

Suppose that the state of a system at time t is described by an Itô process X_t of the form

$$dX_t = dX_t^u = b(t, X_t, u_t)dt + \sigma(t, X_t, u_t)dB_t , \qquad (11.1.1)$$

where $X_t \in \mathbf{R}^n$, $b \colon \mathbf{R} \times \mathbf{R}^n \times U \to \mathbf{R}^n$, $\sigma \colon \mathbf{R} \times \mathbf{R}^n \times U \to \mathbf{R}^{n \times m}$ and B_t is m-dimensional Brownian motion. Here $u_t \in U \subset \mathbf{R}^k$ is a parameter whose value we can choose in the given Borel set U at any instant t in order to control the process X_t. Thus $u_t = u(t, \omega)$ is a stochastic process. Since our decision at time t must be based upon what has happened up to time t, the function $\omega \to u(t, \omega)$ must (at least) be measurable w.r.t. $\mathcal{F}_t^{(m)}$, i.e. the process u_t must be $\mathcal{F}_t^{(m)}$-adapted. Thus the right hand side of (11.1.1) is well-defined as a stochastic integral, under suitable assumptions on the functions b and σ. At the moment we will not specify the conditions on b and σ further, but simply assume that the process X_t satisfying (11.1.1) exists. See further comments on this in the end of this chapter.

Let $\{X_h^{s,x}\}_{h \geq s}$ be the solution of (11.1.1) such that $X_s^{s,x} = x$, i.e.

$$X_h^{s,x} = x + \int\limits_s^h b(r, X_r^{s,x}, u_r)dr + \int\limits_s^h \sigma(r, X_r^{s,x}, u_r)dB_r ; \qquad h \geq s$$

and let the probability law of X_t starting at x for $t = s$ be denoted by $Q^{s,x}$, so that

$$Q^{s,x}[X_{t_1} \in F_1, \ldots, X_{t_k} \in F_k] = P^0[X_{t_1}^{s,x} \in F_1, \ldots, X_{t_k}^{s,x} \in F_k] \qquad (11.1.2)$$

for $s \leq t_i$, $F_i \subset \mathbf{R}^n$; $1 \leq i \leq k$, $k = 1, 2, \ldots$

Let $f : \mathbf{R} \times \mathbf{R}^n \times U \to \mathbf{R}$ (the "profit rate" function) and $g : \mathbf{R} \times \mathbf{R}^n \to \mathbf{R}$ (the "bequest" function) be given continuous functions, let G be a fixed domain in $\mathbf{R} \times \mathbf{R}^n$ and let $\widehat{T}$ be the first exit time after s from G for the process $\{X_r^{s,x}\}_{r \geq s}$, i.e.

$$\widehat{T} = \widehat{T}^{s,x}(\omega) = \inf\{r > s; (r, X_r^{s,x}(\omega)) \notin G\} \leq \infty . \tag{11.1.3}$$

Suppose

$$E^{s,x}\left[\int_s^{\widehat{T}} |f^{u_r}(r, X_r)| dr + |g(\widehat{T}, X_{\widehat{T}})| \mathcal{X}_{\{\widehat{T} < \infty\}} \right] < \infty \quad \text{for all } s, x, u \tag{11.1.4}$$

where $f^u(r, z) = f(r, z, u)$. Then we define the *performance function* $J^u(s, x)$ by

$$J^u(s, x) = E^{s,x}\left[\int_s^{\widehat{T}} f^{u_r}(r, X_r) dr + g(\widehat{T}, X_{\widehat{T}}) \mathcal{X}_{\{\widehat{T} < \infty\}} \right] . \tag{11.1.5}$$

To obtain an easier notation we introduce

$$Y_t = (s + t, X_{s+t}^{s,x}) \quad \text{for } t \geq 0, \, Y_0 = (s, x)$$

and we observe that if we substitute this in (11.1.1) we get the equation

$$dY_t = dY_t^u = b(Y_t, u_t) dt + \sigma(Y_t, u_t) dB_t . \tag{11.1.6}$$

(Strictly speaking, the u, b and σ in (11.1.6) are slightly different from the u, b and σ in (11.1.1).) The probability law of Y_t starting at $y = (s, x)$ for $t = 0$ is (with slight abuse of notation) also denoted by $Q^{s,x} = Q^y$.

Note that

$$\int_s^{\widehat{T}} f^{u_r}(r, X_r) dr = \int_0^{\widehat{T}-s} f^{u_{s+t}}(s + t, X_{s+t}) dt = \int_0^{\tau_G} f^{u_{s+t}}(Y_t) dt ,$$

where

$$\tau_G := \inf\{t > 0; Y_t \notin G\} = \widehat{T} - s . \tag{11.1.7}$$

Moreover,

$$g(\widehat{T}, X_{\widehat{T}}) = g(Y_{\widehat{T}-s}) = g(Y_{\tau_G}) .$$

Therefore the performance function may be written in terms of Y as follows, with $y = (s, x)$,

$$J^u(y) = E^y\left[\int_0^{\tau_G} f^{u_t}(Y_t) dt + g(Y_{\tau_G}) \mathcal{X}_{\{\tau_G < \infty\}} \right] . \tag{11.1.8}$$

(Strictly speaking this u_t is a time shift of the u_t in (11.1.6).)

The problem is – for each $y \in G$ – to find the number $\Phi(y)$ and a control $u^* = u^*(t, \omega) = u^*(y, t, \omega) \in \mathcal{A}$ such that

$$\Phi(y) := \sup_{u(t,\omega)} J^u(y) = J^{u^*}(y) \qquad (11.1.9)$$

where the supremum is taken over a given family $\mathcal{A}$ of *admissible* controls, contained in the set of all $\mathcal{F}_t^{(m)}$-adapted processes $\{u_t\}$ with values in U. Such a control u^* – if it exists – is called an *optimal control* and Φ is called the *optimal performance or the value function*. Examples of types of admissible control functions that may be considered are:

(1) Functions of the form $u(t, \omega) = u(t)$ i.e. not depending on ω. These controls are sometimes called *deterministic* or *open loop controls*.
(2) Processes $\{u_t\}$ which are $\mathcal{M}_t$-adapted, i.e. for each t the function $\omega \to u(t, \omega)$ is $\mathcal{M}_t$-measurable, where $\mathcal{M}_t$ is the σ-algebra generated by $\{X_r^u;\, r \leq t\}$. These controls are called *closed loop* or *feedback* controls.
(3) The controller has only *partial information* of the state of the system. More precisely, to the controller's disposal are only (noisy) observations R_t of X_t, given by an Itô process of the form

$$dR_t = a(t, X_t)dt + \gamma(t, X_t)d\widehat{B}_t \ ,$$

where $\widehat{B}$ is a Brownian motion (not necessarily related to B). Hence the control process $\{u_t\}$ must be adapted w.r.t. the σ-algebra $\mathcal{N}_t$ generated by $\{R_s;\, s \leq t\}$. In this situation the stochastic control problem is linked to the filtering problem (Chapter 6). In fact, if the equation (11.1.1) is linear and the performance function is integral quadratic (i.e. f and g are quadratic) then the stochastic control problem splits into a linear filtering problem and a corresponding deterministic control problem. This is called the Separation Principle. See Example 11.2.4.
(4) Functions $u(t, \omega)$ of the form $u(t, \omega) = u_0(t, X_t(\omega))$ for some function $u_0: \mathbf{R}^{n+1} \to U \subset \mathbf{R}^k$. In this case we assume that u does not depend on the starting point $y = (s, x)$: The value we choose at time t only depends on the state of the system at this time. These are called *Markov controls*, because with such u the corresponding process X_t becomes an Itô diffusion, in particular a Markov process. In the following we will not distinguish between u and u_0. Thus we will identify a function $u: \mathbf{R}^{n+1} \to U$ with the Markov control $u(Y) = u(t, X_t)$ and simply call such functions Markov controls.

11.2 The Hamilton-Jacobi-Bellman Equation

Let us first consider only *Markov controls*

$$u = u(t, X_t(\omega)) .$$

Introducing $Y_t = (s + t, X_{s+t})$ (as explained earlier) the system equation becomes

$$dY_t = b(Y_t, u(Y_t))dt + \sigma(Y_t, u(Y_t))dB_t . \tag{11.2.1}$$

For $v \in U$ and $\phi \in C_0^2(\mathbf{R} \times \mathbf{R}^n)$ define

$$(L^v\phi)(y) = \frac{\partial \phi}{\partial s}(y) + \sum_{i=1}^{n} b_i(y, v)\frac{\partial \phi}{\partial x_i} + \sum_{i,j=1}^{n} a_{ij}(y, v)\frac{\partial^2 \phi}{\partial x_i \partial x_j} \tag{11.2.2}$$

where $a_{ij} = \frac{1}{2}(\sigma\sigma^T)_{ij}$, $y = (s, x)$ and $x = (x_1, \dots, x_n)$. Then for each choice of the function u the solution $Y_t = Y_t^u$ is an Itô diffusion with generator A given by

$$(A\phi)(y) = (L^{u(y)}\phi)(y) \qquad \text{for } \phi \in C_0^2(\mathbf{R} \times \mathbf{R}^n) \text{ (see Theorem 7.3.3) .}$$

For $v \in U$ define $f^v(y) = f(y, v)$. The first fundamental result in stochastic control theory is the following:

Theorem 11.2.1 (The Hamilton-Jacobi-Bellman (HJB) equation (I))
Define

$$\Phi(y) = \sup\{J^u(y); u = u(Y) \text{ Markov control}\} .$$

Suppose that $\Phi \in C^2(G) \cap C(\overline{G})$ satisfies

$$E^y\left[|\Phi(Y_\alpha)| + \int_0^\alpha |L^v\Phi(Y_t)|dt\right] < \infty$$

for all bounded stopping times $\alpha \le \tau_G$, all $y \in G$ and all $v \in U$. Moreover, suppose that an optimal Markov control u^* exists and that ∂G is regular for $Y_t^{u^*}$ (Definition 9.2.8). Then

$$\sup_{v \in U}\{f^v(y) + (L^v\Phi)(y)\} = 0 \qquad \text{for all } y \in G \tag{11.2.3}$$

and

$$\Phi(y) = g(y) \qquad \text{for all } y \in \partial G . \tag{11.2.4}$$

The supremum in (11.2.3) is obtained if $v = u^*(y)$ where $u^*(y)$ is optimal. In other words,

$$f(y, u^*(y)) + (L^{u^*(y)}\Phi)(y) = 0 \qquad \text{for all } y \in G . \tag{11.2.5}$$

Proof. The last two statements are easy to prove: Since $u^* = u^*(y)$ is optimal we have

$$\Phi(y) = J^{u^*}(y) = E^y\left[\int_0^{\tau_G} f(Y_s, u^*(Y_s))ds + g(Y_{\tau_G}) \cdot \mathcal{X}_{\{\tau_G < \infty\}}\right] .$$

If $y \in \partial G$ then $\tau_G = 0$ a.s. Q^y (since ∂G is regular) and (11.2.4) follows. By the solution of the Dirichlet-Poisson problem (Theorem 9.3.3)

$$(L^{u^*(y)}\Phi)(y) = -f(y, u^*(y)) \qquad \text{for all } y \in G,$$

which is (11.2.5). We proceed to prove (11.2.3). Fix $y = (s, x) \in G$ and choose a Markov control u. Let $\alpha \leq \tau_G$ be a bounded stopping time.
Since

$$J^u(y) = E^y\left[\int_0^{\tau_G} f^u(Y_r)dr + g(Y_{\tau_G}) \cdot \mathcal{X}_{\{\tau_G < \infty\}}\right],$$

we get by the strong Markov property (7.2.5), combined with (7.2.6) and (9.3.7)

$$E^y[J^u(Y_\alpha)] = E^y\left[E^{Y_\alpha}\left[\int_0^{\tau_G} f^u(Y_r)dr + g(Y_{\tau_G})\mathcal{X}_{\{\tau_G < \infty\}}\right]\right]$$

$$= E^y\left[E^y\left[\theta_\alpha\left(\int_0^{\tau_G} f^u(Y_r)dr + g(Y_{\tau_G})\mathcal{X}_{\{\tau_G < \infty\}}\right)\Big|\mathcal{F}_\alpha\right]\right]$$

$$= E^y\left[E^y\left[\int_\alpha^{\tau_G} f^u(Y_r)dr + g(Y_{\tau_G})\mathcal{X}_{\{\tau_G < \infty\}}\Big|\mathcal{F}_\alpha\right]\right]$$

$$= E^y\left[\int_0^{\tau_G} f^u(Y_r)dr + g(Y_{\tau_G})\mathcal{X}_{\{\tau_G < \infty\}} - \int_0^\alpha f^u(Y_r)dr\right]$$

$$= J^u(y) - E^y\left[\int_0^\alpha f^u(Y_r)dr\right].$$

So

$$J^u(y) = E^y\left[\int_0^\alpha f^u(Y_r)dr\right] + E^y[J^u(Y_\alpha)]. \tag{11.2.6}$$

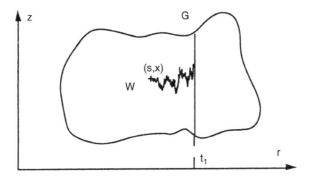

Now let $W \subset G$ be of the form $W = \{(r, z) \in G;\ r < t_1\}$ where $s < t_1$. Put $\alpha = \inf\{t \geq 0;\ Y_t \notin W\}$. Suppose an optimal control $u^*(y) = u^*(r, z)$ exists and choose

$$u(r, z) = \begin{cases} v & \text{if } (r, z) \in W \\ u^*(r, z) & \text{if } (r, z) \in G \setminus W \end{cases}$$

where $v \in U$ is arbitrary. Then

$$\Phi(Y_\alpha) = J^{u^*}(Y_\alpha) = J^u(Y_\alpha) \tag{11.2.7}$$

and therefore, combining (11.2.6) and (11.2.7) we obtain

$$\Phi(y) \geq J^u(y) = E^y\left[\int_0^\alpha f^v(Y_r)dr\right] + E^y[\Phi(Y_\alpha)]. \tag{11.2.8}$$

Since $\Phi \in C^2(G)$ we get by Dynkin's formula

$$E^y[\Phi(Y_\alpha)] = \Phi(y) + E^y\left[\int_0^\alpha (L^u\Phi)(Y_r)dr\right],$$

which substituted in (11.2.8) gives

$$\Phi(y) \geq E^y\left[\int_0^\alpha f^v(Y_r)dr\right] + \Phi(y) + E^y\left[\int_0^\alpha (L^v\Phi)(Y_r)dr\right]$$

or

$$E^y\left[\int_0^\alpha (f^v(Y_r) + (L^v\Phi)(Y_r))dr\right] \leq 0.$$

So

$$\frac{E^y\left[\int_0^\alpha (f^v(Y_r) + (L^v\Phi)(Y_r))dr\right]}{E^y[\alpha]} \leq 0 \qquad \text{for all such } W.$$

Letting $t_1 \downarrow s$ we obtain, since $f^v(\cdot)$ and $(L^v\Phi)(\cdot)$ are continuous at y, that $f^v(y) + (L^v\Phi)(y) \leq 0$, which combined with (11.2.5) gives (11.2.3). That completes the proof.   □

Remark. The HJB (I) equation states that if an optimal control u^* exists, then we know that its value v at the point y is a point v where the function

$$v \to f^v(y) + (L^v\Phi)(y);\qquad v \in U$$

attains its maximum (and this maximum is 0). Thus the original stochastic control problem is associated to the easier problem of finding the maximum of a real function in $U \subset \mathbf{R}^k$. However, the HJB (I) equation only states that it is *necessary* that $v = u^*(y)$ is the maximum of this function. It is just as important to know if this is also *sufficient*: If at each point y we have found $v = u_0(y)$ such that $f^v(y) + (L^v\Phi)(y)$ is maximal and this maximum is 0, will $u_0(Y)$ be an optimal control? The next result states that (under some conditions) this is actually the case:

Theorem 11.2.2 (The HJB (II) equation – a verification theorem)

Let ϕ be a function in $C^2(G) \cap C(\overline{G})$ such that, for all $v \in U$,

$$f^v(y) + (L^v\phi)(y) \leq 0 ; \qquad y \in G \qquad\qquad (11.2.9)$$

with boundary values

$$\lim_{t \to \tau_G} \phi(Y_t) = g(Y_{\tau_G}) \cdot \mathcal{X}_{\{\tau_G < \infty\}} \quad a.s. \ Q^y \qquad (11.2.10)$$

and such that

$$\{\phi^-(Y_\tau); \tau \text{ stopping time}, \tau \leq \tau_G\} \text{ is uniformly } Q^y\text{-integrable}$$
$$\text{for all Markov controls } u \text{ and all } y \in G . \qquad (11.2.11)$$

Then

$$\phi(y) \geq J^u(y) \quad \text{for all Markov controls } u \text{ and all } y \in G . \qquad (11.2.12)$$

Moreover, if for each $y \in G$ we have found $u_0(y)$ such that

$$f^{u_0(y)}(y) + (L^{u_0(y)}\phi)(y) = 0 \qquad\qquad (11.2.13)$$

and

$$\{\phi(Y_\tau^{u_0}); \quad \tau \text{ stopping time}, \tau \leq \tau_G\} \text{ is uniformly}$$
$$Q^y\text{-integrable for all } y \in G \qquad\qquad (11.2.14)$$

then $u_0 = u_0(y)$ is a Markov control such that

$$\phi(y) = J^{u_0}(y)$$

and hence if u_0 is admissible then u_0 must be an optimal control and $\phi(y) = \Phi(y)$.

Proof. Assume that ϕ satisfies (11.2.9) and (11.2.10) above. Let u be a Markov control. Since $L^u\phi \leq -f^u$ in G we have by Dynkin's formula

$$E^y[\phi(Y_{T_R})] = \phi(y) + E^y\left[\int_0^{T_R} (L^u\phi)(Y_r)dr\right]$$

$$\leq \phi(y) - E^y\left[\int_0^{T_R} f^u(Y_r)dr\right]$$

where

$$T_R = \min\{R, \tau_G, \inf\{t > 0; |Y_t| \geq R\}\} \tag{11.2.15}$$

for all $R < \infty$. This gives, by (11.1.4), (11.2.10), (11.2.11) and the Fatou lemma,

$$\phi(y) \geq \lim_{R\to\infty} E^y\left[\int_0^{T_R} f^u(Y_r)dr + \phi(Y_{T_R})\right]$$

$$\geq E^y\left[\int_0^{\tau_G} f^u(Y_r)dr + g(Y_{\tau_G})\mathcal{X}_{\{\tau_G<\infty\}}\right] = J^u(y)$$

which proves (11.2.12). If u_0 is such that (11.2.13) and (11.2.14) hold, then the calculations above give equality and the proof is complete.   □

The HJB equations (I), (II) provide a nice solution to the stochastic control problem in the case where only Markov controls are considered. One might feel that considering only Markov controls is too restrictive, but fortunately one can always obtain as good performance with a Markov control as with an arbitrary $\mathcal{F}_t^{(m)}$-adapted control, at least if some extra conditions are satisfied:

Theorem 11.2.3 *Let*

$$\Phi_M(y) = \sup\{J^u(y); u = u(Y) \text{ Markov control}\}$$

and

$$\Phi_a(y) = \sup\{J^u(y); u = u(t,\omega) \ \mathcal{F}_t^{(m)}\text{-adapted control}\} .$$

Suppose there exists an optimal Markov control $u_0 = u_0(Y)$ for the Markov control problem (i.e. $\Phi_M(y) = J^{u_0}(y)$ for all $y \in G$) such that all the boundary points of G are regular w.r.t. $Y_t^{u_0}$ and that Φ_M is a bounded function in $C^2(G) \cap C(\overline{G})$ satisfying

$$E^y\left[|\Phi_M(Y_\alpha)| + \int_0^\alpha |L^u\Phi_M(Y_t)|dt\right] < \infty \tag{11.2.16}$$

for all bounded stopping times $\alpha \leq \tau_G$, all adapted controls u and all $y \in G$. Then

$$\Phi_M(y) = \Phi_a(y) \qquad \text{for all } y \in G .$$

Proof. Let ϕ be a bounded function in $C^2(G) \cap C(\overline{G})$ satisfying (11.2.16) and

$$f^v(y) + (L^v\phi)(y) \leq 0 \qquad \text{for all } y \in G, v \in U \tag{11.2.17}$$

and

$$\phi(y) = g(y) \qquad \text{for all } y \in \partial G .\qquad (11.2.18)$$

Let $u_t(\omega) = u(t, \omega)$ be an $\mathcal{F}_t^{(m)}$-adapted control. Then Y_t is an Itô process given by

$$dY_t = b(Y_t, u_t)dt + \sigma(Y_t, u_t)dB_t$$

so by Lemma 7.3.2, with T_R as in (11.2.15),

$$E^y[\phi(Y_{T_R})] = \phi(y) + E^y\left[\int_0^{T_R} (L^{u(t,\omega)}\phi)(Y_t)dt\right],$$

where

$$(L^{u(t,\omega)}\phi)(y) =$$
$$= \frac{\partial \phi}{\partial t}(y) + \sum_{i=1}^n b_i(y, u(t,\omega))\frac{\partial \phi}{\partial x_i}(y) + \sum_{i,j=1}^n a_{ij}(y, u(t,\omega))\frac{\partial^2 \phi}{\partial x_i \partial x_j}(y) ,$$

with $a_{ij} = \frac{1}{2}(\sigma\sigma^T)_{ij}$. Thus by (11.2.17) and (11.2.18) this gives

$$E^y[\phi(Y_{T_R})] \le \phi(y) - E^y\left[\int_0^{T_R} f(Y_t, u(t,\omega))dt\right].$$

Letting $R \to \infty$ we obtain

$$\phi(y) \ge J^u(y) .\qquad (11.2.19)$$

But by Theorem 11.2.1 the function $\phi(y) = \Phi_M(y)$ satisfies (11.2.17) and (11.2.18). So by (11.2.19) we have $\Phi_M(y) \ge \Phi_a(y)$ and Theorem 11.2.3 follows.

$\square$

Remark. The theory above also applies to the corresponding *minimum* problem

$$\Psi(y) = \inf_u J^u(y) = J^{u^*}(y) .\qquad (11.2.20)$$

To see the connection we note that

$$\Psi(y) = -\sup_u\{-J^u(y)\} = -\sup_u\left\{E^y\left[\int_0^{\tau_G} -f^u(Y_t)dt - g(Y_{\tau_G})\cdot \mathcal{X}_{\{\tau_G<\infty\}}\right]\right\}$$

so $-\Psi$ coincides with the solution Φ of the problem (11.1.9), but with f replaced by $-f$ and g replaced by $-g$. Using this, we see that the HJB equations apply to Ψ also but with reverse inequalities. For example, equation (11.2.3) for Φ gets for Ψ the form

$$\inf_{v \in U}\{f^v(y) + (L^v\Psi)(y)\} = 0 \qquad \text{for all } y \in G .\qquad (11.2.21)$$

We now illustrate the results by some examples:

Example 11.2.4 (The linear stochastic regulator problem)
Suppose that the state X_t of the system at time t is given by a linear stochastic differential equation:

$$dX_t = (H_t X_t + M_t u_t)dt + \sigma_t dB_t , \qquad t \geq s ; \; X_s = x \qquad (11.2.22)$$

and the cost is of the form

$$J^u(s,x) = E^{s,x}\left[\int\limits_0^{t_1-s} \{X_t^T C_t X_t + u_t^T D_t u_t\} dt + X_{t_1-s}^T R X_{t_1-s}\right], \quad s \leq t_1 \quad (11.2.23)$$

where all the coefficients $H_t \in \mathbf{R}^{n\times n}$, $M_t \in \mathbf{R}^{n\times k}$, $\sigma_t \in \mathbf{R}^{n\times m}$, $C_t \in \mathbf{R}^{n\times n}$, $D_t \in \mathbf{R}^{k\times k}$ and $R \in \mathbf{R}^{n\times n}$ are t-continuous and deterministic. We assume that C_t and R are symmetric, nonnegative definite and D_t is symmetric, positive definite, for all t. We also assume that t_1 is a deterministic time.

The problem is then to choose the control $u = u(t, X_t) \in \mathbf{R}^k$ such that it minimizes $J^u(s,x)$. We may interpret this as follows: The aim is to find a control u which makes $|X_t|$ small fast and such that the energy used ($\sim u^T D u$) is small. The sizes of C_t and R reflect the cost of having large values of $|X_t|$, while the size of D_t reflects the cost (energy) of applying large values of $|u_t|$.

In this case the HJB-equation for $\Psi(s,x) = \inf_u J^u(s,x)$ becomes

$$
\begin{aligned}
0 &= \inf_v \{f^v(s,x) + (L^v\Psi)(s,x)\} \\
&= \frac{\partial \Psi}{\partial s} + \inf_v \left\{ x^T C_s x + v^T D_s v + \sum_{i=1}^n (H_s x + M_s v)_i \frac{\partial \Psi}{\partial x_i} \right. \\
&\qquad \left. + \frac{1}{2} \sum_{i,j=1}^n (\sigma_s \sigma_s^T)_{ij} \frac{\partial^2 \Psi}{\partial x_i \partial x_j} \right\} \quad \text{for } s < t_1
\end{aligned}
\qquad (11.2.24)
$$

and

$$\Psi(t_1, x) = x^T R x . \qquad (11.2.25)$$

Let us try to find a solution ψ of (11.2.24)–(11.2.25) of the form

$$\psi(t,x) = x^T S_t x + a_t \qquad (11.2.26)$$

where $S(t) = S_t \in \mathbf{R}^{n\times n}$ is symmetric, nonnegative definite, $a_t \in \mathbf{R}$ and both a_t and S_t are continuously differentiable w.r.t. t (and deterministic). In order to use Theorem 11.2.2 we need to determine S_t and a_t such that

$$\inf_v \{f^v(t,x) + (L^v\psi)(t,x)\} = 0 \qquad \text{for } t < t_1 \qquad (11.2.27)$$

and

$$\psi(t_1, x) = x^T R x . \qquad (11.2.28)$$

To obtain (11.2.28) we put

$$S_{t_1} = R \tag{11.2.29}$$

$$a_{t_1} = 0 . \tag{11.2.30}$$

Using (11.2.26) we get

$$f^v(t,x) + (L^v\psi)(t,x) = x^T S_t' x + a_t' + x^T C_t x + v^T D_t v +$$
$$+ (H_t x + M_t v)^T (S_t x + S_t^T x) + \sum_{i,j} (\sigma_t \sigma_t^T)_{ij} S_{ij} , \tag{11.2.31}$$

where $S_t' = \frac{d}{dt} S_t$, $a_t' = \frac{d}{dt} a_t$. The minimum of this expression is obtained when

$$\frac{\partial}{\partial v_i} (f^v(t,x) + (L^v\psi)(t,x)) = 0 ; \qquad i = 1, \dots, k$$

i.e. when

$$2D_t v + 2M_t^T S_t x = 0$$

i.e. when

$$v = -D_t^{-1} M_t^T S_t x . \tag{11.2.32}$$

We substitute this value of v in (11.2.31) and obtain

$$f^v(t,x) + (L^v\psi)(t,x) =$$
$$= x^T S_t' x + a_t' + x^T C_t x + x^T S_t M_t D_t^{-1} D_t D_t^{-1} M_t^T S_t x$$
$$+ (H_t x - M_t D_t^{-1} M_t^T S_t x)^T 2 S_t x + tr(\sigma\sigma^T S)_t$$
$$= x^T (S_t' + C_t - S_t M_t D_t^{-1} M_t^T S_t + 2 H_t^T S_t) x + a_t' + tr(\sigma\sigma^T S)_t ,$$

where tr denotes the (matrix) trace. We obtain that this is 0 if we choose S_t such that

$$S_t' = -2 H_t^T S_t + S_t M_t D_t^{-1} M_t^T S_t - C_t ; \qquad t < t_1 \tag{11.2.33}$$

and a_t such that

$$a_t' = -tr(\sigma\sigma^T S)_t ; \qquad t < t_1 . \tag{11.2.34}$$

We recognize (11.2.33) as a Riccati type equation from linear filtering theory (see (6.3.4)). Equation (11.2.33) with boundary condition (11.2.29) determines S_t uniquely. Combining (11.2.34) with the boundary condition (11.2.30) we obtain

$$a_t = \int_t^{t_1} tr(\sigma\sigma^T S)_s ds . \tag{11.2.35}$$

With such a choice of S_t and a_t we see that (11.2.27) and (11.2.28) hold, so by Theorem 11.2.2 we conclude that

$$u^*(t,x) = -D_t^{-1}M_t^T S_t x , \qquad t < t_1 \tag{11.2.36}$$

is an optimal control and the minimum cost is

$$\Psi(s,x) = x^T S_s x + \int_s^{t_1} tr(\sigma\sigma^T S)_t dt , \qquad s < t_1 . \tag{11.2.37}$$

This formula shows that the extra cost due to the noise in the system is given by

$$a_s = \int_s^{t_1} tr(\sigma\sigma^T S)_t dt .$$

The Separation Principle (see Davis (1977), Davis and Vinter (1985) or Fleming and Rishel (1975)) states that if we had only partial knowledge of the state X_t of the system, i.e. if we only had noisy observations

$$dZ_t = \theta_t X_t dt + \gamma_t d\widetilde{B}_t \tag{11.2.38}$$

to our disposal, then the optimal control $u^*(t,\omega)$ (required to be $\mathcal{G}_t$-adapted, where $\mathcal{G}_t$ is the σ-algebra generated by $\{Z_r; r \leq t\}$), would be given by

$$u^*(t,\omega) = -D_t^{-1}M_t^T S_t \widehat{X}_t(\omega) , \tag{11.2.39}$$

where $\widehat{X}_t$ is the filtered estimate of X_t based on the observations $\{Z_r; r \leq t\}$, given by the Kalman-Bucy filter (6.3.3). Comparing with (11.2.36) we see that the stochastic control problem in this case splits into a linear filtering problem and a deterministic control problem.

An important field of applications of the stochastic control theory is economics and finance. Therefore we illustrate the results above by applying them to a simple case of optimal portfolio diversification. This problem has been considered in more general settings by many authors, see for example Markowitz (1976), Merton (1971), Harrison and Pliska (1981), Aase (1984), Karatzas, Lehoczky and Shreve (1987) and the survey article Duffie (1994) and the references therein.

Example 11.2.5 (An optimal portfolio selection problem)
Let X_t denote the wealth of a person at time t. Suppose that the person has the choice of two different investments. The price $X_1(t)$ at time t of one of the assets is assumed to satisfy the equation

$$\frac{dX_1(t)}{dt} = X_1(t)[a + \alpha W_t] \tag{11.2.40}$$

where W_t denotes white noise and a, $\alpha > 0$ are constants measuring the average relative rate of change of $X_1(t)$ and the size of the noise, respectively.

As we have discussed earlier we interpret (11.2.40) as the (Itô) stochastic differential equation

$$dX_1(t) = X_1(t)adt + X_1(t)\alpha dB_t \ . \tag{11.2.41}$$

This investment is called *risky*, since $\alpha > 0$. We assume that the price $X_0(t)$ of the other asset satisfies a similar equation, but with no noise:

$$dX_0(t) = X_0(t)bdt \ . \tag{11.2.42}$$

This investment is called *risk free*. So it is natural to assume $b < a$. At each instant t the person can choose how big fraction $u(t)$ of his wealth he will invest in the risky asset, thereby investing the fraction $1 - u(t)$ in the safe one. This gives the following stochastic differential equation for the wealth $Z_t = Z_t^u$:

$$\begin{aligned} dZ_t &= u(t)Z_t adt + u(t)Z_t \alpha dB_t + (1 - u(t))Z_t bdt \\ &= Z_t(au(t) + b(1 - u(t)))dt + \alpha u(t)Z_t dB_t \ . \end{aligned} \tag{11.2.43}$$

Suppose that, starting with the wealth $Z_s = x > 0$ at time s, the person wants to maximize the expected utility of the wealth at some future time $t_0 > s$. If we do not allow any *borrowing* (i.e. require $u(t) \le 1$) and we do not allow any *shortselling* (i.e. require $u(t) \ge 0$) and we are given a utility function $N : [0, \infty) \to [0, \infty)$, $N(0) = 0$ (usually assumed to be increasing and concave) the problem is to find $\Phi(s, x)$ and a (Markov) control $u^* = u^*(t, Z_t)$, $0 \le u^* \le 1$, such that

$$\begin{aligned} \Phi(s, x) = \sup\{J^u(s, x); \ u \ \text{Markov control}, \ 0 \le u \le 1\} = J^{u^*}(s, x) \ , \\ \text{where} \ \ J^u(s, x) = E^{s,x}[N(Z^u_{\tau_G})] \end{aligned} \tag{11.2.44}$$

and τ_G is the first exit time from the region $G = \{(r, z); \ r < t_0, z > 0\}$. This is a performance criterion of the form (11.1.6)/(11.1.8) with $f = 0$ and $g = N$. The differential operator L^v has the form (see (11.2.2))

$$(L^v\phi)(t, x) = \frac{\partial\phi}{\partial t} + x(av + b(1 - v))\frac{\partial\phi}{\partial x} + \tfrac{1}{2}\alpha^2 v^2 x^2 \frac{\partial^2\phi}{\partial x^2} \ . \tag{11.2.45}$$

The HJB equation becomes

$$\sup_v\{(L^v\Phi)(t, x)\} = 0 \ , \qquad \text{for } (t, x) \in G \ ; \tag{11.2.46}$$

and

$$\Phi(t, x) = N(x) \quad \text{for } t = t_0 \ , \qquad \Phi(t, 0) = N(0) \quad \text{for } t < t_0 \ . \tag{11.2.47}$$

Therefore, for each (t, x) we try to find the value $v = u(t, x)$ which maximizes the function

$$\eta(v) = L^v\Phi = \frac{\partial\Phi}{\partial t} + x(b + (a-b)v)\frac{\partial\Phi}{\partial x} + \frac{1}{2}\alpha^2 v^2 x^2 \frac{\partial^2\Phi}{\partial x^2} \, . \qquad (11.2.48)$$

If $\Phi_x := \frac{\partial\Phi}{\partial x} > 0$ and $\Phi_{xx} := \frac{\partial^2\Phi}{\partial x^2} < 0$, the solution is

$$v = u(t,x) = -\frac{(a-b)\Phi_x}{x\alpha^2\Phi_{xx}} \, . \qquad (11.2.49)$$

If we substitute this into the HJB equation (11.2.48) we get the following nonlinear boundary value problem for Φ :

$$\Phi_t + bx\Phi_x - \frac{(a-b)^2\Phi_x^2}{2\alpha^2\Phi_{xx}} = 0 \qquad \text{for } t < t_0, \, x > 0 \qquad (11.2.50)$$

$$\Phi(t,x) = N(x) \qquad \text{for } t = t_0 \text{ or } x = 0 \, . \qquad (11.2.51)$$

The problem (11.2.50), (11.2.51) is hard to solve for general N. Important examples of increasing and concave functions are the power functions

$$N(x) = x^\gamma \qquad \text{where } 0 < \gamma < 1 \text{ is a constant} \, . \qquad (11.2.52)$$

If we choose such a utility function N, we try to find a solution of (11.2.50), (11.2.51) of the form

$$\phi(t,x) = f(t)x^\gamma \, .$$

Substituting we obtain

$$\phi(t,x) = e^{\lambda(t_0 - t)}x^\gamma \, , \qquad (11.2.53)$$

where $\lambda = b\gamma + \frac{(a-b)^2\gamma}{2\alpha^2(1-\gamma)}$.

Using (11.2.49) we obtain the optimal control

$$u^*(t,x) = \frac{a-b}{\alpha^2(1-\gamma)} \, . \qquad (11.2.54)$$

If $\frac{a-b}{\alpha^2(1-\gamma)} \in (0,1)$ this is the solution to the problem, in virtue of Theorem 11.2.2. Note that u^* is in fact constant.

Another interesting choice of the utility function is $N(x) = \log x$, called the Kelly criterion. As noted by Aase (1984) (in a more general setting) we may in this case obtain the optimal control directly by evaluating $E^{s,x}[\log(X_T)]$ using Dynkin's formula:

$$E^{s,x}[\log(Z_{\tau_G})] =$$

$$= \log x + E^{s,x}\left[\int_s^{\tau_G}\{au(t,Z_t) + b(1 - u(t,Z_t)) - \frac{1}{2}\alpha^2 u^2(t,Z_t)\}dt\right]$$

since $L^v(\log x) = av + b(1-v) - \frac{1}{2}\alpha^2 v^2$.

So it is clear that $J^u(s,x) = E^{s,x}[\log(Z_{\tau_G})]$ is maximal if we for all t, z choose $u(t,z)$ to have the value of v which maximizes

$$av + b(1-v) - \tfrac{1}{2}\alpha^2 v^2$$

i.e. we choose

$$v = u(t, Z_t) = \frac{a-b}{\alpha^2} \qquad \text{for all } t, \omega . \tag{11.2.55}$$

So this is the optimal control if the Kelly criterion is used. Similarly, this direct method also gives the optimal control when $N(x) = x^r$ (See Exercise 11.8).

Example 11.2.6 Finally we include an example which shows that even quite simple – and apparently innocent – stochastic control problems can lead us beyond the reach of the theory developed in this chapter:

Suppose the system is a 1-dimensional Itô integral

$$dX_t = dX_t^u = u(t,\omega)dB_t , \qquad t \geq s; \; X_s = x > 0 \tag{11.2.56}$$

and consider the stochastic control problem

$$\Phi(t,x) = \sup_u E^{t,x}[K(X_{\tau_G}^u)] , \tag{11.2.57}$$

where τ_G is the first exit time from $G = \{(r,z); r \leq t_1, z > 0\}$ for $Y_t = (s+t, X_{s+t}^{s,x})$ and K is a given bounded continuous function.

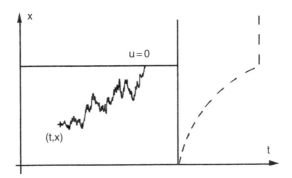

Intuitively, we can think of the system as the state of a game which behaves like an "excited" Brownian motion, where we can control the size u of the excitation at every instant. The purpose of the control is to maximize the expected payoff $K(X_{t_1})$ of the game at a fixed future time t_1.

Assuming that $\Phi \in C^2$ and that u^* exists we get by the HJB (I) equation

$$\sup_{v \in \mathbf{R}} \left\{ \frac{\partial \Phi}{\partial t} + \tfrac{1}{2}v^2 \frac{\partial^2 \Phi}{\partial x^2} \right\} = 0 \qquad \text{for } t < t_1, \; \Phi(t_1, x) = K(x) . \tag{11.2.58}$$

From this we see that we necessarily have

$$\frac{\partial^2 \Phi}{\partial x^2} \leq 0 , \quad v^* \frac{\partial^2 \Phi}{\partial x^2} = 0 \quad \text{and} \quad \frac{\partial \Phi}{\partial t} = 0 \quad \text{for } t < t_1 , \tag{11.2.59}$$

where v^* is the value of $v \in \mathbf{R}$ which gives the supremum in (11.2.58). But if $\frac{\partial \Phi}{\partial t} = 0$, then $\Phi(t, x) = \Phi(t_1, x) = K(x)$. However, this cannot possibly be the solution in general, because we have not assumed that $\frac{\partial^2 K}{\partial x^2} \leq 0$ – in fact, K was not even assumed to be differentiable.

What went wrong? Since the conclusion of the HJB (I) equation was wrong, the assumptions cannot hold. So either Φ is not C^2 or u^* does not exist, or both.

To simplify the problem assume that

$$K(x) = \begin{cases} x^2 ; & 0 \leq x \leq 1 \\ 1 ; & x > 1 . \end{cases}$$

Then considering the figure above and using some intuition we see that it is optimal to excite as much as possible if X_t is in the strip $0 < x < 1$ to avoid exiting from G in the interval $\{t_1\} \times (0, 1)$. Using that X_t is just a time change of Brownian motion (see Chapter 8) we conclude that this optimal control leads to a process X^* which jumps immediately to the value 1 with probability x and to the value 0 with probability $1 - x$, if the starting point is $x \in (0, 1)$. If the starting point is $x \in [1, \infty)$ we simply choose our control to be zero. In other words, heuristically we should have

$$u^*(t, x) = \begin{cases} \infty & \text{if} \quad x \in (0, 1) \\ 0 & \text{if} \quad x \in [1, \infty) \end{cases} \tag{11.2.60}$$

with corresponding expected payoff

$$\phi^*(s, x) = E^{s,x}[K(X_{t_1}^*)] = \begin{cases} x & \text{if} \quad 0 \leq x \leq 1 \\ 1 & \text{if} \quad x > 1 . \end{cases} \tag{11.2.61}$$

Thus we see that our candidate u^* for optimal control is not continuous (not even finite!) and the corresponding optimal process X_t^* is not an Itô diffusion (it is not even continuous). So to handle this case mathematically it is necessary to enlarge the family of admissible controls (and the family of corresponding processes). For example, one can prove an extended version of Theorem 11.2.2 which allows us to conclude that our choice of u^* above does indeed give at least as good performance as any other Markov control u and that ϕ^* given by (11.2.61) does coincide with the maximal expected payoff Φ defined by (11.2.57).

This last example illustrates the importance of the question of *existence* in general, both of the optimal control u^* and of the corresponding solution

X_t of the stochastic differential equation (11.1.1). We briefly outline some results in this direction:

With certain conditions on $b, \sigma, f, g, \partial G$ and assuming that the set of control values is compact, one can show, using general results from nonlinear partial differential equations, that a smooth function ϕ exists such that

$$\sup_{v}\{f^v(y) + (L^v\phi)(y)\} = 0 \qquad \text{for } y \in G$$

and

$$\phi(y) = g(y) \qquad \text{for } y \in \partial G .$$

Then by a measurable selection theorem one can find a (measurable) function $u^*(y)$ such that

$$f^{u^*}(y) + (L^{u^*}\phi)(y) = 0 , \tag{11.2.62}$$

for a.a. $y \in G$ w.r.t. Lebesgue measure in $\mathbf{R}^{n+1}$. Even if u^* is only known to be measurable, one can show that the corresponding solution $X_t = X_t^{u^*}$ of (11.1.1) exists (see Stroock and Varadhan (1979) for general results in this direction). Then by inspecting the proof of Theorem 11.2.2 one can see that it suffices to have (11.2.62) satisfied outside a subset of G with Green measure 0 (see Definition 9.3.4). Under suitable conditions on b and σ one can in fact show that the Green measure is absolutely continuous w.r.t. Lebesgue measure. Thus by (11.2.62) (and a strengthened Theorem 11.2.2) u^* is an optimal control. We refer the reader to Fleming and Rishel (1975), Bensoussan and Lions (1978), Dynkin and Yushkevich (1979) and Krylov (1980) for details and further studies.

11.3 Stochastic control problems with terminal conditions

In many applications there are constraints on the types of Markov controls u to be considered, for example in terms of the probabilistic behaviour of Y_t^u at the terminal time $t = T$. Such problems can often be handled by applying a kind of "Lagrange multiplier" method, which we now describe:

Consider the problem of finding $\Phi(y)$ and $u^*(y)$ such that

$$\Phi(y) = \sup_{u \in \mathcal{K}} J^u(y) = J^{u^*}(y) \tag{11.3.1}$$

where

$$J^u(y) = E^y\left[\int_0^{\tau_G} f^u(Y_t^u)dt + g(Y_{\tau_G}^u)\right] , \tag{11.3.2}$$

and where the supremum is taken over the space $\mathcal{K}$ of all Markov controls $u: \mathbf{R}^{n+1} \to U \subset \mathbf{R}^k$ such that

$$E^y[M_i(Y^u_{\tau_G})] = 0 \,, \qquad i = 1, 2, \ldots, l \,, \tag{11.3.3}$$

where $M = (M_1, \ldots, M_l) \colon \mathbf{R}^{n+1} \to \mathbf{R}^l$ is a given continuous function,

$$E^y[|M(Y^u_{\tau_G})|] < \infty \qquad \text{for all } y, u \,, \tag{11.3.4}$$

and we interpret $g(Y_{\tau_G}(\omega))$ as 0 if $\tau_G(\omega) = \infty$.

Now we introduce a related, but unconstrained problem as follows:

For each $\lambda \in \mathbf{R}^l$ and each Markov control u define

$$J^u_\lambda(y) = E^y\left[\int_0^{\tau_G} f^u(Y^u_t)dt + g(Y^u_{\tau_G}) + \lambda \cdot M(Y^u_{\tau_G})\right] \tag{11.3.5}$$

where $\cdot$ denotes the inner product in $\mathbf{R}^l$. Find $\Phi_\lambda(y)$ and $u^*_\lambda(y)$ such that

$$\Phi_\lambda(y) = \sup_u J^u_\lambda(y) = J^{u^*_\lambda}_\lambda(y) \,, \tag{11.3.6}$$

without terminal conditions.

Theorem 11.3.1 *Suppose that we for all $\lambda \in \Lambda \subset \mathbf{R}^l$ can find $\Phi_\lambda(y)$ and u^*_λ solving the (unconstrained) stochastic control problem (11.3.5)–(11.3.6). Moreover, suppose that there exists $\lambda_0 \in \Lambda$ such that*

$$E^y[M(Y^{u^*_{\lambda_0}}_{\tau_G})] = 0 \,. \tag{11.3.7}$$

Then $\Phi(y) := \Phi_{\lambda_0}(y)$ and $u^* := u^*_{\lambda_0}$ solves the constrained *stochastic control problem (11.3.1)–(11.3.3).*

Proof. Let u be a Markov control, $\lambda \in \Lambda$. Then by the definition of u^*_λ we have

$$E^y\left[\int_0^{\tau_G} f^{u^*_\lambda}(Y^{u^*_\lambda}_t)dt + g(Y^{u^*_\lambda}_{\tau_G}) + \lambda \cdot M(Y^{u^*_\lambda}_{\tau_G})\right] = J^{u^*_\lambda}_\lambda(y)$$

$$\geq J^u_\lambda(y) = E^y\left[\int_0^{\tau_G} f^u(Y^u_t)dt + g(Y^u_{\tau_G}) + \lambda \cdot M(Y^u_{\tau_G})\right] \,. \tag{11.3.8}$$

In particular, if $\lambda = \lambda_0$ and $u \in \mathcal{K}$ then

$$E^y[M(Y^{u^*_{\lambda_0}}_{\tau_G})] = 0 = E^y[M(Y^u_{\tau_G})]$$

and hence by (11.3.8)

$$J^{u^*_{\lambda_0}}(y) \geq J^u(y) \qquad \text{for all } u \in \mathcal{K} \,.$$

Since $u^*_{\lambda_0} \in \mathcal{K}$ the proof is complete.   □

For an application of this result, see Exercise 11.11.

Exercises

11.1.   Write down the HJB equation for the problem

$$\Psi(s,x) = \inf_u E^{s,x}\left[\int_0^\infty e^{-\alpha(s+t)}(\theta(X_t) + u_t^2)dt\right]$$

where
$$dX_t = u_t dt + dB_t \; ; \qquad X_t, u_t, B_t \in \mathbf{R} \; ,$$

$\alpha > 0$ is a constant and $\theta: \mathbf{R} \to \mathbf{R}$ is a given bounded, continuous function. Show that if Ψ satisfies the conditions of Theorem 11.2.1 and u^* exists then

$$u^*(t,x) = -\tfrac{1}{2}e^{\alpha t}\frac{\partial \Psi}{\partial x} \; .$$

11.2.   Consider the stochastic control problem

$$\Psi_0(s,x) = \inf_u E^{s,x}\left[\int_s^\infty e^{-\rho t}f_0(u_t, X_t)dt\right] \; ,$$

where
$$dX_t = dX_t^u = b(u_t, X_t)dt + \sigma(u_t, X_t)dB_t$$
$$X_t \in \mathbf{R}^n, \, u_t \in \mathbf{R}^k, \, B_t \in \mathbf{R}^m \; ,$$

f_0 is a given bounded continuous real function, $\rho > 0$ and the inf is taken over all *time-homogeneous* Markov controls u, i.e. controls u of the form $u = u(X_t)$. Prove that

$$\Psi_0(s,x) = e^{-\rho s}\xi(x) \; , \qquad \text{where } \xi(x) = \Psi(0,x) \; .$$

(Hint: By definition of $E^{s,x}$ we have

$$E^{s,x}\left[\int_s^\infty e^{-\rho t}f_0(u(X_t), X_t)dt\right] = E\left[\int_0^\infty e^{-\rho(s+t)}f_0(u(X_{s+t}^{s,x}), X_{s+t}^{s,x})dt\right]$$

where E denotes expectation w.r.t. P.)

11.3.   Define

$$dX_t = ru_t X_t dt + \alpha u_t X_t dB_t \; ; \qquad X_t, u_t, B_t \in \mathbf{R}$$

and

$$\Phi(s, x) = \sup_u E^{s,x} \left[\int_0^\infty e^{-\rho(s+t)} f_0(X_t) dt \right] ,$$

where r, α, ρ are constants, $\rho > 0$ and f is a bounded continuous real function.

Assume that Φ satisfies the conditions of Theorem 11.2.1 and that an optimal Markov control u^* exists.

a)   Show that

$$\sup_{v \in \mathbf{R}} \left\{ e^{-\rho t} f(x) + \frac{\partial \Phi}{\partial t} + rvx \frac{\partial \Phi}{\partial x} + \tfrac{1}{2} \alpha^2 v^2 x^2 \frac{\partial^2 \Phi}{\partial x^2} \right\} = 0 .$$

Deduce that

$$\frac{\partial^2 \Phi}{\partial x^2} \leq 0 .$$

b)   Assume that $\frac{\partial^2 \Phi}{\partial x^2} < 0$. Prove that

$$u^*(t, x) = -\frac{r \frac{\partial \Phi}{\partial x}}{\alpha^2 x \frac{\partial^2 \Phi}{\partial x^2}}$$

and that

$$2\alpha^2 \left(e^{-\rho t} f_0(x) + \frac{\partial \Phi}{\partial t} \right) \frac{\partial^2 \Phi}{\partial x^2} - r^2 \left(\frac{\partial \Phi}{\partial x} \right)^2 = 0 .$$

c)   Assume that $\frac{\partial^2 \Phi}{\partial x^2} = 0$. Prove that $\frac{\partial \Phi}{\partial x} = 0$ and

$$e^{-\rho t} f_0(x) + \frac{\partial \Phi}{\partial t} = 0 .$$

d)   Assume that $u_t^* = u^*(X_t)$ and that b) holds. Prove that $\Phi(t, x) = e^{-\rho t} \xi(x)$ and

$$2\alpha^2 (f_0 - \rho \xi) \xi'' - r^2 (\xi')^2 = 0 .$$

(See Exercise 11.2)

11.4.   The assumptions in Theorem 11.2.1 often fail (see e.g. Exercise 11.10), so it is useful to have results in such cases also. For example, if we define Φ_a as in Theorem 11.2.3 then, without assuming that u^* exists and without smoothness conditions on Φ, we have the *Bellman principle* (compare with (11.2.6)–(11.2.7))

$$\Phi_a(y) = \sup_u E^y \left[\int_0^\alpha f^u(Y_r^u)dr + \Phi_a(Y_\alpha^u) \right]$$

for all $y \in G$ and all stopping times $\alpha \leq \tau_G$, the sup being taken over all $\mathcal{F}_t^{(m)}$-adapted controls u. (See Krylov (1980, Th. 6, p. 150).) Deduce that if $\Phi_a \in C^2(G)$ then

$$f^v(y) + L^v\Phi_a(y) \leq 0 \qquad \text{for all } y \in G, v \in U \,.$$

11.5.   Assume that $f = 0$ in (11.1.8) and that an optimal Markov control u^* exists. Prove that the function Φ is *superharmonic* in G w.r.t. the process Y_t^u, for any Markov control u. (Hint: See (11.2.6)–(11.2.7).)

11.6.*   Let X_t denote your wealth at time t. Suppose that at any time t you have a choice between two investments:

1)   A risky investment where the unit price $X_1(t) = X_1(t, \omega)$ satisfies the equation

$$dX_1(t) = a_1 X_1(t)dt + \sigma_1 X_1(t)dB_t \,.$$

2)   A safe investment where the unit price $X_0(t) = X_0(t, \omega)$ satisfies

$$dX_0(t) = a_0 X_0(t)dt + \sigma_0 X_0(t)d\tilde{B}_t$$

where a_i, σ_i are constants such that

$$a_1 > a_0 \,, \qquad \sigma_1 > \sigma_0$$

and $B_t, \tilde{B}_t$ are independent 1-dimensional Brownian motions.

a)   Let $u(t, \omega)$ denote the fraction of the fortune $Z_t(\omega)$ which is placed in the riskier investment at time t. Show that

$$dZ_t = dZ_t^{(u)} = Z_t(a_1 u(t) + a_0(1 - u(t)))dt$$
$$+ Z_t(\sigma_1 u(t)dB_t + \sigma_0(1 - u(t))d\tilde{B}_t) \,.$$

b)   Assuming that u is a Markov control, $u = u(t, Z_t^u)$, find the generator A^u of (t, Z_t^u).

c)   Write down the HJB equation for the stochastic control problem

$$\Phi(s, x) = \sup_u E^{s,x}\left[(Z_{\hat{\tau}}^{(u)})^\gamma \right]$$

where $\hat{\tau} = \min(t_1 - s, \tau_0)$, $\tau_0 = \inf\{t > 0; X_t = 0\}$ and t_1 is a given future time (constant), $\gamma \in (0,1)$ is a constant.

d)   Find the optimal control u^* for the problem in c).

11.7.*   Consider the stochastic control problem

(system) $dX_t = au(t)dt + u(t)dB_t$; $X_0 = x > 0$

where $B_t \in \mathbf{R}$, $u(t) \in \mathbf{R}$ and $a \in \mathbf{R}$ is a given constant, and

(performance) $\Phi(s, x) = \sup_u E^{s,x}[(X_{\hat{\tau}})^\gamma]$,

where $0 < \gamma < 1$ is constant and

$$\hat{\tau} = \inf\{t > 0; X_t = 0\} \wedge (T - s) ,$$

T being a given future time (constant).
Show that this problem has the optimal control

$$u^*(t, x) = \frac{ax}{1 - \gamma}$$

with corresponding optimal performance

$$\Phi(s, x) = x^\gamma \exp\left(\frac{a^2(T - s)\gamma}{2(1 - \gamma)}\right) .$$

11.8.   Use Dynkin's formula to prove directly that

$$u^*(t, x) = \min\left(\frac{a - b}{a^2(1 - \gamma)}, 1\right)$$

is the optimal control for the problem in Example 11.2.5, with utility function $N(x) = x^\gamma$. (Hint: See the argument leading to (11.2.55).)

11.9.   In Beneš (1974) the following stochastic control problem is considered:

$$\Psi(s, x) = \inf_u E^{s,x}\left[\int_0^\infty e^{-\rho(s+t)} X_t^2 dt\right] ,$$

where
$$dX_t = dX_t^{(u)} = au_t dt + dB_t ;   X_t, B_t \in \mathbf{R}$$

and a, ρ are (known) constants, $\rho > 0$. Here the controls u are restricted to take values in $U = [-1, 1]$.

a)   Show that the HJB equation for this problem is

$$\inf_{v \in [-1,1]} \left\{ e^{-\rho s} x^2 + \frac{\partial \Psi}{\partial s} + a v \frac{\partial \Psi}{\partial x} + \frac{1}{2} \cdot \frac{\partial^2 \Psi}{\partial x^2} \right\} = 0 \ .$$

b) If $\Psi \in C^2$ and u^* exists, show that

$$u^*(x) = -\text{sign}(ax) \ ,$$

where

$$\text{sign } z = \begin{cases} 1 & \text{if } z > 0 \\ -1 & \text{if } z \leq 0 \ . \end{cases}$$

(Hint: Explain why $x > 0 \Rightarrow \frac{\partial \Psi}{\partial x} > 0$ and $x < 0 \Rightarrow \frac{\partial \Psi}{\partial x} < 0$.)

11.10. Let

$$f_0(x) = \begin{cases} x^2 & \text{for } 0 \leq x \leq 1 \\ \sqrt{x} & \text{for } x > 1 \end{cases}$$

and put

$$J^u(s,x) = E^{s,x} \left[\int_0^{\tau_0} e^{-\rho(s+t)} f_0(X_t^u) dt \right] \ , \quad \Phi(s,x) = \sup_u J^u(s,x)$$

where

$$dX_t^u = u_t dB_t \ ; \qquad t \geq 0$$

with control values $u_t \in \mathbf{R}$, $B_t \in \mathbf{R}$ and

$$\tau_0 = \inf\{t > 0; X_t^u \leq 0\} \ .$$

a) Define

$$\phi(s,x) = \frac{1}{\rho} e^{-\rho s} \widehat{f}(x) \qquad \text{for } x \geq 0, \ s \in \mathbf{R}$$

where

$$\widehat{f}(x) = \begin{cases} x & \text{for } 0 \leq x \leq 1 \\ \sqrt{x} & \text{for } x > 1 \ . \end{cases}$$

Prove that

$$J^u(s,x) \leq \phi(s,x)$$

for all s, x and all (finite) Markov controls u.
(Hint: Put $\phi_1(s,x) = \frac{1}{\rho} e^{-\rho s} x$ for all s, x and $\phi_2(s,x) = \frac{1}{\rho} e^{-\rho s} \sqrt{x}$ for all s, x. Then

$$J^u(s,x) \leq \phi_i(s,x) \qquad \text{for } i = 1, 2$$

by Theorem 11.2.2.)

b) Show that

$$\Phi(s, x) = \phi(s, x) \ .$$

(Hint: Consider $J^{u_k}(s, x)$, where

$$u_k(x) = \begin{cases} k & \text{for } 0 \leq x < 1 \\ 0 & \text{for } x \geq 1 \end{cases}$$

and let $k \to \infty$).

Thus u^* does not exist and Φ is not a C^2 function. Hence both conditions for the HJB (I) equation fail in this case.

11.11.* Consider a 1-dimensional version of the stochastic linear regulator problem of Example 11.2.4:

$$\Psi(s, x) = \inf_{u \in \mathcal{K}} E^{s,x} \left[\int_0^{t_1 - s} ((X_r^u)^2 + \theta u_r^2) dr \right] \tag{11.3.9}$$

where

$$dX_t^u = u_t dt + \sigma dB_t \ ; \qquad \text{for } t \geq 0, \ X_0 = x \ ,$$

$u_t, B_t \in \mathbf{R}$, σ, θ and t_1 constants, $\theta > 0$, $t_1 > 0$. The infimum is taken over the space $\mathcal{K}$ of all Markov controls u satisfying

$$E^{s,x}[(X_{t_1 - s}^u)^2] = m^2 \ , \qquad \text{where } m \text{ is a constant} \ . \tag{11.3.10}$$

Solve this problem by using Theorem 11.3.1.

(Hint: Solve for each $\lambda \in \mathbf{R}$ the unconstrained problem

$$\Psi_\lambda(s, x) = \inf_u E^{s,x} \left[\int_0^{t_1 - s} ((X_r^u)^2 + \theta u_r^2) dr + \lambda (X_{t_1 - s}^u)^2 \right]$$

with optimal control u_λ^*. Then try to find λ_0 such that

$$E^{s,x} \left[\left(X_{t_1 - s}^{u_{\lambda_0}^*} \right)^2 \right] = m^2 \ .)$$

11.12.* Solve the stochastic control problem

$$\Psi(s, x) = \inf_u J^u(s, x) = J^{u^*}(s, x)$$

where

$$J^u(s, x) = E^{s,x} \left[\int_0^\infty e^{-\rho(s+t)} (X_t^2 + \theta u_t^2) dt \right]$$

and

$$dX_t = u_t dt + \sigma dB_t \ ,$$

with $u_t, B_t \in \mathbf{R}$ and $\sigma \in \mathbf{R}$, $\rho > 0$, $\theta > 0$ are constants. (Hint: Try $\psi(s, x) = e^{-\rho s}(ax^2 + b)$ for suitable constants a, b and apply Theorem 11.2.2.)

11.13.* Consider the stochastic control problem

$$\Phi(s, x) = \sup_{u} E^{s,x}\left[\int_0^\tau e^{-\rho(s+t)} u_t \, dt \right]$$

where the (1-dimensional) system X_t is given by

$$dX_t = dX_t^u = (1 - u_t)dt + \sigma \, dB_t \,,$$

and $\rho > 0$, $\sigma \neq 0$ are constants. The control $u_t = u_t(\omega)$ can assume any value in $U = [0, 1]$ and

$$\tau = \inf\{t > 0; \ X_t^u \leq 0\} \qquad \text{(the time of bankruptcy)} \,.$$

Show that if $\rho \geq \frac{2}{\sigma^2}$ then the optimal control is

$$u_t^* = 1 \qquad \text{for all } t$$

and the corresponding value function is

$$\Phi(s, x) = e^{-\rho s} \frac{1}{\rho}\left(1 - \exp\left(-\sqrt{\frac{2}{\sigma^2}} \, x \right) \right); \qquad x \geq 0 \,.$$

11.14. The following problem is an infinite horizon version of Example 11.2.5, with consumption added.

Suppose we have a market with two investment possibilities:

(i)   a bond/bank account, where the price $X_0(t)$ at time t is given by

$$dX_0(t) = \rho \, X_0(t)dt \, ; \qquad X_0(0) = 1, \ \rho \geq 0 \text{ constant}$$

(ii)   a stock, where the price $X_1(t)$ at time t is given by

$$dX_1(t) = \mu \, X_1(t)dt + \sigma \, X_1(t)dB(t); \qquad X_1(0) = x > 0,$$

where μ, σ are constants, $\sigma \neq 0$.

Let $Y_0(t), Y_1(t)$ denote the amount of money that an agent at time t has invested in the bonds and in the stocks, respectively. Assume that the agent at any time can choose her consumption rate $c(t) = c(t, \omega) \geq 0$ and the ratio

$$u(t) = u(t, \omega) = \frac{Y_1(t)}{Y_0(t) + Y_1(t)}$$

of her total wealth invested in the stocks. We assume that $c(t)$ and $u(t)$ are $\mathcal{F}_t$-adapted processes and that the dynamics of the total wealth $Z(t) = Y_1(t) + Y_2(t)$ is given by

$$dZ(t) = Z(t)\big[\{\rho(1 - u(t)) + \mu\, u(t)\}dt + \sigma\, u(t)dB(t)\big] - c(t)dt;$$
$$Z(0) = z > 0 \, .$$

Consider the problem to find Φ, c^* and u^* such that

$$\Phi(s, z) = \sup_{c,u} J^{c,u}(s, z) = J^{c^*,u^*}(s, z) \, ,$$

where

$$J^{c,u}(s, z) = E^{s,z}\left[\int\limits_0^{\tau_0} e^{-\delta(s+t)} \frac{c^\gamma(t)}{\gamma} dt\right]$$

where $\delta > 0$ and $\gamma \in (0, 1)$ are constants and

$$\tau_0 = \inf\{t > 0; Z(t) \le 0\} \le \infty \quad \text{(time of bankruptcy)}$$

Use Theorem 11.2.2 to prove that, under some conditions,

$$\Phi(s, z) = K\, e^{-\delta s} z^\gamma$$

for a certain value of the constant K. Find this value of K and hence find the optimal consumption rate $c^*(t)$ and the optimal portfolio $u^*(t)$.

Chapter 12
Application to Mathematical Finance

12.1 Market, portfolio and arbitrage

In this chapter we describe how the concepts, methods and results in the previous chapters can be applied to give a rigorous mathematical model of finance. We will concentrate on the most fundamental issues and those topics which are most closely related to the theory in this book. We emphasize that this chapter only intends to give a brief introduction to this exciting subject, which has developed very fast during the last years and shows no signs of slowing down. For a more comprehensive treatment see for example Benth (2004), Bingham and Kiesel (1998), Elliott and Kopp (2005), Duffie (1996), Karatzas (1997), Karatzas and Shreve (1998), Shreve (2004), Lamberton and Lapeyre (1996), Musiela and Rutkowski (1997), Kallianpur and Karandikar (2000), Merton (1990), Shiryaev (1999) and the references therein.

First we give the mathematical definitions of some fundamental finance concepts. We point out that other mathematical models are also possible and in fact actively investigated. Other models include more general (possibly discontinuous) semimartingales (see e.g. Barndorff-Nielsen (1998), Eberlein and Keller (1995), Schoutens (2003), Cont and Tankov (2004), Øksendal and Sulem (2007)) and even non-semimartingale models. See e.g. Cutland, Kopp and Willinger (1995), Lin (1995), Rogers (1997), Hu and Øksendal (2003), Elliott and Van der Hoek (2003), Biagini and Øksendal (2005), Øksendal (2005), Di Nunno et al. (2006), Biagini et al. (2007).

Definition 12.1.1 a)   *A market is an $\mathcal{F}_t^{(m)}$-adapted $(n+1)$-dimensional Itô process $X(t) = (X_0(t), X_1(t), \ldots, X_n(t))$; $0 \le t \le T$ which we will assume has the form*

$$dX_0(t) = \rho(t, \omega)X_0(t)dt \; ; \qquad X_0(0) = 1 \qquad (12.1.1)$$

and

$$dX_i(t) = \mu_i(t, \omega)dt + \sum_{j=1}^{m} \sigma_{ij}(t, \omega)dB_j(t) \qquad (12.1.2)$$

$$= \mu_i(t,\omega)dt + \sigma_i(t,\omega)dB(t) ; \qquad X_i(0) = x_i ,$$

where σ_i is row number i of the $n \times m$ matrix $[\sigma_{ij}]$; $1 \leq i \leq n \in \mathbf{N}$.

b)   The market $\{X(t)\}_{t\in[0,T]}$ is called normalized if $X_0(t) \equiv 1$.

c)   A portfolio in the market $\{X(t)\}_{t\in[0,T]}$ is an $(n+1)$-dimensional (t,ω)-measurable and $\mathcal{F}_t^{(m)}$-adapted stochastic process

$$\theta(t,\omega) = (\theta_0(t,\omega), \theta_1(t,\omega), \ldots, \theta_n(t,\omega)); \qquad 0 \leq t \leq T . \qquad (12.1.3)$$

d)   The value at time t of a portfolio $\theta(t)$ is defined by

$$V(t,\omega) = V^\theta(t,\omega) = \theta(t) \cdot X(t) = \sum_{i=0}^n \theta_i(t)X_i(t) \qquad (12.1.4)$$

where $\cdot$ denotes inner product in $\mathbf{R}^{n+1}$.

e)   The portfolio $\theta(t)$ is called self-financing if

$$\int_0^T \left\{ |\theta_0(s)\rho(s)X_0(s) + \sum_{i=1}^n \theta_i(s)\mu_i(s)| + \sum_{j=1}^m \left[\sum_{i=1}^n \theta_i(s)\sigma_{ij}(s) \right]^2 \right\} ds < \infty \quad a.s.$$

$$(12.1.5)$$

and

$$dV(t) = \theta(t) \cdot dX(t) \qquad (12.1.6)$$

i.e.

$$V(t) = V(0) + \int_0^t \theta(s) \cdot dX(s) \qquad \text{for } t \in [0,T]. \qquad (12.1.7)$$

Comments to Definition 12.1.1.

a)   We think of $X_i(t) = X_i(t,\omega)$ as the price of security/asset number i at time t. The assets number $1, \ldots, n$ are called risky because of the presence of their diffusion terms. They can for example represent stock investments. The asset number 0 is called risk free because of the absence of a diffusion term (although $\rho(t,\omega)$ may depend on ω). This asset can for example represent a bank investment. For simplicity we will assume that $\rho(t,\omega)$ is bounded.

b)   Note that we can always make the market normalized by defining

$$\overline{X}_i(t) = X_0(t)^{-1}X_i(t); \qquad 1 \leq i \leq n . \qquad (12.1.8)$$

The market

$$\overline{X}(t) = (1, \overline{X}_1(t), \ldots, \overline{X}_n(t))$$

is called the normalization of $X(t)$.

Thus normalization corresponds to regarding the price $X_0(t)$ of the safe investment as the unit of price (the numeraire) and computing the other prices in terms of this unit. Since

$$X_0(t) = \exp\left(\int_0^t \rho(s,\omega)ds\right)$$

we have

$$\xi(t) := X_0^{-1}(t) = \exp\left(-\int_0^t \rho(s,\omega)ds\right) > 0 \quad \text{for all } t \in [0,T] \qquad (12.1.9)$$

and

$$d\overline{X}_i(t) = d(\xi(t)X_i(t)) = \xi(t)[(\mu_i - \rho X_i)dt + \sigma_i dB(t)]; \quad 1 \le i \le n \qquad (12.1.10)$$

or

$$d\overline{X}(t) = \xi(t)[dX(t) - \rho(t)X(t)dt] . \qquad (12.1.11)$$

c) The components $\theta_0(t,\omega), \ldots, \theta_n(t,\omega)$ represent the *number of units* of the securities number $0, \ldots, n$, respectively, which an investor holds at time t.

d) This is simply the total value of all investments held at time t.

e) Note that condition (12.1.5) is required to make (12.1.7) well-defined. See Definition 3.3.2.

This part e) of Definition 12.1.1 represents a subtle point in the mathematical model. According to Itô's formula the equation (12.1.4) would lead to

$$dV(t) = \theta(t) \cdot dX(t) + X(t) \cdot d\theta(t) + d\theta(t) \cdot dX(t)$$

if $\theta(t)$ was also an Itô process. However, the requirement (12.1.6) stems from the corresponding discrete time model: If investments $\theta(t_k)$ are made at discrete times $t = t_k$, then the increase in the wealth $\Delta V(t_k) = V(t_{k+1}) - V(t_k)$ should be given by

$$\Delta V(t_k) = \theta(t_k) \cdot \Delta X(t_k) \qquad (12.1.12)$$

where $\Delta X(t_k) = X(t_{k+1}) - X(t_k)$ is the change in prices, provided that no money is brought in or taken out from the system i.e. provided the portfolio is *self-financing*. If we consider our continuous time model as a limit of the discrete time case as $\Delta t_k = t_{k+1} - t_k$ goes to 0, then (12.1.6) (with the Itô interpretation of the integral) follows from (12.1.12).

f) Note that if θ is self-financing for $X(t)$ and

$$\overline{V}^\theta(t) = \theta(t) \cdot \overline{X}(t) = \xi(t)V^\theta(t) \qquad (12.1.13)$$

is the value process of the *normalized* market, then by Itô's formula and (12.1.11) we have

$$
\begin{aligned}
d\overline{V}^{\theta}(t) &= \xi(t)dV^{\theta}(t) + V^{\theta}(t)d\xi(t) \\
&= \xi(t)\theta(t)dX(t) - \rho(t)\xi(t)V^{\theta}(t)dt \\
&= \xi(t)\theta(t)[dX(t) - \rho(t)X(t)dt] \\
&= \theta(t)d\overline{X}(t) \ .
\end{aligned} \tag{12.1.14}
$$

Hence θ is also self-financing for the normalized market.

Remark. Note that by combining (12.1.4) and (12.1.6) we get

$$
\theta_0(t)X_0(t) + \sum_{i=1}^{n} \theta_i(t)X_i(t)
$$

$$
= V^{\theta}(0) + \int_0^t \theta_0(s)dX_0(s) + \sum_{i=1}^{n} \int_0^t \theta_i(s)dX_i(s) \ . \tag{12.1.15}
$$

Assume that $\theta_0(t)$ is an Itô process. Then, if we put

$$
Y_0(t) = \theta_0(t)X_0(t) \ ,
$$

we get

$$
dY_0(t) = \rho(t)Y_0(t)dt + dA(t) \ ,
$$

where

$$
A(t) = \sum_{i=1}^{n} \left(\int_0^t \theta_i(s)dX_i(s) - \theta_i(t)X_i(t) \right) . \tag{12.1.16}
$$

This equation has the solution

$$
\xi(t)Y_0(t) = \theta_0(0) + \int_0^t \xi(s)dA(s)
$$

or

$$
\theta_0(t) = \theta_0(0) + \int_0^t \xi(s)dA(s) \ .
$$

Using integration by parts we may rewrite this as

$$
\theta_0(t) = \theta_0(0) + \xi(t)A(t) - A(0) - \int_0^t A(s)d\xi(s)
$$

or

$$
\theta_0(t) = V^{\theta}(0) + \xi(t)A(t) + \int_0^t \rho(s)A(s)\xi(s)ds \ . \tag{12.1.17}
$$

This argument goes both ways, in the sense that if we define $\theta_0(t)$ by (12.1.17) (which is an Itô process), then (12.1.15) holds.

Therefore, if $\theta_1(t), \ldots, \theta_n(t)$ are chosen, we can always make the portfolio $\theta(t) = (\theta_0(t), \theta_1(t), \ldots, \theta_n(t))$ self-financing by choosing $\theta_0(t)$ according to (12.1.17). Moreover, we are free to choose the initial value $V^\theta(0)$ of the portfolio.

We now make the following definition

Definition 12.1.2 *A portfolio $\theta(t)$ which satisfies (12.1.5) and which is self-financing is called* admissible *if the corresponding value process $V^\theta(t)$ is (t, ω) a.s. lower bounded, i. e. there exists $K = K(\theta) < \infty$ such that*

$$V^\theta(t, \omega) \geq -K \qquad \text{for a.a. } (t, \omega) \in [0, T] \times \Omega . \tag{12.1.18}$$

This is the analogue of a *tame* portfolio in the context of Karatzas (1996). The restriction (12.1.18) reflects a natural condition in real life finance: There must be a limit to how much debt the creditors can tolerate. See Example 12.1.4.

Definition 12.1.3 *An admissible portfolio $\theta(t)$ is called an* arbitrage *(in the market $\{X_t\}_{t \in [0,T]}$) if the corresponding value process $V^\theta(t)$ satisfies $V^\theta(0) = 0$ and*

$$V^\theta(T) \geq 0 \quad \text{a.s. and} \quad P[V^\theta(T) > 0] > 0 .$$

In other words, $\theta(t)$ is an arbitrage if it gives an increase in the value from time $t = 0$ to time $t = T$ a.s., and a strictly positive increase with positive probability. So $\theta(t)$ generates a profit without any risk of losing money.

Intuitively, the existence of an arbitrage is a sign of lack of equilibrium in the market: No real market equilibrium can exist in the long run if there are arbitrages there. Therefore it is important to be able to determine if a given market allows an arbitrage or not. Not surprisingly, this question turns out to be closely related to what conditions we pose on the portfolios that should be allowed to use. We have defined our admissible portfolios in Definition 12.1.2 above, where condition (12.1.18) was motivated from a modelling point of view. One could also obtain a mathematically sensible theory with other conditions instead, for example with the L^2-conditions

$$E_Q \left[\int_0^T \sum_{i=1}^n |\theta_i(t)\sigma_i(t)|^2 dt \right] < \infty \tag{12.1.19}$$

where Q is the probability measure defined in Lemma 12.2.3.

In any case, *some* additional conditions are required on the self-financial portfolios: If we only require the portfolio to be self-financing (and satisfying (12.1.5)) we can generate virtually any final value $V(T)$, as the next example illustrates:

Example 12.1.4 Consider the following market

$$dX_0(t) = 0, \quad dX_1(t) = dB(t), \qquad 0 \le t \le T = 1 \,.$$

Let

$$Y(t) = \int_0^t \frac{dB(s)}{\sqrt{1-s}} \qquad \text{for } 0 \le t < 1 \,.$$

By Corollary 8.5.5 there exists a Brownian motion $\widehat{B}(t)$ such that

$$Y(t) = \widehat{B}(\beta_t) \,,$$

where

$$\beta_t = \int_0^t \frac{ds}{1-s} = \ln\left(\frac{1}{1-t}\right) \qquad \text{for } 0 \le t < 1 \,.$$

Let $M < \infty$ be a given constant and define

$$\tau := \tau_M := \inf\{t > 0; \widehat{B}(t) = M\}$$

and

$$\alpha := \alpha_M := \inf\{t > 0; Y(t) = M\} \,.$$

Then

$$\tau < \infty \quad \text{a.s.} \quad \text{(Exercise 7.4a))}$$

and

$$\tau = \ln\left(\frac{1}{1-\alpha}\right), \quad \text{so } \alpha < 1 \text{ a.s.}$$

Define

$$\theta_1(t) = \begin{cases} \frac{1}{\sqrt{1-t}} & \text{for } 0 \le t < \alpha \\ 0 & \text{for } \alpha \le t \le 1 \end{cases}$$

and choose $\theta_0(t)$ according to (12.1.17) and such that $V(0) = 0$. Then $\theta(t) = (\theta_0(t), \theta_1(t))$ is self-financing and the corresponding value process is given by

$$V(t) = \int_0^{t \wedge \alpha} \frac{dB(s)}{\sqrt{1-s}} = Y(t \wedge \alpha) \qquad \text{for } 0 \le t \le 1 \,.$$

In particular,

$$V(1) = Y(\alpha) = M \quad \text{a.s.}$$

so this portfolio generates the terminal wealth M a.s. although the initial wealth is 0. In this case condition (12.1.5) reduces to

$$\int\limits_0^1 \theta_1^2(s)ds < \infty \quad \text{a.s.}$$

Now

$$\int\limits_0^1 \theta_1^2(s)ds = \int\limits_0^\alpha \frac{ds}{1-s} = \ln\left(\frac{1}{1-\alpha}\right) = \tau < \infty \quad \text{a.s.} ,$$

so (12.1.5) holds. But $\theta(t)$ is *not* admissible, because $V(t) = Y(t \wedge \alpha) = \widehat{B}(\ln(\frac{1}{1-t\wedge\alpha}))$ is not (t,ω)-a.s. lower bounded for $(t,\omega) \in [0,1] \times \Omega$. Note that $\theta(t)$ does not satisfy (12.1.19) either, because in this case $Q = P$ and

$$E\left[\int\limits_0^1 \theta_1^2(s)ds\right] = E[\tau] = \infty$$

(Exercise 7.4b).

This example illustrates that with portfolios only required to be self-financing and satisfy (12.1.5) one can generate virtually any terminal value $V(T,\omega)$ from $V(0) = 0$, even when the risky price process $X_1(t)$ is Brownian motion. This clearly contradicts the real life situation in finance, so a realistic mathematical model must put stronger restrictions than (12.1.5) on the portfolios allowed. One such natural restriction is (12.1.18), as we have adopted.

To emphasize the phenomenon illustrated by this example, we state the following striking result, which is due to Dudley (1977):

Theorem 12.1.5 Let F be an $\mathcal{F}_T^{(m)}$-measurable random variable and let $B(t)$ be m-dimensional Brownian motion. Then there exists $\phi \in \mathcal{W}^m$ such that

$$F(\omega) = \int\limits_0^T \phi(t,\omega)dB(t) . \tag{12.1.20}$$

Note that ϕ is not unique. See Exercise 3.4.22 in Karatzas and Shreve (1991). See also Exercise 12.4.

This implies that for *any* constant z there exists $\phi \in \mathcal{W}^m$ such that

$$F(\omega) = z + \int\limits_0^T \phi(t,\omega)dB(t) .$$

Thus, if we let $m = n$ and interpret $B_1(t) = X_1(t), \ldots, B_n(t) = X_n(t)$ as prices, and put $X_0(t) \equiv 1$, this means that we can, with *any* initial fortune z, generate *any* $\mathcal{F}_T^{(m)}$-measurable final value $F = V(T)$, as long as we are allowed to choose the portfolio ϕ freely from $\mathcal{W}^m$. This again underlines the need for some extra restriction on the family of portfolios allowed, like condition (12.1.18).

How can we decide if a given market $\{X(t)\}_{t \in [0,T]}$ allows an arbitrage or not? The following simple result is useful:

Lemma 12.1.6 *Suppose there exists a measure Q on $\mathcal{F}_T^{(m)}$ such that $P \sim Q$ and such that the normalized price process $\{\overline{X}(t)\}_{t \in [0,T]}$ is a local martingale w.r.t. Q. Then the market $\{X(t)\}_{t \in [0,T]}$ has no arbitrage.*

Proof. Suppose $\theta(t)$ is an arbitrage for $\{\overline{X}(t)\}_{t \in [0,T]}$. Let $\overline{V}^\theta(t)$ be the corresponding value process for the normalized market with $\overline{V}^\theta(0) = 0$. Then $\overline{V}^\theta(t)$ is a lower bounded local martingale w.r.t. Q, by (12.1.14). Therefore $\overline{V}^\theta(t)$ is a *supermartingale* w.r.t. Q, by Exercise 7.12. Hence

$$E_Q[\overline{V}^\theta(T)] \leq V^\theta(0) = 0 . \tag{12.1.21}$$

But since $\overline{V}^\theta(T, \omega) \geq 0$ a.s. P we have $\overline{V}^\theta(T, \omega) \geq 0$ a.s. Q (because $Q \ll P$) and since $P[\overline{V}^\theta(T) > 0] > 0$ we have $Q[\overline{V}^\theta(T) > 0] > 0$ (because $P \ll Q$). This implies that

$$E_Q[\overline{V}^\theta(T)] > 0 ,$$

which contradicts (12.1.21). Hence arbitrages do not exist for the normalized price process $\{\overline{X}(t)\}$. It follows that $\{X(t)\}$ has no arbitrage. (Exercise 12.1).
□

Definition 12.1.7 *A measure $Q \sim P$ such that the normalized process $\{\overline{X}(t)\}_{t \in [0,T]}$ is a (local) martingale w.r.t. Q is called* an equivalent (local) martingale measure.

Thus Lemma 12.1.6 states that if there exists an equivalent local martingale measure then the market has no arbitrage. In fact, then the market also satisfies the stronger condition "no free lunch with vanishing risk" (NFLVR). Conversely, if the market satisfies the NFLVR condition, then there exists an equivalent martingale measure. See Delbaen and Schachermayer (1994), (1995), (1997), Leventals and Skorohod (1995) and the references therein. Here we will settle with a weaker result, which nevertheless is good enough for many applications:

Theorem 12.1.8 a)   *Suppose there exists a process $u(t, \omega) \in \mathcal{V}^m(0, T)$ such that, with $\widehat{X}(t, \omega) = (X_1(t, \omega), \dots, X_n(t, \omega))$,*

$$\sigma(t, \omega)u(t, \omega) = \mu(t, \omega) - \rho(t, \omega)\widehat{X}(t, \omega) \qquad \text{for a.a. } (t, \omega) \quad (12.1.22)$$

and such that

$$E\left[\exp\left(\tfrac{1}{2}\int_0^T u^2(t,\omega)dt\right)\right] < \infty .\qquad(12.1.23)$$

Then the market $\{X(t)\}_{t\in[0,T]}$ has no arbitrage.

b)   *(Karatzas (1996), Th. 0.2.4)*
   Conversely, if the market $\{X(t)\}_{t\in[0,T]}$ has no arbitrage, then there exists an $\mathcal{F}_t^{(m)}$-adapted, (t,ω)-measurable process $u(t,\omega)$ such that

$$\sigma(t,\omega)u(t,\omega) = \mu(t,\omega) - \rho(t,\omega)\widehat{X}(t,\omega)$$

   for a.a. (t,ω).

Proof. a) We may assume that $\{X(t)\}$ is normalized, i.e. that $\rho = 0$ (Exercise 12.1). Define the measure $Q = Q_u$ on $\mathcal{F}_T^{(m)}$ by

$$dQ(\omega) = \exp\left(-\int_0^T u(t,\omega)dB(t) - \tfrac{1}{2}\int_0^T u^2(t,\omega)dt\right)dP(\omega) .\qquad(12.1.24)$$

Then $Q \sim P$ and by the Girsanov theorem II (Theorem 8.6.4) the process

$$\widetilde{B}(t) := \int_0^t u(s,\omega)ds + B(t)\qquad(12.1.25)$$

is a Q-Brownian motion and in terms of $\widetilde{B}(t)$ we have

$$dX_i(t) = \mu_i dt + \sigma_i dB(t) = \sigma_i d\widetilde{B}(t); \qquad 1 \le i \le n .$$

Hence $X(t)$ is a local Q-martingale and the conclusion follows from Lemma 12.1.6.

b) Conversely, assume that the market has no arbitrage and is normalized. For $t \in [0,T]$, $\omega \in \Omega$ let

$$\begin{aligned}
F_t &= \{\omega;\ \text{the equation (12.1.22) has no solution}\} \\
&= \{\omega;\ \mu(t,\omega)\ \text{does not belong to the linear span of the columns} \\
&\qquad \text{of } \sigma(t,\omega)\} \\
&= \{\omega;\ \exists v = v(t,\omega)\ \text{with}\ \sigma^T(t,\omega)v(t,\omega) = 0\ \text{and} \\
&\qquad v(t,\omega)\cdot\mu(t,\omega) \ne 0\} .
\end{aligned}$$

Define
$$\theta_i(t,\omega) = \begin{cases} \operatorname{sign}(v(t,\omega)\cdot\mu(t,\omega))v_i(t,\omega) & \text{for } \omega \in F_t \\ 0 & \text{for } \omega \notin F_t \end{cases}$$

for $1 \leq i \leq n$ and $\theta_0(t, \omega)$ according to (12.1.17) and such that $V^\theta(0) = 0$. Since $\sigma(t, \omega)$, $\mu(t, \omega)$ are $\mathcal{F}_t^{(m)}$-adapted and (t, ω)-measurable, it follows that we can choose $\theta(t, \omega)$ to be $\mathcal{F}_t^{(m)}$-adapted and (t, ω)-measurable also. Moreover, $\theta(t, \omega)$ is self-financing and it generates the following terminal value

$$
V^\theta(T, \omega) = \int_0^T \sum_{i=1}^n \theta_i(s, \omega) dX_i(s)
$$

$$
= \int_0^T \mathcal{X}_{F_s}(\omega) |v(s, \omega) \cdot \mu(s, \omega)| ds + \int_0^T \sum_{j=1}^m \left(\sum_{i=1}^n \theta_i(s, \omega) \sigma_{ij}(s, \omega) \right) dB_j(s)
$$

$$
= \int_0^T \mathcal{X}_{F_s}(\omega) |v(s, \omega) \cdot \mu(s, \omega)| ds
$$

$$
+ \int_0^T \mathrm{sign}(v(s, \omega) \cdot \mu(s, \omega)) \mathcal{X}_{F_s}(\omega) \sigma^T(s, \omega) v(s, \omega) dB(s)
$$

$$
= \int_0^T \mathcal{X}_{F_s}(\omega) |v(s, \omega) \cdot \mu(s, \omega)| ds \geq 0 \qquad \text{a.s. .}
$$

Since the market has no arbitrage we must have that

$$
\mathcal{X}_{F_t}(\omega) = 0 \qquad \text{for a.a. } (t, \omega)
$$

i.e. that (12.1.22) has a solution for a.a. (t, ω).   □

Example 12.1.9 a) Consider the price process $X(t)$ given by

$$
dX_0(t) = 0, \quad dX_1(t) = 2dt + dB_1(t), \quad dX_2(t) = -dt + dB_1(t) + dB_2(t) .
$$

In this case we have

$$
\mu = \begin{bmatrix} 2 \\ -1 \end{bmatrix}, \quad \sigma = \begin{bmatrix} 1 & 0 \\ 1 & 1 \end{bmatrix}
$$

and the system $\sigma u = \mu$ has the unique solution

$$
u = \begin{bmatrix} u_1 \\ u_2 \end{bmatrix} = \begin{bmatrix} 2 \\ -3 \end{bmatrix} .
$$

From Theorem 12.1.8a) we conclude that $X(t)$ has no arbitrage.

b) Next, consider the price process $Y(t)$ given by

$$
dY_0(t) = 0 , \quad dY_1(t) = 2dt + dB_1(t) + dB_2(t) ,
$$
$$
dY_2(t) = -dt - dB_1(t) - dB_2(t) .
$$

Here the system of equations $\sigma u = \mu$ gets the form

$$\begin{bmatrix} 1 & 1 \\ -1 & -1 \end{bmatrix} \begin{bmatrix} u_1 \\ u_2 \end{bmatrix} = \begin{bmatrix} 2 \\ -1 \end{bmatrix}$$

which has no solutions. So the market has an arbitrage, according to Theorem 12.1.8 b). Indeed, if we choose

$$\theta(t) = (\theta_0(t), 1, 1)$$

we get

$$V^\theta(T) = V^\theta(0) + \int_0^T 2dt + dB_1(t) + dB_2(t) - dt - dB_1(t) - dB_2(t)$$
$$= V^\theta(0) + T .$$

In particular, if we choose $\theta_0(t)$ according to (12.1.17) and such that $V^\theta(0) = 0$ then θ will be an arbitrage.

12.2 Attainability and Completeness

We start this section by stating without proof the following useful result, which is a special case of Proposition 17.1 in Yor (1997):

Lemma 12.2.1 *Suppose a process* $u(t, \omega) \in \mathcal{V}^m(0, T)$ *satisfies the condition*

$$E\left[\exp\left(\tfrac{1}{2} \int_0^T u^2(s, \omega) ds \right) \right] < \infty . \qquad (12.2.1)$$

Define the measure $Q = Q_u$ *on* $\mathcal{F}_T^{(m)}$ *by*

$$dQ(\omega) = \exp\left(- \int_0^T u(t, \omega) dB(t) - \tfrac{1}{2} \int_0^T u^2(t, \omega) dt \right) dP(\omega) . \qquad (12.2.2)$$

Then

$$\tilde{B}(t) := \int_0^t u(s, \omega) ds + B(t) \qquad (12.2.3)$$

is an $\mathcal{F}_t^{(m)}$ *-martingale (and hence an* $\mathcal{F}_t^{(m)}$ *-Brownian motion) w.r.t.* Q *and any* $F \in L^2(\mathcal{F}_T^{(m)}, Q)$ *has a unique representation*

$$F(\omega) = E_Q[F] + \int_0^T \phi(t,\omega)d\widetilde{B}(t) ,\qquad (12.2.4)$$

where $\phi(t,\omega)$ is an $\mathcal{F}_t^{(m)}$-adapted, (t,ω)-measurable $\mathbf{R}^m$-valued process such that

$$E_Q\left[\int_0^T \phi^2(t,\omega)dt\right] < \infty .\qquad (12.2.5)$$

Remark.   a) Note that the filtration $\{\widetilde{\mathcal{F}}_t^{(m)}\}$ generated by $\{\widetilde{B}(t)\}$ is *contained* in $\{\mathcal{F}_t^{(m)}\}$ (by (12.2.3)), but not necessarily equal to $\{\mathcal{F}_t^{(m)}\}$. Therefore the representation (12.2.4) is not a consequence of the Itô representation theorem (Theorem 4.3.3) or the Dudley theorem (Theorem 12.1.5), which in this setting would require that F be $\widetilde{\mathcal{F}}_T^{(m)}$-measurable.

b) To prove that $\widetilde{B}(t)$ is an $\mathcal{F}_t^{(m)}$-martingale w.r.t Q, we apply Itô's formula to the process

$$Y(t) := Z(t)\widetilde{B}(t) ,$$

where

$$Z(t) = \exp\left(-\int_0^t u(s,\omega)dB(s) - \tfrac{1}{2}\int_0^t u^2(s,\omega)ds\right) ,$$

and use the Bayes formula, Lemma 8.6.2. The details are left to the reader. (Exercise 12.5.)

Next we make the following simple, but useful observation:

Lemma 12.2.2 *Let* $\overline{X}(t) = \xi(t)X(t)$ *be the normalized price process, as in (12.1.8)–(12.1.11). Suppose* $\theta(t)$ *is an admissible portfolio for the market* $\{X(t)\}$ *with value process*

$$V^\theta(t) = \theta(t) \cdot X(t) .\qquad (12.2.6)$$

Then $\theta(t)$ *is also an admissible portfolio for the normalized market* $\{\overline{X}(t)\}$ *with value process*

$$\overline{V}^\theta(t) := \theta(t) \cdot \overline{X}(t) = \xi(t)V^\theta(t)\qquad (12.2.7)$$

and vice versa.
   In other words,

$$V^\theta(t) = V^\theta(0) + \int_0^t \theta(s)dX(s) ;\quad 0 \le t \le T\qquad (12.2.8)$$

$$\Updownarrow$$

$$\xi(t)V^\theta(t) = V^\theta(0) + \int_0^t \theta(s)d\overline{X}(s) ;\ \ 0 \le t \le T\qquad (12.2.9)$$

Proof. Note that $\overline{V}^\theta(t)$ is lower bounded if and only if $V^\theta(t)$ is lower bounded (since $\rho(t)$ is bounded). Consider first the market consisting of the price process $X(t)$. Let $\theta(t)$ be an admissible portfolio for this market with value process $V^\theta(t)$. Then

$$\overline{V}^\theta(t) = \theta(t) \cdot \overline{X}(t) = \xi(t)V^\theta(t) \tag{12.2.10}$$

and since $\theta(t)$ is self-financing for the market $\{X(t)\}$ we have, by (12.1.14),

$$d\overline{V}^\theta(t) = \theta(t)d\overline{X}(t) . \tag{12.2.11}$$

Hence $\theta(t)$ is also admissible for $\{\overline{X}(t)\}$ and $\overline{V}^\theta(t) = V^\theta(0) + \int_0^t \theta(s)d\overline{X}(s)$, which shows that (12.2.8) implies (12.2.9).

The argument goes both ways, so the lemma is proved. $\square$

Before we proceed we note the following useful result:

Lemma 12.2.3 *Suppose there exists an m-dimensional process $u(t,\omega) \in \mathcal{V}^m(0,T)$ such that, with $\widehat{X}(t,\omega) = (X_1(t,\omega),\ldots,X_n(t,\omega))$,*

$$\sigma(t,\omega)u(t,\omega) = \mu(t,\omega) - \rho(t,\omega)\widehat{X}(t,\omega) \qquad \textit{for a.a. } (t,\omega) \tag{12.2.12}$$

and

$$E\left[\exp\left(\tfrac{1}{2}\int_0^T u^2(s,\omega)ds\right)\right] < \infty . \tag{12.2.13}$$

Define the measure $Q = Q_u$ and the process $\widetilde{B}(t)$ as in (12.2.2), (12.2.3), respectively. Then $\widetilde{B}$ is a Brownian motion w.r.t. Q and in terms of $\widetilde{B}$ we have the following representation of the normalized market $\overline{X}(t) = \xi(t)X(t)$:

$$d\overline{X}_0(t) = 0 \tag{12.2.14}$$
$$d\overline{X}_i(t) = \xi(t)\sigma_i(t)d\widetilde{B}(t) ; \qquad 1 \le i \le n . \tag{12.2.15}$$

In particular, if $\int_0^T E_Q[\xi^2(t)\sigma_i^2(t)]dt < \infty$, then Q is an equivalent martingale measure (Definition 12.1.7).

In any case the normalized value process $\overline{V}^\theta(t)$ of an admissible portfolio θ is a local Q-martingale given by

$$d\overline{V}^\theta(t) = \xi(t)\sum_{i=1}^n \theta_i(t)\sigma_i(t)d\widetilde{B}(t) = \xi(t)\widehat{\theta}(t)\sigma(t)d\widetilde{B}(t) , \tag{12.2.16}$$

where $\widehat{\theta}(t) = (\theta_1(t),\ldots,\theta_n(t)) \in \mathbf{R}^{1\times n}$.

Proof. The first statement follows from the Girsanov theorem. To prove the representation (12.2.15) we compute

$$
\begin{aligned}
d\overline{X}_i(t) &= d(\xi(t)X_i(t)) = \xi(t)dX_i(t) + X_i(t)d\xi(t) \\
&= \xi(t)[(\mu_i(t) - \rho(t)X_i(t))dt + \sigma_i(t)dB(t)] \\
&= \xi(t)[(\mu_i(t) - \rho(t)X_i(t))dt + \sigma_i(t)(d\widetilde{B}(t) - u_i(t)dt)] \\
&= \xi(t)\sigma_i(t)d\widetilde{B}(t) \ .
\end{aligned}
$$

In particular, if $\int\limits_0^T E_Q[\xi^2(t)\sigma_i^2(t)]dt < \infty$, then $\overline{X}_i(t)$ is a martingale w.r.t. Q
by Corollary 3.2.6.

Finally, the representation (12.2.16) follows from (12.2.11) and (12.2.15).

$\square$

Note. **From now on we assume that there exists a (not necessarily**
unique) process $u(t, \omega) \in \mathcal{V}^m(0, T)$ satisfying (12.2.12) and (12.2.13)
and we let $Q = Q_u$ and $\widetilde{B}$ be as in (12.2.2), (12.2.3), as described
in Lemma 12.2.3. In particular, by Theorem 12.1.8 this guarantees
that the market $\{X(t)\}_{t\in[0,T]}$ has no arbitrage.

Definition 12.2.4

a)   *A (European) contingent T-claim (or just a T-claim or claim) is a lower*
     bounded $\mathcal{F}_T^{(m)}$-measurable random variable $F(\omega) \in L^2(Q)$.

b)   *We say that the claim $F(\omega)$ is attainable (in the market $\{X(t)\}_{t\in[0,T]}$)*
     if there exists an admissible portfolio $\theta(t)$ and a real number z such that

$$
F(\omega) = V_z^\theta(T) := z + \int\limits_0^T \theta(t)dX(t) \quad a.s.
$$

     and such that

$$
\overline{V}_z^\theta(t) = z + \int\limits_0^t \xi(s) \sum_{i=1}^n \theta_i(s)\sigma_i(s)d\widetilde{B}(s) ; \quad 0 \le t \le T \text{ is a } Q\text{-martingale} .
$$

     If such a $\theta(t)$ exists, we call it a replicating *or* hedging *portfolio for F.*
     Note that we necessarily have $z = E_Q[F\xi(T)]$.

c)   *The market $\{X(t)\}_{t\in[0,T]}$ is called* complete *if every T-claim is attain-*
     able.

In other words, a claim $F(\omega)$ is attainable if there exists a real number z
such that if we start with z as our initial fortune we can find an admissible
portfolio $\theta(t)$ which generates a value $V_z^\theta(T)$ at time T which a.s. equals F:

$$
V_z^\theta(T, \omega) = F(\omega) \quad \text{for a.a. } \omega \ .
$$

In addition we require that the corresponding normalized value process $\overline{V}^{\theta}(t)$, which has the representation (12.2.16), is a *martingale* and not just a *local martingale* w.r.t. Q.

Remark. If we drop the martingale condition in Definition 12.2.4b) then the replicating portfolio θ need not be unique. See Exercise 12.4.

   What claims are attainable? Which markets are complete? These are important, but difficult questions in general. We will give some partial answers.
   We are now ready for the main result of this section:

Theorem 12.2.5 *The market* $\{X(t)\}$ *is complete if and only if* $\sigma(t,\omega)$ *has a left inverse* $\Lambda(t,\omega)$ *for a.a.* (t,ω), *i.e. there exists an* $\mathcal{F}_t^{(m)}$-*adapted matrix valued process* $\Lambda(t,\omega) \in \mathbf{R}^{m \times n}$ *such that*

$$\Lambda(t,\omega)\sigma(t,\omega) = I_m \qquad \text{for a.a. } (t,\omega) \, . \qquad (12.2.17)$$

Remark. Note that the property (12.2.17) is equivalent to the property

$$\text{rank } \sigma(t,\omega) = m \qquad \text{for a.a. } (t,\omega) \, . \qquad (12.2.18)$$

Proof of Theorem 12.2.5. (i) Assume that (12.2.17) hold. Let Q and $\tilde{B}$ be as in (12.2.2), (12.2.3). Let F be a T-claim. We want to prove that there exists an admissible portfolio $\theta(t) = (\theta_0(t), \ldots, \theta_n(t))$ and a real number z such that if we put

$$V_z^{\theta}(t) = z + \int_0^t \theta(s)dX(s) \, ; \qquad 0 \leq t \leq T$$

then $\overline{V}_z^{\theta}(t)$ is a Q-martingale and

$$V_z^{\theta}(T) = F(\omega) \quad \text{a.s.}$$

By (12.2.16) this is equivalent to

$$\xi(T)F(\omega) = \overline{V}^{\theta}(T) = z + \int_0^T \xi(t) \sum_{i=1}^n \theta_i(t)\sigma_i(t)d\tilde{B}(t) \, .$$

By Lemma 12.2.1 we have a unique representation

$$\xi(T)F(\omega) = E_Q[\xi(T)F] + \int_0^T \phi(t,\omega)d\tilde{B}(t) = E_Q[\xi(T)F] + \int_0^T \sum_{j=1}^m \phi_j(t,\omega)d\tilde{B}_j(t)$$

for some $\phi(t,\omega) = (\phi_1(t,\omega), \ldots, \phi_m(t,\omega)) \in \mathbf{R}^m$. Hence we put

$$z = E_Q[\xi(T)F]$$

and we choose $\widehat{\theta}(t) = (\theta_1(t), \ldots, \theta_n(t))$ such that

$$\xi(t) \sum_{i=1}^{n} \theta_i(t)\sigma_{ij}(t) = \phi_j(t) ; \qquad 1 \leq j \leq m$$

i.e. such that

$$\xi(t)\widehat{\theta}(t)\sigma(t) = \phi(t) .$$

By (12.2.17) this equation in $\widehat{\theta}(t)$ has the solution

$$\widehat{\theta}(t, \omega) = X_0(t)\phi(t, \omega)\Lambda(t, \omega) .$$

By choosing θ_0 according to (12.1.17) the portfolio becomes self-financing. We can also choose $V^\theta(0) = z$. Moreover, since $\xi(t)V_z^\theta(t) = z + \int\limits_0^t \theta(s)dX(s) = z + \int\limits_0^t \phi(s)d\widetilde{B}(s)$ is a Q-martingale, we get the useful formula

$$\xi(t)V_z^\theta(t) = E_Q[\xi(T)V_z^\theta(T)|\widetilde{\mathcal{F}}_t^{(m)}] = E_Q[\xi(T)F|\widetilde{\mathcal{F}}_t^{(m)}] , \qquad (12.2.19)$$

where $\widetilde{\mathcal{F}}_t^{(m)}$ is the σ-algebra generated by $\widetilde{B}(s)$; $s \leq t$. In particular, $V_z^\theta(t)$ is lower bounded. Hence the market $\{X(t)\}$ is complete.

(ii) Conversely, assume that $\{X(t)\}$ is complete. Then $\{\overline{X}(t)\}$ is complete, so we may assume that $\rho = 0$. The calculation in part a) shows that the value process $V_z^\theta(t)$ generated by an admissible portfolio $\theta(t) = (\theta_0(t), \theta_1(t), \ldots, \theta_n(t))$ with $V_z^\theta(0) = z$ is

$$V_z^\theta(t) = z + \int\limits_0^t \sum_{j=1}^m \left(\sum_{i=1}^n \theta_i\sigma_{ij} \right) d\widetilde{B}_j = z + \int\limits_0^t \widehat{\theta}\sigma \, d\widetilde{B} , \qquad (12.2.20)$$

where $\widehat{\theta}(t) = (\theta_1(t), \ldots, \theta_n(t))$.

Since $\{X(t)\}$ is complete we can hedge any T-claim. Choose an $\mathcal{F}_t^{(m)}$-adapted process $\phi(t, \omega) \in \mathbf{R}^m$ such that $E_Q[\int\limits_0^T \phi^2(t, \omega)dt] < \infty$ and define $F(\omega) := \int\limits_0^T \phi(t, \omega)d\widetilde{B}(t)$. Then $E_Q[F^2] < \infty$ so by completeness there exists for an admissible portfolio $\theta = (\theta_0, \widehat{\theta})$ such that $V^\theta(t) = \int\limits_0^t \widehat{\theta}\sigma d\widetilde{B}$ is a Q-martingale and

$$F(\omega) = V^\theta(T) = \int_0^T \widehat{\theta}\sigma \, d\widetilde{B} \,.$$

But then

$$E_Q[F \mid \widetilde{\mathcal{F}}_t^{(m)}] = \int_0^t \phi \, d\widetilde{B}$$

a.s. for all $t \in [0, T]$, where $\widetilde{\mathcal{F}}_t^{(m)}$ is the σ-algebra generated by $\widetilde{B}(s)$; $s \leq t$.

Hence by uniqueness we have $\phi(t, \omega) = \widehat{\theta}(t, \omega)\sigma(t, \omega)$ for a.a. (t, ω). This implies that $\phi(t, \omega)$ belongs to the linear span of the rows $\{\sigma_i(t, \omega)\}_{i=1}^n$ of $\sigma(t, \omega)$. Since this applies to, e.g., all processes ϕ of the exponential form given in (4.3.1), it follows by Lemma 4.3.2 that the linear span of $\{\sigma_i(t, \omega)\}_{i=1}^n$ is the whole of $\mathbf{R}^m$ for a.a. (t, ω). So $\text{rank}\,\sigma(t, \omega) = m$ and there exists $\Lambda(t, \omega) \in \mathbf{R}^{m \times n}$ such that

$$\Lambda(t, \omega)\sigma(t, \omega) = I_m \,.$$

$\square$

Corollary 12.2.6 (a) *If $n = m$ then the market is complete if and only if $\sigma(t, \omega)$ is invertible for a.a. (t, ω).*

(b) *If the market is complete, then*

$$\text{rank}\,\sigma(t, \omega) = m \qquad \text{for a.a. } (t, \omega) \,.$$

In particular, $n \geq m$.
Moreover, the process $u(t, \omega)$ satisfying (12.2.12) is unique.

Proof. (a) is a direct consequence of Theorem 12.2.5, since the existence of a left inverse implies invertibility when $n = m$. The existence of a left inverse of an $n \times m$ matrix is only possible if the rank is equal to m, which again implies that $n \geq m$. Moreover, the only solution $u(t, \omega)$ of (12.2.12) is given by

$$u(t, \omega) = \Lambda(t, \omega)[\mu(t, \omega) - \rho(t, \omega)\widehat{X}(t, \omega)] \,.$$

This shows (b). $\square$

Example 12.2.7 Define $X_0(t) \equiv 1$ and

$$\begin{bmatrix} dX_1(t) \\ dX_2(t) \\ dX_3(t) \end{bmatrix} = \begin{bmatrix} 1 \\ 2 \\ 3 \end{bmatrix} dt + \begin{bmatrix} 1 & 0 \\ 0 & 1 \\ 1 & 1 \end{bmatrix} \begin{bmatrix} dB_1(t) \\ dB_2(t) \end{bmatrix} \,.$$

Then $\rho = 0$ and the equation (12.2.12) gets the form

$$\sigma u = \begin{bmatrix} 1 & 0 \\ 0 & 1 \\ 1 & 1 \end{bmatrix} \begin{bmatrix} u_1 \\ u_2 \end{bmatrix} = \begin{bmatrix} 1 \\ 2 \\ 3 \end{bmatrix}$$

which has the unique solution $u_1 = 1$, $u_2 = 2$. Since u is constant, it is clear that (12.2.12) and (12.2.13) hold. It is immediate that $\operatorname{rank}\sigma = 2$, so (12.2.18) holds and the market is complete by Theorem 12.2.5.

Since

$$\begin{bmatrix} 1 & 0 & 0 \\ 0 & 1 & 0 \end{bmatrix} \begin{bmatrix} 1 & 0 \\ 0 & 1 \\ 1 & 1 \end{bmatrix} = \begin{bmatrix} 1 & 0 \\ 0 & 1 \end{bmatrix} = I_2 \ ,$$

we see that in this case

$$\Lambda = \begin{bmatrix} 1 & 0 & 0 \\ 0 & 1 & 0 \end{bmatrix}$$

is a left inverse of $\sigma = \begin{bmatrix} 1 & 0 \\ 0 & 1 \\ 1 & 1 \end{bmatrix}$.

Example 12.2.8 Let $X_0(t) \equiv 1$ and

$$dX_1(t) = 2dt + dB_1(t) + dB_2(t) \ .$$

Then $\mu = 2$, $\sigma = (1,1) \in \mathbf{R}^{1 \times 2}$, so $n = 1 < 2 = m$. Hence this market cannot be complete, by Corollary 12.2.6. So there exist T-claims which cannot be hedged. Can we find such a T-claim? Let $\theta(t) = (\theta_0(t), \theta_1(t))$ be an admissible portfolio. Then the corresponding value process $V_z^\theta(t)$ is given by (see (12.2.20))

$$V_z^\theta(t) = z + \int\limits_0^t \theta_1(s)(d\widetilde{B}_1(s) + d\widetilde{B}_2(s)) \ .$$

So if θ hedges a T-claim $F(\omega)$ we have that $V_z^\theta(t)$ is a Q-martingale and

$$F(\omega) = z + \int\limits_0^T \theta_1(s)(d\widetilde{B}_1(s) + d\widetilde{B}_2(s)) \ . \qquad (12.2.21)$$

Choose $F(\omega) = g(\widetilde{B}_1(T))$, where $g: \mathbf{R} \to \mathbf{R}$ is bounded. Then by the Itô representation theorem applied to the 2-dimensional Brownian motion $\widetilde{B}(t) = (\widetilde{B}_1(t), \widetilde{B}_2(t))$ there is a unique $\phi(t, \omega) = (\phi_1(t, \omega), \phi_2(t, \omega))$ such that $E_Q\left[\int_0^T (\phi_1^2(s) + \phi_2^2(s))ds\right] < \infty$ and

$$g(\widetilde{B}_1(T)) = E_Q[g(\widetilde{B}_1(T))] + \int\limits_0^T \phi_1(s)d\widetilde{B}_1(s) + \phi_2(s)d\widetilde{B}_2(s)$$

and by the Itô representation theorem applied to $\widetilde{B}_1(t)$, we must have $\phi_2 = 0$, i.e.

$$g(\widetilde{B}_1(T)) = E_Q[g(\widetilde{B}_1(T))] + \int_0^T \phi_1(s)d\widetilde{B}_1(s)$$

Comparing this with (12.2.21) and using the martingale property we get

$$z + \int_0^t \theta_1(s)(d\widetilde{B}_1(s) + d\widetilde{B}_2(s)) = E_Q[F|\mathcal{F}_t] = E_Q[F] + \int_0^t \phi_1(s)d\widetilde{B}_1(s) ,$$

which implies both that $\theta_1(s) = \phi_1(s)$ and that $\theta_1(s) = 0$ for a.a. (s, ω). This contradiction proves that $F(\omega) = g(\widetilde{B}_1(T))$ cannot be hedged if $g \neq 0$.

Remark. There is a striking characterization of completeness in terms of equivalent martingale measures, due to Harrison and Pliska (1983) and Jacod (1979):

A market $\{X(t)\}$ is complete if and only if there is *one and only one* equivalent martingale measure for the normalized market $\{\overline{X}(t)\}$.

(Compare this result with the equivalent martingale measure characterization of markets with no arbitrage/NFLVR, stated after Definition 12.1.7!)

12.3 Option Pricing

European Options

Let $F(\omega)$ be a T-claim. A *European option* on the claim F is a guarantee to be paid the amount $F(\omega)$ at time $t = T > 0$. How much would you be willing to pay at time $t = 0$ for such a guarantee? You could argue as follows: If I – the buyer of the option – pay the price y for this guarantee, then I have an initial fortune $-y$ in my investment strategy. With this initial fortune (debt) it must be possible to hedge to time T a value $V_{-y}^\theta(T, \omega)$ which, if the guaranteed payoff $F(\omega)$ is added, gives me a nonnegative result:

$$V_{-y}^\theta(T, \omega) + F(\omega) \geq 0 \quad \text{a.s.}$$

Thus the maximal price $p = p(F)$ the buyer is willing to pay is

(*Buyer's price of the (European) contingent claim F*)   (12.3.1)

$p(F) = \sup\{y;$ There exists an admissible portfolio φ

$$\text{such that } V_{-y}^\varphi(T, \omega) := -y + \int_0^T \varphi(s)dX(s) \geq -F(\omega) \text{ a.s.}\}$$

On the other hand, the seller of this guarantee could argue as follows:

If I – the seller – receive the price z for this guarantee, then I can use this as the initial value in an investment strategy. With this initial fortune it must be possible to hedge to time T a value $V_z^\varphi(T, \omega)$ which is not less than the amount $F(\omega)$ that I have promised to pay to the buyer:

$$V_z^\varphi(T, \omega) \geq F(\omega) \quad \text{a.s.}$$

Thus the minimal price $q = q(F)$ the seller is willing to accept is

(Seller's price of the (European) contingent claim F)   (12.3.2)

$$q(F) = \inf\{z; \text{ There exists an admissible portfolio } \psi$$

$$\text{such that } V_z^\psi(T, \omega) := z + \int_0^T \psi(s)dX(s) \geq F(\omega) \text{ a.s.}\}$$

Definition 12.3.1 *If $p(F) = q(F)$ we call this common value* the price *(at $t = 0$) of the (European) T-contingent claim $F(\omega)$.*

Two important examples of European contingent claims are

a) **the European call**, where

$$F(\omega) = (X_i(T, \omega) - K)^+$$

for some $i \in \{1, 2, \ldots, n\}$ and some $K > 0$. This option gives the owner the right (but not the obligation) to *buy* one unit of security number i at the specified price K (the *exercise* price) at time T. So if $X_i(T, \omega) > K$ then the owner of the option will obtain the payoff $X_i(T, \omega) - K$ at time T, while if $X_i(T.\omega) \leq K$ then the owner will not exercise his option and the payoff is 0.

b) Similarly, **the European put** option gives the owner the right (but not the obligation) to *sell* one unit of security number i at a specified price K at time T. This option gives the owner the payoff

$$F(\omega) = (K - X_i(T, \omega))^+ .$$

Theorem 12.3.2 a) *Suppose (12.2.12) and (12.2.13) hold and let Q be as in (12.2.2). Let F be a (European) T-claim. Then*

$$\operatorname{ess\,inf} F(\omega) \leq p(F) \leq E_Q[\xi(T)F] \leq q(F) \leq \infty . \qquad (12.3.3)$$

b) *Suppose, in addition to the conditions in a), that the market $\{X(t)\}$ is complete. Then the price of the (European) T-claim F is*

$$p(F) = E_Q[\xi(T)F] = q(F) . \qquad (12.3.4)$$

Proof. a) Suppose $y \in \mathbf{R}$ and there exists an admissible portfolio φ such that

$$V_{-y}^{\varphi}(T, \omega) = -y + \int_0^T \varphi(s) dX(s) \geq -F(\omega) \quad \text{a.s.}$$

i.e., using (12.2.7) and Lemma 12.2.2,

$$-y + \int_0^T \sum_{i=1}^n \varphi_i(s) \xi(s) \sigma_i(s) d\widetilde{B}(s) \geq -\xi(T) F(\omega) \quad \text{a.s.} \tag{12.3.5}$$

where $\widetilde{B}$ is defined in (12.2.3).

Since $\int_0^t \sum_{i=1}^n \varphi_i(s) \xi(s) \sigma_i(s) d\widetilde{B}(s)$ is a lower bounded local Q-martingale, it is a supermartingale, by Exercise 7.12. Hence $E_Q[\int_0^t \sum_{i=1}^n \varphi_i(s) \xi(s) \sigma_i(s) d\widetilde{B}(s)] \leq 0$ for all $t \in [0, T]$. Therefore, taking the expectation of (12.3.5) with respect to Q we get

$$y \leq E_Q[\xi(T) F] .$$

Hence

$$p(F) \leq E_Q[\xi(T) F] ,$$

provided such a portfolio φ exists for some $y \in \mathbf{R}$. This proves the second inequality in (12.3.3). Clearly, if $y < F(\omega)$ for a.a. ω, we can choose $\varphi = 0$. Hence the first inequality in (12.3.3) holds.

Similarly, if there exists $z \in \mathbf{R}$ and an admissible portfolio ψ such that

$$z + \int_0^T \psi(s) dX(s) \geq F(\omega) \quad \text{a.s.}$$

then, as in (12.3.5)

$$z + \int_0^T \sum_{i=1}^n \psi_i(s) \xi(s) \sigma_i(s) d\widetilde{B}(s) \geq \xi(T) F(\omega) \quad \text{a.s.}$$

Taking Q-expectations we get

$$z \geq E_Q[\xi(T) F] ,$$

provided such z and ψ exist.

If no such z, ψ exist, then $q(F) = \infty > E_Q[\xi(T) F]$.

b) Next, assume in addition that the market is complete. Then by completeness we can find (unique) $y \in \mathbf{R}$ and θ such that

$$-y + \int_0^T \theta(s)dX(s) = -F(\omega) \quad \text{a.s.}$$

i.e. (by (12.2.7) and Lemma 12.2.2)

$$-y + \int_0^T \sum_{i=1}^n \theta_i(s)\xi(s)\sigma_i(s)d\widetilde{B}(s) = -\xi(T)F(\omega) \quad \text{a.s.} ,$$

which gives, since $\int_0^t \sum_{i=1}^n \theta_i(s)\xi(s)\sigma_i(s)d\widetilde{B}(s)$ is a Q-martingale (Definition 12.2.4c)),

$$y = E_Q[\xi(T)F] .$$

Hence

$$p(F) \geq E_Q[\xi(T)F]$$

Combined with a) this gives

$$p(F) = E_Q[\xi(T)F] .$$

A similar argument gives that

$$q(F) = E_Q[\xi(T)F] .$$

$\square$

How to Hedge an Attainable Claim

We have seen that if $V_z^\theta(t)$ is the value process of an admissible portfolio $\theta(t)$ for the market $\{X(t)\}$, then $\overline{V}_z^\theta(t) := \xi(t)V_z^\theta(t)$ is the value process of $\theta(t)$ for the normalized market $\{\overline{X}(t)\}$ (Lemma 12.2.2). Hence we have

$$\xi(t)V_z^\theta(t) = z + \int_0^t \theta(s)d\overline{X}(s) . \tag{12.3.6}$$

If (12.2.12) and (12.2.13) hold, then – if Q, $\widetilde{B}$ are defined as before ((12.2.2) and (12.2.3)) – we can rewrite this as (see Lemma 12.2.3)

$$\xi(t)V_z^\theta(t) = z + \int_0^t \sum_{i=1}^n \theta_i(s)\xi(s) \sum_{j=1}^m \sigma_{ij}(s)d\widetilde{B}_j(s) . \tag{12.3.7}$$

Therefore, the portfolio $\theta(t) = (\theta_0(t), \ldots, \theta_n(t))$ needed to hedge a given T-claim F is given by

$$\xi(t,\omega)(\theta_1(t), \ldots, \theta_n(t))\sigma(t,\omega) = \phi(t,\omega) , \qquad (12.3.8)$$

i.e.

$$\theta(t) = X_0(t)\phi(t)\Lambda(t) ,$$

where $\phi(t,\omega) \in \mathbf{R}^m$ is such that

$$\xi(T)F(\omega) = z + \int_0^T \phi(t,\omega)d\widetilde{B}(t) \qquad (12.3.9)$$

(and $\theta_0(t)$ is given by (12.1.17)).

In view of this it is of interest to find explicitly the integrand $\phi(t,\omega)$ when F is given. One way of doing this is by using a generalized version of the Clark-Ocone theorem from the Malliavin calculus. See Karatzas and Ocone (1991). A survey containing their result is in Øksendal (1996)). See also Di Nunno et al (2009). In the Markovian case, however, there is a simpler method, which we now describe. It is a modification of a method used by Hu (1995).

Let $Y(t)$ be an Itô diffusion in $\mathbf{R}^k$ of the form

$$dY(t) = b(Y(t))dt + \sigma(Y(t))d\widetilde{B}(t), \qquad Y(0) = y \qquad (12.3.10)$$

where $b \colon \mathbf{R}^k \to \mathbf{R}^k$ and $\sigma \colon \mathbf{R}^k \to \mathbf{R}^{k \times \ell}$ are given Lipschitz continuous functions. Assume that $Y(t)$ is *uniformly elliptic*, i.e. that there exists a constant $c > 0$ such that

$$x^T \sigma(y)\sigma^T(y)x \geq c|x|^2 \qquad (12.3.11)$$

for all $x \in \mathbf{R}^k$, $y \in \mathbf{R}^k$.

Let $h \in C_0^2(\mathbf{R}^k)$ and define

$$g(t,y) = E_Q^y[h(Y(T-t))]; \qquad t \in [0,T], \quad y \in \mathbf{R}^k . \qquad (12.3.12)$$

Then by uniform ellipticity it is known that $g(t,y) \in C^{1,2}((0,T) \times \mathbf{R}^k)$ (see Dynkin 1965 II, Theorem 13.18 p. 53 and Dynkin 1965 I, Theorem 5.11 p. 162) and hence Kolmogorov's backward equation (Theorem 8.1.1) gives that

$$\frac{\partial g}{\partial t} + \sum_{i=1}^k b_i(y)\frac{\partial g}{\partial y_i} + \frac{1}{2}\sum_{i,j=1}^k (\sigma\sigma^T)_{ij}(y)\frac{\partial^2 g}{\partial y_i \partial y_j} = 0 . \qquad (12.3.13)$$

Now put

$$Z(t) = g(t, Y(t)) . \qquad (12.3.14)$$

Then by Itô's formula we have, by (12.3.13),

$$dZ(t) = \frac{\partial g}{\partial t}(t, Y(t))dt + \sum_{i=1} \frac{\partial g}{\partial y_i}(t, Y(t))dY_i(t))$$

$$+ \tfrac{1}{2} \sum_{i,j=1}^{k} (\sigma\sigma^T)_{ij}(Y(t)) \frac{\partial^2 g}{\partial y_i \partial y_j}(t, Y(t))dt$$

$$= \sum_{i=1}^{k} \frac{\partial g}{\partial y_i}(t, Y(t))\sigma_i(Y(t))d\widetilde{B}(t) . \qquad (12.3.15)$$

Moreover,

$$Z(T) = g(T, Y(T)) = E_Q^{Y(T)}[h(Y(0))] = h(Y(T)) \qquad (12.3.16)$$

and

$$Z(0) = g(0, Y(0)) = E_Q^y[h(Y(T))] . \qquad (12.3.17)$$

Combining (12.3.15)–(12.3.17) we get

$$h(Y(T)) = E_Q^y[h(Y(T))] + \int_0^T \phi(t, \omega)d\widetilde{B}(t) , \qquad (12.3.18)$$

where

$$\phi(t, \omega) = \sum_{i=1}^{k} \frac{\partial}{\partial y_i} E_Q^y[h(Y(T-t))]_{y=Y(t)} \sigma_i(Y(t)) . \qquad (12.3.19)$$

More generally, by approximating more general processes $Y(t)$ by uniformly elliptic processes $Y^{(n)}(t)$ and more general functions $h(y)$ by C_0^2 functions $H^{(m)}(y); n, m = 1, 2, \ldots$ we obtain the following conclusion:

Theorem 12.3.3 *Let $Y(t) \in \mathbf{R}^k$ be an Itô diffusion of the form (12.3.10) and assume that $h : \mathbf{R}^k \to \mathbf{R}$ is a given function such that*

$$\left\{ \frac{\partial}{\partial y_i} E_Q^y[h(Y(T-t))] \right\}_{i=1}^{k} \quad \text{exists} \qquad (12.3.20)$$

and

$$E_Q^y \left[\int_0^T \phi^2(t, \omega)dt \right] < \infty , \qquad (12.3.21)$$

where

$$\phi(t, \omega) = \sum_{i=1}^{k} \frac{\partial}{\partial y_i} E_Q^y[h(Y(T-t))]_{y=Y(t)} \sigma_i(Y(t)) . \qquad (12.3.22)$$

Then we have the Itô representation formula

$$h(Y(T)) = E_Q[h(Y(T))] + \int_0^T \phi(t,\omega)d\widetilde{B}(t) . \qquad (12.3.23)$$

To apply this result to find ϕ in (12.3.9) we argue as follows:
Suppose the equation for $X(t)$ is *Markovian*, i.e.

$$dX_0(t) = \rho(X(t))X_0(t)dt; \qquad X_0(0) = 1$$
$$dX_i(t) = \mu_i(X(t))dt + \sigma_i(X(t))dB(t); \qquad i = 1, \ldots, n$$

where σ_i is row number i of the matrix σ. We rewrite this equation for $X(t)$ in terms of $\widetilde{B}(t)$ and get, using (12.2.12) and (12.2.3), the following system

$$dX_0(t) = \rho(X(t))X_0(t)dt \qquad (12.3.24)$$
$$dX_i(t) = \rho(X(t))X_i(t)dt + \sigma_i(X(t))d\widetilde{B}(t) ; \quad 1 \le i \le n . \qquad (12.3.25)$$

Therefore Theorem 12.3.3 gives

Corollary 12.3.4 *Let* $X(t) = (X_0(t), \ldots, X_n(t)) \in \mathbf{R}^{n+1}$ *be given by* (12.3.24)–(12.3.25) *and assume that* $h_0 : \mathbf{R}^{n+1} \to \mathbf{R}$ *is a given function such that*

$$\left\{ \frac{\partial}{\partial x_i} E_Q^x[\xi(T-t)h_0(X(T-t))] \right\}_{i=1}^n \quad exists \qquad (12.3.26)$$

and

$$E_Q^x\left[\int_0^T \phi^2(t,\omega)dt \right] < \infty \qquad (12.3.27)$$

where

$$\phi(t,\omega) := \sum_{i=1}^n \frac{\partial}{\partial x_i} E_Q^x[\xi(T-t)h_0(X(T-t))]_{x=X(t)} \sigma_i(X(t)) . \qquad (12.3.28)$$

Then we have the Itô representation formula

$$\xi(T)h_0(X(T)) = E_Q[\xi(T)h_0(X(T))] + \int_0^T \phi(t,\omega)d\widetilde{B}(t) . \qquad (12.3.29)$$

We summarize our results about pricing and hedging of European T-claims as follows:

Theorem 12.3.5 *Let* $\{X(t)\}_{t \in [0,T]}$ *be a complete market. Suppose* (12.2.12) *and* (12.2.13) *hold and let* $Q, \widetilde{B}$ *be as in* (12.2.2), (12.2.3). *Let* F *be a European* T-*claim such that* $E_Q[\xi(T)F] < \infty$. *Then the price of the claim* F *is*

$$p(F) = E_Q[\xi(T)F] . \qquad (12.3.30)$$

Moreover, to find a replicating (hedging) portfolio $\theta(t) = (\theta_0(t), \ldots, \theta_n(t))$ for the claim F we first find (for example by using Corollary 12.3.4 if possible) an adapted process ϕ such that $E_Q\left[\int_0^T \phi^2(t)dt\right] < \infty$ and

$$\xi(T)F = E_Q[\xi(T)F] + \int_0^T \phi(t,\omega)d\widetilde{B}(t) \,. \tag{12.3.31}$$

Then we choose $\widehat{\theta}(t) = (\theta_1(t), \ldots, \theta_n(t))$ such that

$$\widehat{\theta}(t,\omega)\xi(t,\omega)\sigma(t,\omega) = \phi(t,\omega) \tag{12.3.32}$$

and we choose $\theta_0(t)$ as in (12.1.17).

Proof. (12.3.30) is just part b) of Theorem 12.3.2. The equation (12.3.32) follows from (12.3.8). Note that the equation (12.3.32) has the solution

$$\widehat{\theta}(t,\omega) = X_0(t)\phi(t,\omega)\Lambda(t,\omega) \tag{12.3.33}$$

where $\Lambda(t,\omega)$ is the left inverse of $\sigma(t,\omega)$ (Theorem 12.2.5). $\qquad\square$

Example 12.3.6 Suppose the market is $(X_0(t), X_1(t))$, where $X_0(t) = e^{\rho t}$ and $X_1(t)$ is an Ornstein-Uhlenbeck process

$$dX_1(t) = \alpha X_1(t)dt + \sigma dB(t); \qquad X_1(0) = x_1$$

where $\rho > 0$, α, σ are constants, $\sigma \neq 0$. How do we hedge the claim

$$F(\omega) = \exp(X_1(T)) \,?$$

The portfolio $\theta(t) = (\theta_0(t), \theta_1(t))$ that we seek is given by (12.3.33), i.e.

$$\theta_1(t,\omega) = e^{\rho t}\sigma^{-1}\phi(t,\omega)$$

where $\phi(t,\omega)$ and $V(0) = z$ are uniquely given by (12.3.9), i.e.

$$\xi(T)F(\omega) = z + \int_0^T \phi(t,\omega)d\widetilde{B}(t)$$

or

$$F(\omega) = \exp(X_1(T)) = ze^{\rho T} + \int_0^T \phi_0(t,\omega)d\widetilde{B}(t) \,,$$

where

$$\phi_0(t,\omega) = e^{\rho T}\phi(t,\omega) \,.$$

To find $\phi(t,\omega)$ explicitly we apply Corollary 12.3.4:

Note that in terms of $\widetilde{B}$ we have

$$dX_1(t) = \rho X_1(t)dt + \sigma d\widetilde{B}(t) ; \qquad X_1(0) = x_1 .$$

This equation has the solution (see Exercise 5.5)

$$X_1(t) = x_1 e^{\rho t} + \sigma \int_0^t e^{\rho(t-s)} d\widetilde{B}(s) .$$

Hence, if we choose $h_0(x_1) = \exp(x_1)$ we have

$$E_Q^{x_1}[h_0(X(T-t))] = E_Q^{x_1}[\exp(X_1(T-t))]$$

$$= E_Q\left[\exp\left\{x_1 e^{\rho(T-t)} + \sigma \int_0^{T-t} e^{\rho(T-t-s)} d\widetilde{B}(s)\right\}\right]$$

$$= \exp\left\{x_1 e^{\rho(T-t)} + \frac{\sigma^2}{4\rho}\left(e^{2\rho(T-t)} - 1\right)\right\} \qquad \text{if } \rho \neq 0 .$$

This gives, by (12.3.28),

$$\phi_0(t,\omega) = \frac{d}{dx_1} E_Q^{x_1}[h_0(X(T-t))]_{x_1 = X_1(t)} \sigma$$

$$= \sigma e^{\rho(T-t)} \exp\left\{X_1(t)e^{\rho(T-t)} + \frac{\sigma^2}{4\rho}(e^{2\rho(T-t)} - 1)\right\} \qquad \text{if } \rho \neq 0 .$$

Therefore, by (12.3.33),

$$\theta_1(t) = \begin{cases} \exp\{X_1(t)e^{\rho(T-t)} + \frac{\sigma^2}{4\rho}(e^{2\rho(T-t)} - 1)\} & \text{if } \rho \neq 0 \\ \exp\{X_1(t) + \frac{\sigma^2}{2}(T-t)\} & \text{if } \rho = 0 . \end{cases}$$

The Generalized Black-Scholes Model

Let us now specialize to a situation where the market has just two securities $X_0(t), X_1(t)$ where X_0, X_1 are Itô processes of the form

$$dX_0(t) = \rho(t,\omega)X_0(t)dt \qquad \text{(as before)} \tag{12.3.34}$$

$$dX_1(t) = \alpha(t,\omega)X_1(t)dt + \beta(t,\omega)X_1(t)dB(t) , \tag{12.3.35}$$

where $B(t)$ is 1-dimensional and $\alpha(t,\omega), \beta(t,\omega)$ are 1-dimensional processes in $\mathcal{W}$.

Note that the solution of (12.3.35) is

$$X_1(t) = X_1(0)\exp\left(\int_0^t \beta(s,\omega)dB(s) + \int_0^t (\alpha(s,\omega) - \tfrac{1}{2}\beta^2(s,\omega))ds\right). \quad (12.3.36)$$

The equation (12.2.12) gets the form

$$X_1(t)\beta(t,\omega)u(t,\omega) = X_1(t)\alpha(t,\omega) - X_1(t)\rho(t,\omega)$$

which has the solution

$$u(t,\omega) = \beta^{-1}(t,\omega)[\alpha(t,\omega) - \rho(t,\omega)] \qquad \text{if } \beta(t,\omega) \neq 0 . \quad (12.3.37)$$

So (12.2.13) holds iff

$$E\left[\exp\left(\tfrac{1}{2}\int_0^T \frac{(\alpha(s,\omega) - \rho(s,\omega))^2}{\beta^2(s,\omega)}ds\right)\right] < \infty . \quad (12.3.38)$$

In this case we have an equivalent martingale measure Q given by (12.2.2) and the market has no arbitrage, by Theorem 12.1.8. Moreover, the market is complete by Corollary 12.2.6. Therefore we get by Theorem 12.3.2 that the price at $t = 0$ of a European option with payoff given by a contingent T-claim F is

$$p(F) = q(F) = E_Q[\xi(T)F] , \quad (12.3.39)$$

provided this quantity is finite.

Now suppose that $\rho(t,\omega) = \rho(t)$ and $\beta(t,\omega) = \beta(t)$ are *deterministic* and that the payoff $F(\omega)$ has the form

$$F(\omega) = f(X_1(T,\omega))$$

for some lower bounded function $f: \mathbf{R} \to \mathbf{R}$ such that

$$E_Q[f(X_1(T))] < \infty .$$

Then by (12.3.39) the price $p = p(F) = q(F)$ is, with $x_1 = X_1(0)$,

$$p = \xi(T)E_Q\left[f\left(x_1\exp\left(\int_0^T \beta(s)d\widetilde{B}(s) + \int_0^T (\rho(s) - \tfrac{1}{2}\beta^2(s))ds\right)\right)\right].$$

Under the measure Q the random variable $Y = \int_0^T \beta(s)d\widetilde{B}(s)$ is normally distributed with mean 0 and variance $\delta^2 := \int_0^T \beta^2(t)dt$ and therefore we can write down a more explicit formula for p. The result is the following:

Theorem 12.3.7 (The generalized Black-Scholes formula)
Suppose $X(t) = (X_0(t), X_1(t))$ is given by

$$dX_0(t) = \rho(t)X_0(t)dt \; ; \qquad X_0(0) = 1 \qquad (12.3.40)$$

$$dX_1(t) = \alpha(t,\omega)X_1(t)dt + \beta(t)X_1(t)dB(t) \; ; \qquad X_1(0) = x_1 > 0 \qquad (12.3.41)$$

where $\rho(t)$, $\beta(t)$ are deterministic and

$$E\left[\exp\left(\tfrac{1}{2} \int\limits_0^T \frac{(\alpha(t,\omega) - \rho(t))^2}{\beta^2(t)} dt \right) \right] < \infty \; .$$

a)   *Then the market $\{X(t)\}$ is arbitrage free and complete and the price at time $t = 0$ of the European T-claim $F(\omega) = f(X_1(T,\omega))$, where $E_Q[f(X_1(T,\omega))] < \infty$, is*

$$p = \frac{\xi(T)}{\delta\sqrt{2\pi}} \int\limits_{\mathbf{R}} f\left(x_1 \exp\left[y + \int\limits_0^T (\rho(s) - \tfrac{1}{2}\beta^2(s))ds \right] \right) \exp\left(-\frac{y^2}{2\delta^2} \right) dy$$

$$(12.3.42)$$

where $\xi(T) = \exp(-\int\limits_0^T \rho(s)ds)$ and $\delta^2 = \int\limits_0^T \beta^2(s)ds$.

b)   *If $\rho, \alpha, \beta \neq 0$ are constants and $f \in C^1(\mathbf{R})$, then the self-financing portfolio $\theta(t) = (\theta_0(t), \theta_1(t))$ needed to replicate the T-claim $F(\omega) = f(X_1(T,\omega))$ is given by*

$$\theta_1(t,\omega) = \frac{1}{\sqrt{2\pi(T-t)}} \int\limits_{\mathbf{R}} f'(X_1(t,\omega)\exp\{\beta x + (\rho - \tfrac{1}{2}\beta^2)(T-t)\})$$

$$\cdot \exp\left(\beta x - \frac{x^2}{2(T-t)} - \tfrac{1}{2}\beta^2(T-t) \right) dx \qquad (12.3.43)$$

and $\theta_0(t,\omega)$ is determined by (12.1.17) and $V^\theta(0) = p$.

Proof. Part a) is already proved and part b) follows from Theorem 12.3.4 and Theorem 12.3.5.

b) The portfolio we seek is by (12.3.33) given by

$$\theta_1(t,\omega) = X_0(t)(\beta X_1(t,\omega))^{-1}\phi(t,\omega)$$

where $\phi(t,\omega)$ is given by (12.3.28) with $h(y) = f(y)$ and

$$X_1(t) = x_1 \exp\{\beta\tilde{B}(t) + (\rho - \tfrac{1}{2}\beta^2)t\} \; .$$

Hence

$$\theta_1(t,\omega) = e^{\rho t}(\beta X_1(t,\omega))^{-1}\frac{\partial}{\partial x_1}\big[E_Q^{x_1}[e^{-\rho T}f(X_1(T-t))]\big]_{x_1=X_1(t)}\cdot\beta X_1(t,\omega)$$

$$= e^{\rho(t-T)}\frac{\partial}{\partial x_1}E\big[f(x_1\exp\{\beta B(T-t)+(\rho-\tfrac{1}{2}\beta^2)(T-t)\})\big]_{x_1=X_1(t)}$$

$$= e^{\rho(t-T)}E\big[f'(x_1\exp\{\beta B(T-t)+(\rho-\tfrac{1}{2}\beta^2)(T-t)\})$$
$$\cdot\exp\{\beta B(T-t)+(\rho-\tfrac{1}{2}\beta^2)(T-t)\}\big]_{x_1=X_1(t)}$$

$$= \frac{e^{\rho(t-T)}}{\sqrt{2\pi(T-t)}}\int_{\mathbf{R}}f'(X_1(t,\omega)\exp\{\beta x+(\rho-\tfrac{1}{2}\beta^2)(T-t)\})$$
$$\cdot\exp\{\beta x+(\rho-\tfrac{1}{2}\beta^2)(T-t)\}e^{-\frac{x^2}{2(T-t)}}\,dx\ ,$$

which is (12.3.43).   □

An important special case of Theorem 12.3.7 is the following:

Corollary 12.3.8 (The classical Black-Scholes formula)
a) *Suppose* $X(t)=(X_0(t),X_1(t))$ *is the classical Black-Scholes market*

$$dX_0(t)=\rho X_0(t)dt\ ;\qquad X_0(0)=1$$
$$dX_1(t)=\alpha X_1(t)dt+\beta X_1(t)dB(t)\ ;\qquad X_1(0)=x_1>0$$

where $\rho,\alpha,\beta\neq 0$ *are constants. Then the price* p *at time 0 of the European call option, with payoff*

$$F(\omega)=(X_1(T,\omega)-K)^+ \tag{12.3.44}$$

where $K>0$ *is a constant (the exercise price), is*

$$p=x_1\Phi\big(\eta+\tfrac{1}{2}\beta\sqrt{T}\big)-Ke^{-\rho T}\Phi\big(\eta-\tfrac{1}{2}\beta\sqrt{T}\big) \tag{12.3.45}$$

where

$$\eta=\beta^{-1}T^{-1/2}\Big(\ln\frac{x_1}{K}+\rho T\Big) \tag{12.3.46}$$

and

$$\Phi(y)=\frac{1}{\sqrt{2\pi}}\int_{-\infty}^{y}e^{-\frac{1}{2}x^2}\,dx\ ;\qquad y\in\mathbf{R} \tag{12.3.47}$$

is the standard normal distribution function.

b) *The replicating portfolio* $\theta(t)=(\theta_0(t),\theta_1(t))$ *for this claim* F *in (12.3.44) is given by*

$$\theta_1(t,\omega)=\Phi\Big(\beta^{-1}(T-t)^{-1/2}\Big(\ln\frac{X_1(t)}{K}+\rho(T-t)+\tfrac{1}{2}\beta^2(T-t)\Big)\Big) \tag{12.3.48}$$

with $\theta_0(t,\omega)$ *determined by (12.1.17) and* $V^\theta(0)=p$.

Remark. Note that, in particular, $\theta_1(t,\omega) > 0$ for all $t \in [0,T]$. This means that we can replicate the European call without *shortselling*. For the European *put* the situation is different. See Exercise 12.16.

Proof. a) This follows by applying Theorem 12.3.7a) to the function

$$f(z) = (z - K)^+ .$$

Then the corresponding integral (12.3.42) can be written

$$p = \frac{e^{-\rho T}}{\beta\sqrt{2\pi T}} \int_{\gamma}^{\infty} (x_1 \exp[y + (\rho - \tfrac{1}{2}\beta^2)T] - K) \exp\left(-\frac{y^2}{2\beta^2 T}\right) dy$$

where $\gamma = \ln(\frac{K}{x_1}) - \rho T + \tfrac{1}{2}\beta^2 T$.

This integral splits into two parts, both of which can be reduced to integrals of standard normal distribution type by completing the square. We leave the details to the reader. (Exercise 12.13.)

b) This follows similarly from Theorem 12.3.7b). The function $f(z) = (z - K)^+$ is not C^1, but an approximation argument shows that formula (12.3.43) still holds if we represent f' by

$$f'(z) = \mathcal{X}_{[K,\infty)}(z) .$$

Then the rest follows by completing the square as in a). (Exercise 12.13.) $\square$

American options

The difference between European and American options is that in the latter case the buyer of the option is free to choose any exercise time τ before or at the given expiration time T (and the guaranteed payoff may depend on both τ and ω.) This exercise time τ may be stochastic (depend on ω), but only in such a way that the decision to exercise before or at a time t only depends on the history up to time t. More precisely, we require that for all t we have

$$\{\omega; \tau(\omega) \le t\} \in \mathcal{F}_t^{(m)} .$$

In other words, τ must be an $\mathcal{F}_t^{(m)}$-stopping time (Definition 7.2.1).

Definition 12.3.9 *An* American contingent T-claim *is an* $\mathcal{F}_t^{(m)}$-*adapted,* (t,ω)-*measurable and a.s. lower bounded continuous stochastic process* $F(t) = F(t,\omega)$; $t \in [0,T]$, $\omega \in \Omega$. An American option *on such a claim* $F(t,\omega)$ *gives the owner of the option the right (but not the obligation) to choose* *any stopping time $\tau(\omega) \leq T$ as exercise time for the option, resulting in a payment $F(\tau(\omega), \omega)$ to the owner.*

Let $F(t) = F(t, \omega)$ be an American contingent claim. Suppose you were offered a guarantee to be paid the amount $F(\tau(\omega), \omega)$ at the (stopping) time $\tau(\omega) \leq T$ that you are free to choose. How much would you be willing to pay for such a guarantee? We repeat the argument preceding Definition 12.3.1:

If I – the buyer – pay the price y for this guarantee, then I will have an initial fortune (debt) $-y$ in my investment strategy. With this initial fortune $-y$ it must be possible to find a stopping time $\tau \leq T$ and an admissible portfolio φ such that

$$V_{-y}^{\varphi}(\tau(\omega), \omega) + F(\tau(\omega), \omega) \geq 0 \quad \text{a.s.}$$

Thus the maximal price $p = p_A(F)$ the buyer is willing to pay is

(*Buyer's price of the American contingent claim F*)   (12.3.49)

$p_A(F) = \sup\{y;$ There exists a stopping time $\tau \leq T$

and an admissible portfolio φ such that

$$V_{-y}^{\varphi}(\tau(\omega), \omega) := -y + \int_0^{\tau(\omega)} \varphi(s)dX(s) \geq -F(\tau(\omega), \omega) \text{ a.s.}\}$$

On the other hand, the seller could argue as follows: If I – the seller – receive the price z for such a guarantee, then with this initial fortune z it must be possible to find an admissible portfolio ψ which generates a value process which at any time is not less than the amount promised to pay to the buyer:

$$V_z^{\psi}(t, \omega) \geq F(t, \omega) \quad \text{a.s. for all } t \in [0, T] \ .$$

Thus the minimal price $q = q_A(F)$ the seller is willing to accept is

(*Seller's price of the American contingent claim F*)   (12.3.50)

$q_A(F) = \inf\{z;$ There exists an admissible portfolio ψ

such that for all $t \in [0, T]$ we have

$$V_z^{\psi}(t, \omega) := z + \int_0^t \psi(s)dX(s) \geq F(t, \omega) \text{ a.s.}\}$$

We can now prove a result analogous to Theorem 12.3.2. The result is basically due to Bensoussan (1984) and Karatzas (1988).

Theorem 12.3.10 (Pricing formula for American options)

a)   *Suppose (12.2.12) and (12.2.13) and let Q be as in (12.2.2). Let $F(t) = F(t, \omega); t \in [0, T]$ be an American contingent T-claim such that*

$$\sup_{\tau \leq T} E_Q[\xi(\tau)F(\tau)] < \infty \qquad (12.3.51)$$

Then

$$p_A(F) \leq \sup_{\tau \leq T} E_Q[\xi(\tau)F(\tau)] \leq q_A(F) \leq \infty . \qquad (12.3.52)$$

b) *Suppose, in addition to the conditions in a), that the market $\{X(t)\}$ is complete. Then*

$$p_A(F) = \sup_{\tau \leq T} E_Q[\xi(\tau)F(\tau)] = q_A(F) . \qquad (12.3.53)$$

Proof. a) We proceed as in the proof of Theorem 12.3.2: Suppose $y \in \mathbf{R}$ and there exists a stopping time $\tau \leq T$ and an admissible portfolio φ such that

$$V_{-y}^{\varphi}(\tau, \omega) = -y + \int_0^{\tau} \varphi(s)dX(s) \geq -F(\tau) \quad \text{a.s.}$$

Then as before we get

$$-y + \int_0^{\tau} \sum_{i=1}^{n} \varphi_i(s)\xi(s)\sigma_i(s)d\widetilde{B}(s) = \overline{V}_{-y}^{\varphi}(\tau) = \xi(\tau)V_{-y}^{\varphi}(\tau) \geq -\xi(\tau)F(\tau) \quad \text{a.s.}$$

Taking expectations with respect to Q we get, since $\overline{V}_{-y}(t)$ is a Q-supermartingale

$$y \leq E_Q[\xi(\tau)F(\tau)] \leq \sup_{\tau \leq T} E_Q[\xi(\tau)F(\tau)] .$$

Since this holds for all such y we conclude that

$$p_A(F) \leq \sup_{\tau \leq T} E_Q[\xi(\tau)F(\tau)] . \qquad (12.3.54)$$

Similarly, suppose $z \in \mathbf{R}$ and there exists an admissible portfolio ψ such that

$$V_z^{\psi}(t, \omega) = z + \int_0^{t} \psi(s)dX(s) \geq F(t) \quad \text{a.s. for all } t \in [0, T] .$$

Then, as above, if $\tau \leq T$ is a stopping time we get

$$z + \int_0^{\tau} \sum_{i=1}^{n} \psi_i(s)\xi(s)\sigma_i(s)d\widetilde{B}(s) = \overline{V}_z^{\psi}(\tau) = \xi(\tau)V_z^{\psi}(\tau) \geq \xi(\tau)F(\tau) \quad \text{a.s.}$$

Again, taking expectations with respect to Q and then supremum over $\tau \leq T$ we get

$$z \geq \sup_{\tau \leq T} E_Q[\xi(\tau)F(\tau)] \ .$$

Since this holds for all such z, we get

$$q_A(F) \geq \sup_{\tau \leq T} E_Q[\xi(\tau)F(\tau)] \ . \tag{12.3.55}$$

b) Next, assume in addition that the market is complete. Choose a stopping time $\tau \leq T$. Define

$$F_k(t) = F_k(t,\omega) = \begin{cases} k & \text{if} \quad F(t,\omega) \geq k \\ F(t,\omega) & \text{if} \quad F(t,\omega) < k \end{cases}$$

and put

$$G_k(\omega) = X_0(T)\xi(\tau)F_k(\tau) \ .$$

Then G_k is a bounded T-claim, so by completeness we can find $y_k \in \mathbf{R}$ and a portfolio $\theta^{(k)}$ such that

$$-y_k + \int_0^T \theta^{(k)}(s)dX(s) = -G_k(\omega) \quad \text{a.s.}$$

and such that

$$-y_k + \int_0^t \theta^{(k)}(s)d\overline{X}(s)$$

is a Q-martingale. Then, by (12.2.8)–(12.2.9),

$$-y_k + \int_0^T \theta^{(k)}(s)d\overline{X}(s) = -\xi(T)G_k(\omega) = -\xi(\tau)F_k(\tau)$$

and hence

$$-y_k + \int_0^\tau \theta^{(k)}(s)d\overline{X}(s) = E_Q\left[-y_k + \int_0^T \theta^{(k)}(s)d\overline{X}(s) \mid \mathcal{F}_\tau^{(m)} \right]$$

$$= E_Q[-\xi(\tau)F_k(\tau) \mid \mathcal{F}_\tau^{(m)}] = -\xi(\tau)F_k(\tau) \ .$$

From this we get, again by (12.2.8)–(12.2.9),

$$-y_k + \int_0^\tau \theta^{(k)}(s)dX(s) = -F_k(\tau) \quad \text{a.s.}$$

and

$$y_k = E_Q[\xi(\tau)F_k(\tau)] \ .$$

This shows that any price of the form $E_Q[\xi(\tau)F_k(\tau)]$ for some stopping time $\tau \leq T$ would be acceptable for the buyer of an American option on the claim $F_k(t,\omega)$. Hence

$$p_A(F) \geq p_A(F_k) \geq \sup_{\tau \leq T} E_Q[\xi(\tau)F_k(\tau)] .$$

Letting $k \to \infty$ we obtain by monotone convergence

$$p_A(F) \geq \sup_{\tau \leq T} E_Q[\xi(\tau)F(\tau)] .$$

It remains to show that if we put

$$z = \sup_{0 \leq \tau \leq T} E_Q[\xi(\tau)F(\tau)] \tag{12.3.56}$$

then there exists an admissible portfolio $\theta(s,\omega)$ which *superreplicates* $F(t,\omega)$, in the sense that

$$z + \int_0^t \theta(s,\omega)dX(s) \geq F(t,\omega) \qquad \text{for a.a. } (t,\omega) \in [0,T] \times \Omega . \tag{12.3.57}$$

The details of the proof of this can be found in Karatzas (1997), Theorem 1.4.3. Here we only sketch the proof:
Define the *Snell envelope*

$$S(t) = \sup_{t \leq \tau \leq T} E_Q[\xi(\tau)F(\tau)|\mathcal{F}_t^{(m)}] ; \qquad 0 \leq t \leq T .$$

Then $S(t)$ is a *supermartingale* w.r.t. Q and $\{\mathcal{F}_t^{(m)}\}$, so by the *Doob-Meyer decomposition* we can write

$$S(t) = M(t) - A(t) ; \qquad 0 \leq t \leq T$$

where $M(t)$ is a $Q, \{\mathcal{F}_t^{(m)}\}$-martingale with $M(0) = S(0) = z$ and $A(t)$ is a nondecreasing process with $A(0) = 0$. It is a consequence of Lemma 12.2.1 that we can represent the martingale M as an Itô integral w.r.t $\widetilde{B}$. Hence

$$z + \int_0^t \phi(s,\omega)d\widetilde{B}(s) = M(t) = S(t) + A(t) \geq S(t) ; \qquad 0 \leq t \leq T \tag{12.3.58}$$

for some $\mathcal{F}_t^{(m)}$-adapted process $\phi(s,\omega)$. Since the market is complete, we know by Theorem 12.2.5 that $\sigma(t,\omega)$ has a left inverse $\Lambda(t,\omega)$. So if we define $\widehat{\theta} = (\theta_1, \ldots, \theta_n)$ by

$$\widehat{\theta}(t,\omega) = X_0(t)\phi(t,\omega)\Lambda(t,\omega)$$

then by (12.3.58) and Lemma 12.2.4 we get

$$z + \int_0^t \widehat{\theta}\, d\overline{X} = z + \int_0^t \sum_{i=1}^n \xi\theta_i\sigma_i d\widetilde{B} = z + \int_0^t \phi\, d\widetilde{B} \geq S(t) ; \qquad 0 \leq t \leq T .$$

Hence, by Lemma 12.2.3,

$$z + \int_0^t \theta(s,\omega)dX(s) \geq X_0(t)S(t) \geq X_0(t)\xi(t)F(t) = F(t) ; \qquad 0 \leq t \leq T .$$

$\square$

The Itô Diffusion Case: Connection to Optimal Stopping

Theorem 12.3.8 shows that pricing an American option is an optimal stopping problem. In the general case the solution to this problem can be expressed in terms of the Snell envelope. See e.g. El Karoui (1981) and Fakeev (1970). In the Itô diffusion case we get an optimal stopping problem of the type discussed in Chapter 10. We now consider this case in more detail:

Assume the market is an $(n + 1)$-dimensional Itô *diffusion* $X(t) = (X_0(t), X_1(t), \dots, X_n(t))$; $t \geq 0$ of the form (see (12.1.1)–(12.1.2))

$$dX_0(t) = \rho(t, X(t))X_0(t)dt ; \qquad X_0(0) = x_0 > 0 \qquad (12.3.59)$$

and

$$dX_i(t) = \mu_i(t, X(t))dt + \sum_{j=1}^m \sigma_{ij}(t, X(t))dB_j(t) \qquad (12.3.60)$$

$$= \mu_i(t, X(t))dt + \sigma_i(t, X(t))dB(t) ; \qquad X_i(0) = x_i ,$$

where ρ, μ_i and σ_{ij} are given functions satisfying the conditions of Theorem 5.2.1.

Moreover, assume that the American claim $F(t)$ is Markovian, i.e.

$$F(t) = h(X(t)) \qquad (12.3.61)$$

for some lower bounded function $h : \mathbf{R}^{n+1} \to \mathbf{R}$. Define

$$Y(t) = \begin{bmatrix} s + t \\ X(t) \end{bmatrix} \in \mathbf{R}^{n+2} \qquad (12.3.62)$$

and put

$$g(y) = g(s,x) = x_0^{-1}h(x) ; \qquad x = (x_0, x_1, \ldots, x_n) \in \mathbf{R}^{n+1} . \qquad (12.3.63)$$

Then the problem to find the price

$$p_A(F) = q_A(F) = \sup_{\tau \le T} E_Q[\xi(\tau)F(\tau)] \qquad (12.3.64)$$

in (12.3.53) can be regarded as a special case of the general optimal stopping problem considered in Section 10.4. Indeed, if we put

$$\Phi(y) = \sup_{\tau \le \tau_G} E_Q^y[g(Y(\tau))] , \qquad (12.3.65)$$

where

$$\tau_G = \inf\{t > 0; s + t \ge T\} = T - s \qquad (12.3.66)$$

is the first exit time of $Y(t)$ from the region

$$G = \{(s,x); s < T\} \subset \mathbf{R}^{n+1} , \qquad (12.3.67)$$

then

$$p_A(F) = \Phi(0, 1, x_1, \ldots, x_n) . \qquad (12.3.68)$$

To apply the theory of Section 10.4 to problem (12.3.65) we must rewrite equations (12.3.59)–(12.3.60) in terms of $\widetilde{B}(t)$. Using (12.2.12) and (12.2.3) we get, as in (12.3.24)–(12.3.25),

$$dX_0(t) = \rho(X(t))dt ; \qquad X_0(0) = x_0 > 0 \qquad (12.3.69)$$
$$dX_i(t) = \rho(X(t))X_i(t)dt + \sigma_i(X(t))d\widetilde{B}(t); \quad X_i(0) = x_i; \ 1 \le i \le n \quad (12.3.70)$$

We summarize this as follows:

Theorem 12.3.11   *Suppose the market $\{X(t)\}$ has the Markovian form (12.3.59)–(12.3.60), which is equivalent to (12.3.69)–(12.3.70), and that the American claim $F(t)$ has the Markovian form (12.3.61). Then the price $p_A(F)$ of the American option with this payoff F is the solution of the optimal stopping problem (12.3.65) at $y = (0, 1, x_1, \ldots, x_n)$.*

Example 12.3.12 (The American put)
Consider the Black-Scholes market

$$dX_0(t) = \rho X_0(t)dt ; \qquad X_0(0) = 1$$
$$dX_1(t) = \alpha X_1(t)dt + \beta X_1(t)dB(t) ; \qquad X_1(0) = x_1 > 0$$

where $\rho, \alpha, \beta \ne 0$ are constants. In terms of $\widetilde{B}$ the system becomes

$$dX_0(t) = \rho X_0(t)dt ; \qquad X_0(0) = 1$$
$$dX_1(t) = \rho X_1(t)dt + \beta X_1(t)d\widetilde{B}(t) ; \qquad X_1(0) = x_1 > 0 .$$

Therefore the generator L of the process $Y(t)$ defined in (12.3.62) is given by

$$Lf(s, x_0, x_1) = \frac{\partial f}{\partial s} + \rho x_0 \frac{\partial f}{\partial x_0} + \rho x_1 \frac{\partial f}{\partial x_1} + \tfrac{1}{2}\beta^2 x_1^2 \frac{\partial^2 f}{\partial x_1^2} \qquad (12.3.71)$$

for $f \in C_0^2(\mathbf{R}^3)$.

Therefore, according to Theorem 10.4.1, to find the value function Φ in (12.3.63)–(12.3.65) we look for a C^1 function ϕ such that

$$\phi(s, x_0, x_1) \geq x_0^{-1} h(x_0, x_1) \quad \text{for all } s < T \qquad (12.3.72)$$
$$\phi(T, x_0, x_1) = x_0^{-1} h(x_0, x_1) . \qquad (12.3.73)$$

Define the *continuation region*

$$D = \{(s, x_0, x_1); \phi(s, x_0, x_1) > x_0^{-1} h(x_0, x_1)\} . \qquad (12.3.74)$$

Then we require ϕ to be C^2 outside ∂D and

$$L\phi(s, x_0, x_1) \leq 0 \quad \text{outside } \bar{D} \qquad (12.3.75)$$

and

$$L\phi(s, x_0, x_1) = 0 \quad \text{on } D . \qquad (12.3.76)$$

Suppose
$$F(t) = (K - X_1(t))^+ \qquad \textbf{(the American put)}$$

Then the function g in (12.3.63) is

$$g(s, x_0, x_1) = x_0^{-1}(K - x_1)^+ = e^{-\rho s}(K - x_1)^+ ,$$

so we can disregard the variable x_0 and use s instead. Then the variational inequalities (12.3.72)–(12.3.76) get the form

$$\phi(s, x_1) \geq e^{-\rho s}(K - x_1)^+ \qquad \text{for all } s < T \qquad (12.3.77)$$
$$\phi(T, x_1) = e^{-\rho T}(K - x_1)^+ \qquad (12.3.78)$$
$$D = \{(s, x_1); \phi(s, x_1) > e^{-\rho s}(K - x_1)^+\} \qquad (12.3.79)$$
$$\frac{\partial \phi}{\partial s} + \rho x_1 \frac{\partial \phi}{\partial x_1} + \tfrac{1}{2}\beta^2 x_1^2 \frac{\partial^2 \phi}{\partial x_1^2} \leq 0 \qquad \text{outside } \bar{D} \qquad (12.3.80)$$
$$\frac{\partial \phi}{\partial s} + \rho x_1 \frac{\partial \phi}{\partial x_1} + \tfrac{1}{2}\beta^2 x_1^2 \frac{\partial^2 \phi}{\partial x_1^2} = 0 \qquad \text{in } D . \qquad (12.3.81)$$

In this case we cannot factor out $e^{-\rho s}$ as we often did in Chapter 10, because that would lead to a conflict between (12.3.78) and (12.3.81).

If such a ϕ is found, and the additional assumptions of Theorem 10.4.1 hold, then we can conclude that

$$\phi(s, x_1) = \Phi(s, x_1)$$

and hence $p_A(F) = \phi(0, x_1)$ is the option price at time $t = 0$. Moreover,

$$\tau^* = \tau_D = \inf\{t > 0; (s + t, X_1(t)) \notin D\}$$

is the corresponding optimal stopping time, i.e. the optimal time to exercise the American option. Unfortunately, even in this case it seems that an explicit analytic solution is very hard (possibly impossible) to find. However, there are interesting partial results and good approximation procedures. See e.g. Barles et al (1995), Bather (1997), Jacka (1991), Karatzas (1997), Musiela and Rutkowski (1997) and the references therein. For example, it is known (see Jacka (1991)) that the continuation region D has the form

$$D = \{(t, x_1) \in (-\infty, T) \times \mathbf{R}, \ x_1 > f(t)\},$$

i.e. D is the region above the graph of f, for some continuous, increasing function $f: (0, T) \to \mathbf{R}$.

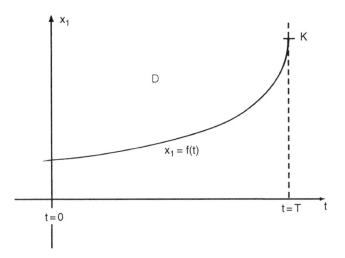

Figure 12.1. The shape of the continuation region for the American put.

Thus the problem is to find the function f. In Barles et al (1995) it is shown that

$$f(t) \sim K - \beta K \sqrt{(T - t)|\ln(T - t)|} \qquad \text{as } t \to T^-,$$

in the sense that

$$\frac{f(t) - K}{-\beta K \sqrt{(T - t)|\ln(T - t)|}} \to 1 \qquad \text{as } t \to T^-.$$

This indicates that the continuation region has the shape shown in the Figure 12.1. But its exact form is still unknown.

For the corresponding American *call* option the situation is much simpler. See Exercise 12.14.

Exercises

12.1. a) Prove that the price process $\{X(t)\}_{t\in[0,T]}$ has an arbitrage iff the normalized price process $\{\overline{X}(t)\}_{t\in[0,T]}$ has an arbitrage.

   b) Suppose $\{X(t)\}_{t\in[0,T]}$ is normalized. Prove that $\{X(t)\}_{t\in[0,T]}$ has an arbitrage iff there exists an admissible portfolio θ such that

$$V^\theta(0) \le V^\theta(T) \quad \text{a.s. and} \quad P[V^\theta(T) > V^\theta(0)] > 0 . \quad (12.3.82)$$

   In other words, in normalized markets it is not essential that we require $V^\theta(0) = 0$ for an arbitrage θ, only that the *gains* $V^\theta(T) - V^\theta(0)$ is nonnegative a.s. and positive with positive probability.
   (Hint: If θ is as in (12.3.82) define $\widetilde{\theta}(t) = (\widetilde{\theta}_0(t), \ldots, \widetilde{\theta}_n(t))$ as follows:
   Let $\widetilde{\theta}_i(t) = \theta(t)$ for $i = 1, \ldots, n$; $t \in [0, T]$. Then choose $\widetilde{\theta}_0(0)$ such that $V^{\widetilde{\theta}}(0) = 0$ and define $\widetilde{\theta}_0(t)$ according to (12.1.15) to make $\widetilde{\theta}$ self-financing. Then

$$V^{\widetilde{\theta}}(t) = \widetilde{\theta}(t) \cdot X(t) = \int_0^t \widetilde{\theta}(s)dX(s) = \int_0^t \theta(s)dX(s) = V^\theta(t) - V^\theta(0).)$$

12.2. Let $\theta(t) = (\theta_0, \ldots, \theta_n)$ be a *constant* portfolio. Prove that θ is self-financing.

12.3. Suppose $\{X(t)\}$ is a normalized market and that (12.2.12) and (12.2.13) hold. Suppose $n = m$ and that σ is invertible with a bounded inverse. Suppose every *bounded* claim is attainable. Show that then any lower bounded claim F such that

$$E_Q[F^2] < \infty$$

is attainable.
(Hint: Choose bounded T-claims F_k such that

$$F_k \to F \quad \text{in } L^2(Q) \quad \text{and} \quad E[F_k] = E[F] .$$

By assumption there exist admissible portfolios $\theta^{(k)} = (\theta_0^{(k)}, \ldots, \theta_n^{(k)})$ and constants $V_k(0)$ such that

$$F_k(\omega) = V_k(0) + \int_0^T \theta^{(k)}(s)dX(s) = V_k(0) + \int_0^T \widehat{\theta}^{(k)}(s)\sigma(s)d\widetilde{B}(s)$$

where $\widehat{\theta}^{(k)} = (\theta_1^{(k)}, \ldots, \theta_n^{(k)})$. It follows that $V_k(0) = E_Q[F_k] \to E_Q[F]$ as $k \to \infty$.

By the Itô isometry the sequence $\{\widehat{\theta}^{(k)}\sigma\}_k$ is a Cauchy sequence in $L^2(\lambda \times Q)$ and hence converges in this space. Conclude that there exists an admissible θ such that

$$F(\omega) = E_Q[F] + \int_0^T \theta(s)dX(s) \ .)$$

12.4. Let $B(t)$ be 1-dimensional Brownian motion. Show that there exist $\theta_1(t, \omega), \theta_2(t, \omega) \in \mathcal{W}$ such that if we define

$$V_1(t) = 1 + \int_0^t \theta_1(s, \omega)dB(s), \quad V_2(t) = 2 + \int_0^t \theta_2(s, \omega)dB(s); \quad t \in [0, 1]$$

then

$$V_1(1) = V_2(1) = 0$$

and

$$V_1(t) \geq 0, \quad V_2(t) \geq 0$$

for a.a. (t, ω).

Therefore both $\theta_1(t, \omega)$ and $\theta_2(t, \omega)$ are admissible portfolios for the claim $F(\omega) = 0$ in the normalized market with $n = 1$ and $X_1(t) = B(t)$. In particular, if we drop the martingale condition in Definition 12.2.4b) we have no uniqueness of replicating portfolios, even if we require the portfolio to be admissible. (Note, however, that we have uniqueness if we require that $\theta \in \mathcal{V}(0, 1)$, by Theorem 4.3.3). (Hint: Use Example 12.1.4 with $M = -1$ and with $M = -2$. Then define, for $i = 1, 2$,

$$\theta_i(t) = \begin{cases} \frac{1}{\sqrt{1-t}} & \text{for } 0 \leq t < \alpha_{-i} \\ 0 & \text{for } \alpha_{-i} \leq t \leq 1 \end{cases}$$

and

$$V_i(t) = i + \int_0^t \theta_i(s)dB(s) = i + Y(t \wedge \alpha_{-i}) ; \qquad 0 \le t \le 1 .)$$

12.5. Prove the first part of Lemma 12.2.1, i.e. that $\tilde{B}(t)$ given by (12.2.3) is an $\mathcal{F}_t^{(m)}$-martingale (see the Remark b) following this lemma).

12.6.* Determine if the following normalized markets $\{X(t)\}_{t \in [0,T]}$ allow an arbitrage. If so, find one.

a) $(n = m = 2)$
$dX_1(t) = 3dt + dB_1(t) + dB_2(t),$
$dX_2(t) = -dt + dB_1(t) - dB_2(t).$

b) $(n = 2, m = 3)$
$dX_1(t) = dt + dB_1(t) + dB_2(t) - dB_3(t)$
$dX_2(t) = 5dt - dB_1(t) + dB_2(t) + dB_3(t)$

c) $(n = 2, m = 3)$
$dX_1(t) = dt + dB_1(t) + dB_2(t) - dB_3(t)$
$dX_2(t) = 5dt - dB_1(t) - dB_2(t) + dB_3(t)$

d) $(n = 2, m = 3)$
$dX_1(t) = dt + dB_1(t) + dB_2(t) - dB_3(t)$
$dX_2(t) = -3dt - 3dB_1(t) - 3dB_2(t) + 3dB_3(t)$

e) $(n = 3, m = 2)$
$dX_1(t) = dt + dB_1(t) + dB_2(t)$
$dX_2(t) = 2dt + dB_1(t) - dB_2(t)$
$dX_3(t) = 3dt - dB_1(t) + dB_2(t)$

f) $(n = 3, m = 2)$
$dX_1(t) = dt + dB_1(t) + dB_2(t)$
$dX_2(t) = 2dt + dB_1(t) - dB_2(t)$
$dX_3(t) = -2dt - dB_1(t) + dB_2(t)$

12.7.* Determine which of the nonarbitrage markets $\{X(t)\}_{t \in [0,T]}$ of Exercise 12.6 a)–f) are complete. For those which are not complete, find a T-claim which is not attainable.

12.8. Let B_t be 1-dimensional Brownian motion. Use Theorem 12.3.3 to find $z \in \mathbf{R}$ and $\phi(t, \omega) \in \mathcal{V}(0, T)$ such that

$$F(\omega) = z + \int_0^T \phi(t, \omega)dB(t)$$

in the following cases:

(i) $F(\omega) = B^2(T, \omega)$
(ii) $F(\omega) = B^3(T, \omega)$

(iii) $F(\omega) = \exp B(T, \omega)$.
(Compare with the methods you used in Exercise 4.14.)

12.9.*   Let B_t be n-dimensional Brownian motion. Use Theorem 12.3.3 to find $z \in \mathbf{R}$ and $\phi(t, \omega) \in \mathcal{V}^n(0, T)$ such that

$$F(\omega) = z + \int_0^T \phi(t, \omega) dB(t)$$

in the following cases

(i) $F(\omega) = B^2(T, \omega)$ $(= B_1^2(T, \omega) + \cdots + B_n^2(T, \omega))$
(ii) $F(\omega) = \exp(B_1(T, \omega) + \cdots + B_n(T, \omega))$.

12.10.   Let $X(t)$ be a geometric Brownian motion given by

$$dX(t) = \alpha X(t) dt + \beta X(t) dB(t) ,$$

where α and β are constants. Use Theorem 12.3.3 to find $z \in \mathbf{R}$ and $\phi(t, \omega) \in \mathcal{V}(0, T)$ such that

$$X(T, \omega) = z + \int_0^T \phi(t, \omega) dB(t) .$$

12.11.   Suppose the market is given by

$$dX_0(t) = \rho X_0(t) dt ; \qquad X_0(0) = 1$$
$$dX_1(t) = (m - X_1(t)) dt + \sigma dB(t) ; \qquad X_1(0) = x_1 > 0 .$$

(the mean-reverting Ornstein-Uhlenbeck process) where $\rho > 0$, $m > 0$ and $\sigma \neq 0$ are constants.

a)   Find the price $E_Q[\xi(T)F]$ of the European T-claim

$$F(\omega) = X_1(T, \omega) .$$

b)   Find the replicating portfolio $\theta(t) = (\theta_0(t), \theta_1(t))$ for this claim. (Hint: Use Theorem 12.3.5, as in Example 12.3.6.)

12.12.*   Consider a market $(X_0(t), X_1(t)) \in \mathbf{R}^2$ where

$$dX_0(t) = \rho X_0(t) dt ; \qquad X_0(0) = 1 \quad \rho > 0 \text{ constant}) .$$

Find the price $E_Q[\xi(T)F]$ of the European T-claim

$$F(\omega) = B(T, \omega)$$

and find the corresponding replicating portfolio $\theta(t) = (\theta_0(t), \theta_1(t))$ in the following cases

a) $dX_1(t) = \alpha X_1(t)dt + \beta X_1(t)dB(t)$; α, β constants, $\beta \neq 0$
b) $dX_1(t) = c\, dB(t)$; $c \neq 0$ constant
c) $dX_1(t) = \alpha X_1(t)dt + \sigma\, dB(t)$; α, σ constants, $\sigma \neq 0$.

12.13. (The classical Black-Scholes formula).
Complete the details in the proof of the Black-Scholes option pricing formula (12.3.45) and the corresponding replicating portfolio formula (12.3.48) of Corollary 12.3.8.

12.14. (The American call)
Let $X(t) = (X_0(t), X_1(t))$ be as in Exercise 12.13. If the American T-claim is given by

$$F(t, \omega) = (X_1(t, \omega) - K)^+ , \qquad 0 \leq t \leq T ,$$

then the corresponding option is called *the American call*.
According to Theorem 12.3.10 the price of an American call is given by

$$p_A(F) = \sup_{\tau \leq T} E_Q[e^{-\rho\tau}(X_1(\tau) - K)^+] .$$

Prove that

$$p_A(F) = e^{-\rho T} E_Q[(X_1(T) - K)^+] ,$$

i.e. that it is always optimal to exercise the American call at the terminal time T, if at all. Hence the price of an American call option coincides with that of a European call option.
(Hint: Define

$$Y(t) = e^{-\rho t}(X_1(t) - K) .$$

a) Prove that $Y(t)$ is a Q-submartingale (Appendix C), i.e.

$$Y(t) \leq E_Q[Y(s)|\mathcal{F}_t] \qquad \text{for } s > t .$$

b) Then use the Jensen inequality (Appendix B) to prove that

$$Z(t) := e^{-\rho t}(X_1(t) - K)^+$$

is also a Q-submartingale.
c) Complete the proof by using Doob's optional sampling theorem (see the proof of Lemma 10.1.3 e)).

12.15.* (The perpetual American put)
Solve the optimal stopping problem

$$\Phi(s, x) = \sup_{\tau \geq 0} E^x [e^{-\rho(s+\tau)} (K - X(\tau))^+]$$

where

$$dX(t) = \alpha X(t)dt + \beta X(t)dB(t) \ ; \qquad X(0) = x > 0 \ .$$

Here $\rho > 0$, $K > 0$, α and $\beta \neq 0$ are constants.

If $\alpha = \rho$ then $\Phi(s, x)$ gives the price of the American put option with infinite horizon $(T = \infty)$. (Hint: Proceed as in Example 10.4.2.)

12.16.* Let

$$dX_0(t) = \rho X_0(t)dt \ ; \qquad X_0(0) = 1$$
$$dX_1(t) = \alpha X_1(t)dt + \beta X_1(t)dB(t) \ ; \qquad X_1(0) = x_1 > 0$$

be the classical Black-Scholes market. Find the replicating portfolio $\theta(t) = (\theta_0(t), \theta_1(t))$ for the following European T-claims:

a) $F(\omega) = (K - X_1(T, \omega))^+$ (the European put)

b) $F(\omega) = (X_1(T, \omega))^2$.

Remark. Note that in case a) we get $\theta_1(t) < 0$ for all $t \in [0, T]$, which means that we have to *shortsell* in order to replicate the European put. This is not necessary if we want to replicate the European *call*. See (12.3.48).

Appendix A: Normal Random Variables

Here we recall some basic facts which are used in the text.

Definition A.13 *Let* $(\Omega, \mathcal{F}, P)$ *be a probability space. A random variable* $X \colon \Omega \to \mathbf{R}$ *is* normal *if the distribution of* X *has a density of the form*

$$p_X(x) = \frac{1}{\sigma\sqrt{2\pi}} \cdot \exp\left(-\frac{(x-m)^2}{2\sigma^2}\right), \qquad (A.1)$$

where $\sigma > 0$ *and* m *are constants. In other words,*

$$P[X \in G] = \int_G p_X(x)dx, \qquad \text{for all Borel sets } G \subset \mathbf{R}.$$

If this is the case, then

$$E[X] = \int_\Omega X dP = \int_R x p_X(x)dx = m \qquad (A.2)$$

and

$$\text{var}[X] = E[(X-m)^2] = \int_R (x-m)^2 p_X(x)dx = \sigma^2. \qquad (A.3)$$

More generally, a random variable $X \colon \Omega \to \mathbf{R}^n$ is called *(multi-)normal* $\mathcal{N}(m, C)$ if the distribution of X has a density of the form

$$p_X(x_1, \cdots, x_n) = \frac{\sqrt{|A|}}{(2\pi)^{n/2}} \cdot \exp\left(-\tfrac{1}{2} \cdot \sum_{j,k}(x_j - m_j)a_{jk}(x_k - m_k)\right) \quad (A.4)$$

where $m = (m_1, \cdots, m_n) \in \mathbf{R}^n$ and $C^{-1} = A = [a_{jk}] \in \mathbf{R}^{n \times n}$ is a symmetric positive definite matrix.

If this is the case then

$$E[X] = m \qquad (A.5)$$

and

$$A^{-1} = C = [c_{jk}] \quad \text{is the covariance matrix of } X, \text{ i.e.}$$
$$c_{jk} = E[(X_j - m_j)(X_k - m_k)] \, . \tag{A.6}$$

Definition A.14 *The* characteristic function *of a random variable*
$X: \Omega \rightarrow \mathbf{R}^n$ *is the function* $\phi_X: \mathbf{R}^n \rightarrow \mathbf{C}$ *(where* $\mathbf{C}$ *denotes the complex numbers) defined by*

$$\phi_X(u_1, \cdots, u_n) = E[\exp(i(u_1 X_1 + \cdots + u_n X_n))] = \int_{\mathbf{R}^n} e^{i\langle u, x \rangle} \cdot P[X \in dx] \, , \tag{A.7}$$

where $\langle u, x \rangle = u_1 x_1 + \cdots + u_n x_n$ *(and* $i \in \mathbf{C}$ *is the imaginary unit). In other words,* ϕ_X *is the Fourier transform of* X *(or, more precisely, of the measure* $P[X \in dx]$*). Therefore we have*

Theorem A.15 *The characteristic function of* X *determines the distribution of* X *uniquely.*

It is not hard to verify the following:

Theorem A.16 *If* $X: \Omega \rightarrow \mathbf{R}^n$ *is normal* $\mathcal{N}(m, C)$*, then*

$$\phi_X(u_1, \cdots, u_n) = \exp\left(-\tfrac{1}{2} \sum_{j,k} u_j c_{jk} u_k + i \sum_j u_j m_j \right) . \tag{A.8}$$

Theorem A.4 is often used as a basis for an extended concept of a normal random variable: We *define* $X: \Omega \rightarrow \mathbf{R}^n$ to be normal (in the extended sense) if ϕ_X satisfies (A.8) for some symmetric non-negative definite matrix $C = [c_{jk}] \in \mathbf{R}^{n \times n}$ and some $m \in \mathbf{R}^n$. So by this definition it is not required that C be invertible. From now on we will use this extended definition of normality. In the text we often use the following result:

Theorem A.17 *Let* $X_j: \Omega \rightarrow \mathbf{R}$ *be random variables;* $1 \leq j \leq n$*. Then*

$$X = (X_1, \cdots, X_n) \quad \text{is normal}$$

if and only if

$$Y = \lambda_1 X_1 + \cdots + \lambda_n X_n \quad \text{is normal for all } \lambda_1, \ldots, \lambda_n \in \mathbf{R} \, .$$

Proof. If X is normal, then

$$E[\exp(iu(\lambda_1 X_1 + \cdots + \lambda_n X_n))] = \exp\left(-\tfrac{1}{2} \sum_{j,k} u\lambda_j c_{jk} u\lambda_k + i \sum_j u\lambda_j m_j \right)$$

$$= \exp\left(-\tfrac{1}{2} u^2 \sum_{j,k} \lambda_j c_{jk} \lambda_k + iu \sum_j \lambda_j m_j \right) ,$$

so Y is normal with $E[Y] = \sum \lambda_j m_j$, $\mathrm{var}[Y] = \sum \lambda_j c_{jk} \lambda_k$.

Conversely, if $Y = \lambda_1 X_1 + \cdots + \lambda_n X_n$ is normal with $E[Y] = m$ and $\mathrm{var}[Y] = \sigma^2$, then

$$E[\exp(iu(\lambda_1 X_1 + \cdots + \lambda_n X_n))] = \exp(-\tfrac{1}{2}u^2\sigma^2 + ium) ,$$

where

$$m = \sum_j \lambda_j E[X_j], \sigma^2 = E\left[\left(\sum_j \lambda_j X_j - \sum_j \lambda_j E[X_j]\right)^2\right]$$

$$= E\left[\left(\sum_j \lambda_j (X_j - m_j)\right)^2\right] = \sum_{j,k} \lambda_j \lambda_k E[(X_j - m_j)(X_k - m_k)] ,$$

where $m_j = E[X_j]$. Hence X is normal.

Theorem A.18 *Let $Y_0, Y_1, \ldots, Y_n$ be real, random variables on Ω. Assume that $X = (Y_0, Y_1, \ldots, Y_n)$ is normal and that Y_0 and Y_j are uncorrelated for each $j \geq 1$, i.e*

$$E[(Y_0 - E[Y_0])(Y_j - E[Y_j])] = 0 ; \qquad 1 \leq j \leq n .$$

Then Y_0 is independent of $\{Y_1, \cdots, Y_n\}$.

Proof. We have to prove that

$$P[Y_0 \in G_0, Y_1 \in G_1, \ldots, Y_n \in G_n] = P[Y_0 \in G_0] \cdot P[Y_1 \in G_1, \ldots, Y_n \in G_n] ,$$
$$(A.9)$$

for all Borel sets $G_0, G_1, \ldots, G_n \subset \mathbf{R}$.

We know that in the first line (and the first column) of the covariance matrix $c_{jk} = E[(Y_j - E[Y_j])(Y_k - E[Y_k])]$ only the first entry $c_{00} = \mathrm{var}[Y_0]$, is non-zero. Therefore the characteristic function of X satisfies

$$\phi_X(u_0, u_1, \ldots, u_n) = \phi_{Y_0}(u_o) \cdot \phi_{(Y_1, \ldots, Y_n)}(u_1, \ldots, u_n)$$

and this is equivalent to (A.9).

Finally we establish the following:

Theorem A.19 *Suppose $X_k \colon \Omega \to \mathbf{R}^n$ is normal for all k and that $X_k \to X$ in $L^2(\Omega)$, i.e.*
$$E[|X_k - X|^2] \to 0 \qquad as \ k \to \infty .$$

Then X is normal.

Proof. Since $|e^{i\langle u, x \rangle} - e^{i\langle u, y \rangle}| < |u| \cdot |x - y|$, we have

$$E[\{\exp(i\langle u, X_k \rangle) - \exp(i\langle u, X \rangle)\}^2] \leq |u|^2 \cdot E[|X_k - X|^2] \to 0 \quad as \ k \to \infty .$$

Therefore

$$E[\exp(i\langle u, X_k\rangle)] \to E[\exp(i\langle u, X\rangle)] \quad \text{as } k \to \infty .$$

So X is normal, with mean $E[X] = \lim E[X_k]$ and covariance matrix $C = \lim C_k$, where C_k is the covariance matrix of X_k.

Appendix B: Conditional Expectation

Let $(\Omega, \mathcal{F}, P)$ be a probability space and let $X : \Omega \to \mathbf{R}^n$ be a random variable such that $E[\|X\|] < \infty$. If $\mathcal{H} \subset \mathcal{F}$ is a σ-algebra, then *the conditional expectation of X given $\mathcal{H}$*, denoted by $E[X|\mathcal{H}]$, is defined as follows:

Definition B.1 $E[X|\mathcal{H}]$ *is the (a.s. unique) function from Ω to $\mathbf{R}^n$ satisfying:*

(1) $E[X|\mathcal{H}]$ *is $\mathcal{H}$-measurable*
(2) $\int\limits_H E[X|\mathcal{H}]dP = \int\limits_H X dP$, *for all $H \in \mathcal{H}$.*

The existence and uniqueness of $E[X|\mathcal{H}]$ comes from the Radon-Nikodym theorem: Let μ be the measure on $\mathcal{H}$ defined by

$$\mu(H) = \int\limits_H X dP \; ; \qquad H \in \mathcal{H} \,.$$

Then μ is absolutely continuous w.r.t. $P|\mathcal{H}$, so there exists a $P|\mathcal{H}$-unique $\mathcal{H}$-measurable function F on Ω such that

$$\mu(H) = \int\limits_H F dP \qquad \text{for all } H \in \mathcal{H} \,.$$

Thus $E[X|\mathcal{H}] := F$ does the job and this function is unique a.s. w.r.t. the measure $P|\mathcal{H}$.

Note that (2) is equivalent to

(2)' $$\int\limits_\Omega Z \cdot E[X|\mathcal{H}]dP = \int\limits_\Omega Z \cdot X dP \qquad \text{for all } \mathcal{H}\text{-measurable } Z \,.$$

We list some of the basic properties of the conditional expectation:

Theorem B.2 *Suppose $Y : \Omega \to \mathbf{R}^n$ is another random variable with $E[\|Y\|] < \infty$ and let $a, b \in \mathbf{R}$. Then*

a) $E[aX + bY|\mathcal{H}] = aE[X|\mathcal{H}] + bE[Y|\mathcal{H}]$
b) $E[E[X|\mathcal{H}]] = E[X]$
c) $E[X|\mathcal{H}] = X$ *if* X *is* $\mathcal{H}$-*measurable*
d) $E[X|\mathcal{H}] = E[X]$ *if* X *is independent of* $\mathcal{H}$
e) $E[Y \cdot X|\mathcal{H}] = Y \cdot E[X|\mathcal{H}]$ *if* Y *is* $\mathcal{H}$-*measurable, where* $\cdot$ *denotes the usual inner product in* $\mathbf{R}^n$.

Proof. d):   If X is independent of $\mathcal{H}$ we have for $H \in \mathcal{H}$

$$\int_H X dP = \int_\Omega X \cdot \mathcal{X}_H dP = \int_\Omega X dP \cdot \int_\Omega \mathcal{X}_H dP = E[X] \cdot P(H) \,,$$

so the constant $E[X]$ satisfies (1) and (2).

e):   We first establish the result in the case when $Y = \mathcal{X}_H$ (where $\mathcal{X}$ denotes the indicator function), for some $H \in \mathcal{H}$.
Then for all $G \in \mathcal{H}$

$$\int_G Y \cdot E[X|\mathcal{H}] dP = \int_{G \cap H} E[X|\mathcal{H}] dP = \int_{G \cap H} X dP = \int_G Y X dP \,,$$

so $Y \cdot E[X|\mathcal{H}]$ satisfies both (1) and (2). Similarly we obtain that the result is true if Y is a simple function

$$Y = \sum_{j=1}^m c_j \mathcal{X}_{H_j} \,, \qquad \text{where } H_j \in \mathcal{H} \,.$$

The result in the general case then follows by approximating Y by such simple functions.   □

Theorem B.3 *Let* $\mathcal{G}, \mathcal{H}$ *be* σ-*algebras such that* $\mathcal{G} \subset \mathcal{H}$. *Then*

$$E[X|\mathcal{G}] = E[E[X|\mathcal{H}]|\mathcal{G}] \,.$$

Proof. If $G \in \mathcal{G}$ then $G \in \mathcal{H}$ and therefore

$$\int_G E[X|\mathcal{H}] dP = \int_G X dP \,.$$

Hence $E[E[X|\mathcal{H}]|\mathcal{G}] = E[X|\mathcal{G}]$ by uniqueness.   □

The following useful result can be found in Chung (1974), Theorem 9.1.4:

Theorem B.4 (The Jensen inequality)
If $\phi \colon \mathbf{R} \to \mathbf{R}$ *is convex and* $E[|\phi(X)|] < \infty$ *then*

$$\phi(E[X|\mathcal{H}]) \leq E[\phi(X)|\mathcal{H}] \,.$$

Corollary B.5 (i) $|E[X|\mathcal{H}]| \leq E[|X| \mid \mathcal{H}]$
(ii) $|E[X|\mathcal{H}]|^2 \leq E[|X|^2 \mid \mathcal{H}]$.

Corollary B.6 *If $X_n \to X$ in L^2 then $E[X_n \mid \mathcal{H}] \to E[X \mid \mathcal{H}]$ in L^2.*

Appendix C: Uniform Integrability and Martingale Convergence

We give a brief summary of the definitions and results which are the background for the applications in this book. For proofs and more information we refer to Doob (1984), Liptser and Shiryaev (1977), Meyer (1966) or Williams (1979).

Definition C.1 *Let* $(\Omega, \mathcal{F}, P)$ *be a probability space. A family* $\{f_j\}_{j \in J}$ *of real, measurable functions* f_j *on* Ω *is called* uniformly integrable *if*

$$\lim_{M \to \infty} \left(\sup_{j \in J} \left\{ \int_{\{|f_j| > M\}} |f_j| dP \right\} \right) = 0 \; .$$

One of the most useful tests for uniform integrability is obtained by using the following concept:

Definition C.2 *A function* $\psi \colon [0, \infty) \to [0, \infty)$ *is called a u.i. (uniform integrability)* test function *if* ψ *is increasing, convex (i.e.* $\psi(\lambda x + (1 - \lambda)y) \leq \lambda \psi(x) + (1 - \lambda)\psi(y)$ *for all* $x, y \in [0, \infty)$, $\lambda \in [0, 1]$) *and*

$$\lim_{x \to \infty} \frac{\psi(x)}{x} = \infty \; .$$

So for example $\psi(x) = x^p$ *is a u.i. test function if* $p > 1$, *but not if* $p = 1$.

The justification for the name in Definition C.2 is the following:

Theorem C.3 *The family* $\{f_j\}_{j \in J}$ *is uniformly integrable if and only if there is a u.i. test function* ψ *such that*

$$\sup_{j \in J} \left\{ \int \psi(|f_j|) dP \right\} < \infty \; .$$

One major reason for the usefulness of uniform integrability is the following result, which may be regarded as the ultimate generalization of the various convergence theorems in integration theory:

Theorem C.4 *Suppose* $\{f_k\}_{k=1}^{\infty}$ *is a sequence of real measurable functions on* Ω *such that*

$$\lim_{k \to \infty} f_k(\omega) = f(\omega) \qquad \text{for a.a. } \omega .$$

Then the following are equivalent:
1) $\{f_k\}$ *is uniformly integrable*
2) $f \in L^1(P)$ *and* $f_k \to f$ *in* $L^1(P)$, *i.e.* $\int |f_k - f| dP \to 0$ *as* $k \to \infty$.

Note that 2) implies that

$$\int f_k \, dP \to \int f \, dP \qquad \text{as } k \to \infty .$$

An important application of uniform integrability is within the convergence theorems for martingales:

Let $(\Omega, \mathcal{N}, P)$ be a probability space and let $\{\mathcal{N}_t\}_{t \geq 0}$ be an increasing family of σ-algebras, $\mathcal{N}_t \subset \mathcal{N}$ for all t. A stochastic process $N_t \colon \Omega \to \mathbf{R}$ is called a *supermartingale* (w.r.t. $\{\mathcal{N}_t\}$) if N_t is $\mathcal{N}_t$-adapted, $E[|N_t|] < \infty$ for all t and

$$N_t \geq E[N_s | \mathcal{N}_t] \qquad \text{for all } s > t . \tag{C.1}$$

Similarly, if (C.1) holds with the inequality reversed for all $s > t$, then N_t is called a *submartingale*. And if (C.1) holds with equality then N_t is called a *martingale*.

As is customary we will assume that each $\mathcal{N}_t$ contains all the null sets of $\mathcal{N}$, that $t \to N_t(\omega)$ is right continuous for a.a.ω and that $\{\mathcal{N}_t\}$ is right continuous, in the sense that $\mathcal{N}_t = \bigcap_{s > t} \mathcal{N}_s$ for all $t \geq 0$.

Theorem C.5 (Doob's martingale convergence theorem I)
Let N_t be a right continuous supermartingale with the property that

$$\sup_{t > 0} E[N_t^-] < \infty ,$$

where $N_t^- = \max(-N_t, 0)$. *Then the pointwise limit*

$$N(\omega) = \lim_{t \to \infty} N_t(\omega)$$

exists for a.a. ω *and* $E[N^-] < \infty$.

Note, however, that the convergence need not be in $L^1(P)$. In order to obtain this we need uniform integrability:

Theorem C.6 (Doob's martingale convergence theorem II)
Let N_t be a right-continuous supermartingale. Then the following are equivalent:

1) $\{N_t\}_{t \geq 0}$ *is uniformly integrable*

2)   *There exists $N \in L^1(P)$ such that $N_t \to N$ a.e. (P) and $N_t \to N$ in $L^1(P)$, i.e. $\int |N_t - N| dP \to 0$ as $t \to \infty$.*

Combining Theorems C.6 and C.3 (with $\psi(x) = x^p$) we get

Corollary C.7 *Let M_t be a continuous martingale such that*

$$\sup_{t>0} E[|M_t|^p] < \infty \qquad for \ some \ p > 1 .$$

Then there exists $M \in L^1(P)$ such that $M_t \to M$ a.e. (P) and

$$\int |M_t - M| dP \to 0 \qquad as \ t \to \infty .$$

Finally, we mention that similar results can be obtained for the analogous discrete time super/sub-martingales $\{N_k, \mathcal{N}_k\}$, $k = 1, 2, \ldots$. Of course, no continuity assumptions are needed in this case. For example, we have the following result, which is used in Chapter 9:

Corollary C.8 *Let M_k; $k = 1, 2, \ldots$ be a discrete time martingale and assume that*

$$\sup_k E[|M_k|^p] < \infty \qquad for \ some \ p > 1 .$$

Then there exists $M \in L^1(P)$ such that $M_k \to M$ a.e. (P) and

$$\int |M_k - M| dP \to 0 \qquad as \ k \to \infty .$$

Corollary C.9 *Let $X \in L^p(P)$ for some $p \in [1, \infty]$, let $\{\mathcal{N}_k\}_{k=1}^{\infty}$ be an increasing family of σ-algebras, $\mathcal{N}_k \subset \mathcal{F}$ and define $\mathcal{N}_\infty$ to be the σ-algebra generated by $\{\mathcal{N}_k\}_{k=1}^{\infty}$. Then*

$$E[X|\mathcal{N}_k] \to E[X|\mathcal{N}_\infty] \qquad as \ k \to \infty ,$$

a.e. P and in $L^p(P)$.

Proof. We give the proof in the case $p = 1$. The process $M_k := E[X|\mathcal{N}_k]$ is a u.i. martingale, so there exists $M \in L^1(P)$ such that $M_k \to M$ a.e. P and in $L^1(P)$, as $k \to \infty$. It remains to prove that $M = E[X|\mathcal{N}_\infty]$: Note that

$$\|M_k - E[M|\mathcal{N}_k]\|_{L^1(P)} = \|E[M_k|\mathcal{N}_k] - E[M|\mathcal{N}_k]\|_{L^1(P)}$$
$$\leq \|M_k - M\|_{L^1(P)} \to 0 \qquad as \ k \to \infty .$$

Hence if $F \in \mathcal{N}_{k_0}$ and $k \geq k_0$ we have

$$\int_F (X - M) dP = \int_F E[X - M|\mathcal{N}_k] dP = \int_F (M_k - E[M|\mathcal{N}_k]) dP \to 0 \quad as \ k \to \infty .$$

Therefore

$$\int_F (X - M)dP = 0 \qquad \text{for all } F \in \bigcup_{k=1}^{\infty} \mathcal{N}_k$$

and hence

$$E[X|\mathcal{N}_\infty] = E[M|\mathcal{N}_\infty] = M \ .$$

$\square$

Appendix D: An Approximation Result

In this Appendix we prove an approximation result which was used in Theorem 10.4.1. We use the notation from that Theorem.

Theorem D.1 *Let $D \subset V \subset \mathbf{R}^n$ be open sets such that*
$$\partial D \quad \text{is a Lipschitz surface} \tag{D.1}$$

and let $\phi: \overline{V} \to \mathbf{R}$ be a function with the following properties
$$\phi \in C^1(V) \cap C(\overline{V}) \tag{D.2}$$

$$\phi \in C^2(V \setminus \partial D) \quad \text{and the second order derivatives} \tag{D.3}$$
$$\text{of } \phi \text{ are locally bounded near } \partial D \,,$$

Then there exists a sequence $\{\phi_j\}_{j=1}^\infty$ of functions $\phi_j \in C^2(V) \cap C(\overline{V})$ such that

$$\phi_j \to \phi \quad \text{uniformly on compact subsets of } \overline{V}, \text{ as } j \to \infty \tag{D.4}$$
$$L\phi_j \to L\phi \quad \text{uniformly on compact subsets of } V \setminus \partial D, \text{ as } j \to \infty \tag{D.5}$$
$$\{L\phi_j\}_{j=1}^\infty \quad \text{is locally bounded on } V \,. \tag{D.6}$$

Proof. We may assume that ϕ is extended to a continuous function on the whole of $\mathbf{R}^n$. Choose a C^∞ function $\eta: \mathbf{R}^n \to [0, \infty)$ with compact support such that

$$\int_{\mathbf{R}^n} \eta(y)dy = 1 \tag{D.7}$$

and put

$$\eta_\epsilon(x) = \epsilon^{-n}\eta\left(\frac{x}{\epsilon}\right) \qquad \text{for } \epsilon > 0, \ x \in \mathbf{R}^n \,. \tag{D.8}$$

Fix a sequence $\epsilon_j \downarrow 0$ and define

$$\phi_j(x) = (\phi * \eta_{\epsilon_j})(x) = \int_{\mathbf{R}^n} \phi(x - z)\eta_{\epsilon_j}(z)dz = \int_{\mathbf{R}^n} \phi(y)\eta_{\epsilon_j}(x - y)dy \,, \tag{D.9}$$

i.e. ϕ_j is the *convolution* of ϕ and η_{ϵ_j}. Then it is well-known that $\phi_j(x) \to \phi(x)$ uniformly on compact subsets of any set in $\overline{V}$ where ϕ is continuous. See e.g. Folland (1984), Theorem 8.14 (c). Note that since η has compact support we need not assume that ϕ is globally bounded, just locally bounded (which follows from continuity).

We proceed to verify (D.4)–(D.6): Let $W \subset V$ be open with a Lipschitz boundary. Put $V_1 = W \cap D$, $V_2 = W \setminus \overline{D}$.

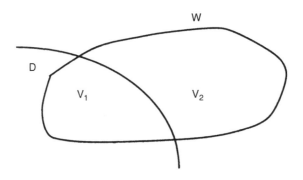

Then V_1, V_2 are Lipschitz domains and integration by parts gives, for $i = 1, 2$ and $x \in W \setminus \partial D$

$$
\int\limits_{V_i} \phi(y) \frac{\partial^2}{\partial y_k \partial y_\ell} \eta_{\epsilon_j}(x - y) dy =
$$

$$
\int\limits_{\partial V_i} \phi(y) \frac{\partial}{\partial y_\ell} \eta_{\epsilon_j}(x - y) n_{ik} d\nu(y) - \int\limits_{V_i} \frac{\partial \phi}{\partial y_k}(y) \frac{\partial}{\partial y_\ell} \eta_{\epsilon_j}(x - y) dy , \quad \text{(D.10)}
$$

where n_{ik} is component number k of the outer unit normal $\mathbf{n}_i$ from V_i at ∂V_i. (This outer normal exists a.e. with respect to the surface measure ν on ∂V_i since ∂V_i is a Lipschitz surface.)

Another integration by parts yields

$$
\int\limits_{V_i} \frac{\partial \phi}{\partial y_k}(y) \frac{\partial}{\partial y_\ell} \eta_{\epsilon_j}(x - y) dy =
$$

$$
\int\limits_{\partial V_i} \frac{\partial \phi}{\partial y_k}(y) \eta_{\epsilon_j}(x - y) n_{i\ell} d\nu(y) - \int\limits_{V_i} \frac{\partial^2 \phi}{\partial y_k \partial y_\ell}(y) \eta_{\epsilon_j}(x - y) dy . \quad \text{(D.11)}
$$

Combining (D.10) and (D.11) we get

$$
\int\limits_{V_i} \phi(y) \frac{\partial^2}{\partial y_k \partial y_\ell} \eta_{\epsilon_j}(x - y) dy =
$$

$$\int_{\partial V_i} \left[\phi(y) \frac{\partial}{\partial y_\ell} \eta_{\epsilon_j}(x - y) n_{ik} - \frac{\partial \phi}{\partial y_k}(y) \eta_{\epsilon_j}(x - y) n_{i\ell} \right] d\nu(y)$$

$$+ \int_{V_i} \frac{\partial^2 \phi}{\partial y_k \partial y_\ell}(y) \eta_{\epsilon_j}(x - y) dy ; \qquad i = 1, 2 . \tag{D.12}$$

Adding (D.12) for $i = 1, 2$ and keeping in mind that the outer unit normal for V_i is the inner unit normal for V_{3-i} on $\partial V_1 \cap \partial V_2$, we get

$$\int_W \phi(y) \frac{\partial^2}{\partial y_k \partial y_\ell} \eta_{\epsilon_j}(x - y) dy =$$

$$\int_{\partial W} \left\{ \phi(y) \frac{\partial}{\partial y_\ell} \eta_{\epsilon_j}(x - y) N_k - \frac{\partial \phi}{\partial y_k}(y) \eta_{\epsilon_j}(x - y) N_\ell \right\} d\nu(y)$$

$$+ \int_W \frac{\partial^2 \phi}{\partial y_k \partial y_\ell}(y) \eta_{\epsilon_j}(x - y) dy , \tag{D.13}$$

where N_k, N_ℓ are components number k, ℓ of the outer unit normal $\mathbf{N}$ from W at ∂W.

If we fix $x \in W \setminus \partial D$ then for j large enough we have $\eta_{\epsilon_j}(x - y) = 0$ for all y outside W and for such j we get from (D.13)

$$\int_{\mathbf{R}^n} \phi(y) \frac{\partial^2}{\partial y_k \partial y_\ell} \eta_{\epsilon_j}(x - y) dy = \int_{\mathbf{R}^n} \frac{\partial^2 \phi}{\partial y_k \partial y_\ell}(y) \eta_{\epsilon_j}(x - y) dy . \tag{D.14}$$

In other words, we have proved that

$$\frac{\partial^2}{\partial x_k \partial x_\ell} \phi_j(x) = \left(\frac{\partial^2 \phi}{\partial y_k \partial y_\ell} * \eta_{\epsilon_j} \right)(x) \qquad \text{for } x \in V \setminus \partial D . \tag{D.15}$$

Similarly, integration by parts applied to W gives, if j is large enough

$$\int_W \phi(y) \frac{\partial}{\partial y_k} \eta_{\epsilon_j}(x - y) dy = - \int_W \frac{\partial \phi}{\partial y_k}(y) \eta_{\epsilon_j}(x - y) dy$$

from which we conclude that

$$\frac{\partial}{\partial x_k} \phi_j(x) = \left(\frac{\partial \phi}{\partial y_k} * \eta_{\epsilon_j} \right)(x) \qquad \text{for } x \in V . \tag{D.16}$$

From (D.15) and (D.16), combined with Theorem 8.14 (c) in Folland (1984), we get that

$$\frac{\partial \phi_j}{\partial x_k} \to \frac{\partial \phi}{\partial x} \quad \text{uniformly on compact subsets of } V \text{ as } j \to \infty \tag{D.17}$$

and

$$\frac{\partial^2 \phi_j}{\partial x_k \partial x_\ell} \to \frac{\partial^2 \phi}{\partial x_k \partial x_\ell} \quad \text{uniformly on compact subsets of } V \setminus \partial D \text{ as } j \to \infty .$$

$$(D.18)$$

Moreover, $\left\{ \frac{\partial \phi_j}{\partial x_k} \right\}_{j=1}^{\infty}$ and $\left\{ \frac{\partial^2 \phi_j}{\partial x_k \partial x_\ell} \right\}_{j=1}^{\infty}$ are locally bounded on V, by (D.15), (D.16) combined with the assumptions (D.2), (D.3).

We conclude that (D.4)–(D.6) hold.   □

Solutions and Additional Hints to Some of the Exercises

2.4. a) Put $A = \{\omega; |X(\omega)| \geq \lambda\}$. Then

$$\int_\Omega |X(\omega)|^p dP(\omega) \geq \int_A |X(\omega)|^p dP(\omega) \geq \lambda^p P(A).$$

□

b) By a) we have (choosing $p = 1$)

$$P[|X| \geq \lambda] = P[\exp(k|X|) \geq \exp(k\lambda)]$$
$$\leq \exp(-k\lambda)E[\exp(k|X|)].$$

□

2.6. $P\left(\bigcap_{m=1}^\infty \bigcup_{k=m}^\infty A_k\right) = \lim_{m\to\infty} P\left(\bigcup_{k=m}^\infty A_k\right) \leq \overline{\lim_{m\to\infty}} \sum_{k=m}^\infty P(A_k) = 0$

since $\sum_{k=1}^\infty P(A_k) < \infty$.

Hence

$$P(\{\omega; \omega \text{ belongs to infinitely many } A_k\text{'s}\})$$
$$= P(\{\omega; (\forall m)(\exists k \geq m) \text{ s.t. } \omega \in A_k\})$$
$$= P\left(\left\{\omega; \omega \in \bigcap_{m=1}^\infty \bigcup_{k=m}^\infty A_k\right\}\right) = 0.$$

□

2.7. a) We must verify that
(i) $\phi \in \mathcal{G}$
(ii) $F \in \mathcal{G} \Rightarrow F^C \in \mathcal{G}$
(iii) $F_1, \ldots, F_k \in \mathcal{G} \Rightarrow F_1 \cup F_2 \cup \cdots \cup F_k \in \mathcal{G}$

Since each element F of $\mathcal{G}$ consists of a union of some of the G_i's, these statements are easily verified.

b) Let $\mathcal{F}$ be a *finite* σ-algebra of subsets of Ω. For each $x \in \Omega$ define

$$F_x = \bigcap \{F \in \mathcal{F}; x \in F\}$$

Since $\mathcal{F}$ is finite we have $F_x \in \mathcal{F}$ and clearly F_x is the smallest set in $\mathcal{F}$ which contains x. We claim that for given $x, y \in \Omega$ the following holds:

(i) Either $F_x \cap F_y = \emptyset$ or $F_x = F_y$.

To prove this we argue by contradiction: Assume that we both have

(ii) $F_x \cap F_y \neq \emptyset$ and $F_x \neq F_y$, e.g. $F_x \setminus F_y \neq \emptyset$.

Then there are at most two possibilities:

a) $x \in F_x \cap F_y$
     Then $F_x \cap F_y$ is a set in $\mathcal{F}$ containing x and $F_x \cap F_y$ is strictly smaller than F_x. This is impossible.
b) $x \in F_x \setminus F_y$
     Then $F_x \setminus F_y$ is a set in $\mathcal{F}$ containing x and $F_x \setminus F_y$ is strictly smaller than F_x. This is again impossible.

Hence the claim (i) is proved. Therefore there exist $x_1, \ldots, x_m \in \Omega$ such that

$$F_{x_1}, F_{x_2}, \ldots, F_{x_m}$$

forms a partition of Ω. Hence any $F \in \mathcal{F}$ is a disjoint union of some of these F_{x_i}'s. Therefore $\mathcal{F}$ is of the form $\mathcal{G}$ described in a).   □

c) Let $X : \Omega \to \mathbf{R}$ be $\mathcal{F}$-measurable. Then for all $x \in \mathbf{R}$ we have

$$\{\omega \in \Omega; X(\omega) = c\} = X^{-1}(\{c\}) \in \mathcal{F}$$

Therefore X has the constant value c on a finite (possibly empty) union of the F_i's, where $F_i = F_{x_i}$ is as in b).   □

2.9.   With
$$X_t(\omega) = \begin{cases} 1 & \text{if } t = \omega \\ 0 & \text{otherwise} \end{cases}$$
and $Y_t(\omega) = 0$ for all $(t, \omega) \in [0, \infty) \times [0, \infty)$ we have

$$P[X_t = Y_t] = P[X_t = 0] = P(\{\omega; \omega \neq t\}) = 1 .$$

Hence X_t is a version of Y_t.

2.13.   Define $D_\rho = \{x \in \mathbf{R}^2; |x| < \rho\}$. Then by (2.2.2) with $n = 2$ we have

$$P^0[B_t \in D_\rho] = (2\pi t)^{-1} \iint_{D_\rho} e^{-\frac{x^2 + y^2}{2t}} \, dx \, dy .$$

Introducing polar coordinates

$$x = r\cos\theta\,;$$
$$y = r\sin\theta\,;\quad 0 \le r \le \rho,\quad 0 \le \theta \le 2\pi$$

we get

$$P^0[B_t \in D_\rho] = (2\pi t)^{-1} \int_0^{2\pi}\int_0^{\rho} re^{-\frac{r^2}{2t}}\,dr\,d\theta$$

$$= (2\pi t)^{-1}\cdot 2\pi t \left[-e^{-\frac{r^2}{2t}}\right]_0^{\rho} = 1 - e^{-\frac{\rho^2}{2t}}\ .$$

2.14. The expected total length of time that B_t spends in the set K is given by

$$E^x\left[\int_0^\infty \mathcal{X}_K(B_t)dt\right] = \int_0^\infty P^x[B_t \in K]dt = \int_0^\infty (2\pi t)^{-n/2}\int_K e^{-\frac{(x-y)^2}{2t}}\,dy = 0,$$

since K has Lebesgue measure 0.

2.15. $P[\widetilde{B}_{t_1} \in F_1,\dots,\widetilde{B}_{t_k} \in F_k] = P[B_{t_1} \in U^{-1}F_1,\dots,B_{t_k} \in U^{-1}F_k]$

$$= \int_{U^{-1}F_1\times\cdots\times U^{-1}F_k} p(t_1,0,x_1)p(t_2-t_1,x_1,x_2)\cdots p(t_k-t_{k-1},x_{k-1},x_k)$$
$$\times dx_1\cdots dx_k$$

$$= \int_{F_1\times\cdots\times F_k} p(t_1,0,y_1)p(t_2-t_1,y_1,y_2)\cdots p(t_k-t_{k-1},y_{k-1},y_k)dy_1\cdots dy_k$$
$$= P[B_{t_1} \in F_1,\dots,B_{t_k} \in F_k]\,,$$

by (2.2.1), using the substitutions $y_j = Ux_j$ and the fact that $|Ux_j - Ux_{j-1}|^2 = |x_j - x_{j-1}|^2$.

2.17. a) $E[(Y_n(t,\cdot) - t)^2] = E\left[\left(\sum_{k=0}^{2^n-1}(\Delta B_k)^2 - \sum_{k=0}^{2^n-1} 2^{-n}t\right)^2\right]$

$$= E\left[\left\{\sum_{k=0}^{2^n-1}((\Delta B_k)^2 - 2^{-n}t)\right\}^2\right]$$

$$= E\left[\sum_{j,k=0}^{2^n-1}((\Delta B_j)^2 - 2^{-n}t)((\Delta B_k)^2 - 2^{-n}t)\right]$$

$$= \sum_{k=0}^{2^n-1} E[((\Delta B_k)^2 - 2^{-n}t)^2]$$

$$= \sum_{k=0}^{2^n-1} E[(\Delta B_k)^4 - 2\cdot 2^{-2n}t^2 + 2^{-2n}t^2]$$

$$= \sum_{k=0}^{2^n-1} 2 \cdot 2^{-2n} t^2 = 2 \cdot 2^{-n} t^2 \to 0 \qquad \text{as } n \to \infty .$$

b)   This follows from the following general result: If the quadratic variation of a real function over an interval is positive, then the total variation of the function over that interval is infinite.

3.1.   Using the notation $\Delta X_j = X_{t_j+1} - X_{t_j}$ we have

$$
\begin{aligned}
tB_t &= \sum_j \Delta(t_j B_j) = \sum_j (t_{j+1} B_{t_j+1} - t_j B_{t_j}) \\
&= \sum_j t_j \, \Delta B_j + \sum_j B_{j+1} \Delta t_j \\
&\to \int_0^t s \, dB_s + \int_0^t B_s ds \quad \text{as } \Delta t_j \to 0 .
\end{aligned}
$$

3.3.   a) Suppose X_t is a martingale with respect to some filtration $\{\mathcal{N}_t\}_{t \geq 0}$. Then $\mathcal{H}_t^{(X)} \subseteq \mathcal{N}_t$ and hence, for $s > t$,

$$E[X_s | \mathcal{H}_t^{(X)}] = E[E[X_s | \mathcal{N}_t] | \mathcal{H}_t^{(X)}] = E[X_t | \mathcal{H}_t^{(X)}] = X_t .$$

b)   If X_t is a martingale with respect to $\mathcal{H}_t^{(X)}$ then

$$E[X_t | \mathcal{H}_0^{(X)}] = X_0$$

i.e.

$$E[X_t] = E[X_0] \quad \text{for all } t \geq 0 .$$

c)   The process $X_t := B_t^3$ satisfies $(*)$, but it is not a martingale. To see this, choose $s > t$ and consider

$$
\begin{aligned}
E[B_s^3 | \mathcal{F}_t] &= E[(B_s - B_t)^3 + 3B_s^2 B_t - 3B_s B_t^2 + B_t^3 | \mathcal{F}_t] \\
&= 0 + 3B_t E[B_s^2 | \mathcal{F}_t] - 3B_t^2 E[B_s | \mathcal{F}_t] + B_t^3 \\
&= 3B_t E[(B_s - B_t)^2 + 2B_s B_t - B_t^2 | \mathcal{F}_t] - 2B_t^3 \\
&= 3B_t(s - t) + 6B_t^3 - 3B_t^3 - 2B_t^3 = 3B_t(s - t) + B_t^3 \neq B_t^3 .
\end{aligned}
$$

3.4.   (i) If $X_t = B_t + t$ then $E[X_t] = B(0) + t \neq E[X_0]$, so X_t does not satisfy $(*)$ of Exercise 3.3 b). Hence X_t cannot be a martingale.

(ii) If $X_t = B_t^2$ then $E[X_t] = nt + B_0^2 \neq E[X_0]$, so X_t cannot be a martingale.

(iii) If $X_t = t^2 B_t - 2 \int_0^t r \, B_r dr$ then, for $s > t$,

$$E[X_s|\mathcal{F}_t] = E[s^2 B_s|\mathcal{F}_t] - 2\int_0^t r\,B_r dr - 2\int_t^s r\,E[B_r|\mathcal{F}_t]dr$$

$$= s^2 B_t - 2\int_0^t r\,B_r dr - 2B_t\int_t^s r\,dr$$

$$= s^2 B_t - 2\int_0^t r\,B_r dr - B_t(s^2 - t^2) = t^2 B_t - 2\int_0^t r\,B_r dr = X_t\ .$$

Hence X_t is a martingale.   □

(iv) If $X_t = B_1(t)B_2(t)$ then

$$E[X_s|\mathcal{F}_t] = E[B_1(s)B_2(s)|\mathcal{F}_t]$$
$$= E[(B_1(s) - B_1(t))(B_2(s) - B_2(t))|\mathcal{F}_t]$$
$$+ E[B_1(t)(B_2(s) - B_2(t))|\mathcal{F}_t]$$
$$+ E[B_2(t)(B_1(s) - B_1(t))|\mathcal{F}_t] + E[B_1(t)B_2(t)|\mathcal{F}_t]$$
$$= E[(B_1(s) - B_1(t)) \cdot (B_2(s) - B_2(t))] + 0 + 0 + B_1(t)B_2(t)$$
$$= E[B_1(s) - B_1(t)] \cdot E[B_2(s) - B_2(t)] + B_1(t)B_2(t)$$
$$= B_1(t)B_2(t) = X_t\ .$$

Hence X_t is a martingale.   □

3.5.   To prove that $M_t := B_t^2 - t$ is a martingale, choose $s > t$ and consider

$$E[M_s\,|\,\mathcal{F}_t] = E[B_s^2 - s\,|\,\mathcal{F}_t] = E[(B_s - B_t)^2 + 2B_s B_t - B_t^2\,|\,\mathcal{F}_t] - s$$
$$= s - t + 2B_t^2 - B_t^2 - s = B_t^2 - t = M_t\ .$$

                                                               □

3.6.   To prove that $N_t := B_t^3 - 3tB_t$ is a martingale, choose $s > t$ and consider

$$E[N_s\,|\,\mathcal{F}_t] = E[(B_s - B_t)^3 + 3B_s^2 B_t - 3B_s B_t^2 + B_t^3\,|\,\mathcal{F}_t] - 3sB_t$$
$$= 3B_t E[B_s^2\,|\,\mathcal{F}_t] - 3B_t^2 B_t + B_t^3 - 3sB_t$$
$$= 3B_t E[(B_s - B_t)^2 + 2B_s B_t - B_t^2\,|\,\mathcal{F}_t] - 2B_t^3 - 3sB_t$$
$$= 3B_t(s - t) + B_t^3 - 3sB_t = B_t^3 - 3tB_t = N_t\ .$$

                                                               □

3.8.   a) If $M_t = E[Y\,|\,\mathcal{F}_t]$ then for $s > t$,

$$E[M_s\,|\,\mathcal{F}_t] = E[E[Y\,|\,\mathcal{F}_s]\,|\,\mathcal{F}_t] = E[Y\,|\,\mathcal{F}_t] = M_t\ .$$

                                                               □

3.9. $\int_0^T B_t \circ dB_t = \frac{1}{2}B_T^2$ if $B_0 = 0$.

3.12. (i) a) $dX_t = (\gamma + \frac{1}{2}\alpha^2)X_t dt + \alpha X_t dB_t$.

 b) $dX_t = \frac{1}{2}\sin X_t[\cos X_t - t^2]dt + (t^2 + \cos X_t)dB_t$.

(ii) a) $dX_t = (r - \frac{1}{2}\alpha^2)X_t dt + \alpha X_t \circ dB_t$.

 b) $dX_t = (2e^{-X_t} - X_t^3)dt + X_t^2 \circ dB_t$.

3.15. Suppose

$$C + \int_S^T f(t,w)dB_t(w) = D + \int_S^T g(t,w)dB_t(w)$$

By taking expectation we get

$$C = D .$$

Hence

$$\int_S^T f(t,w)dB_t(w) = \int_S^T g(t,w)dB_t(w)$$

and therefore, by the Itô isometry,

$$0 = E\Big[\Big(\int_S^T (f(t,w) - g(t,w))dB(t)\Big)^2\Big] = E\Big[\int_S^T (f(t,w) - g(t,w))^2 dt\Big],$$

which implies that

$$f(t,w) = g(t,w) \qquad \text{for a.a. } (t,w) \in [S,T] \times \Omega.$$

□

4.1. a) $dY_t = 2B_t dB_t + dt$.

 b) $dY_t = \big(1 + \frac{1}{2}e^{B_t}\big)dt + e^{B_t}dB_t$.

 c) $dY_t = 2dt + 2B_1 dB_1(t) + 2B_2 dB_2(t)$.

 d) $dY_t = \begin{bmatrix} dt \\ dB_t \end{bmatrix} = \begin{bmatrix} 1 \\ 0 \end{bmatrix}dt + \begin{bmatrix} 0 \\ 1 \end{bmatrix}dB_t$.

 e) $dY_1(t) = dB_1(t) + dB_2(t) + dB_3(t)$

   $dY_2(t) = dt - B_3(t)dB_1(t) + 2B_2(t)dB_2(t) - B_1(t)dB_3(t)$

   or

$$dY_t = \begin{bmatrix} dY_1(t) \\ dY_2(t) \end{bmatrix} = \begin{bmatrix} 0 \\ 1 \end{bmatrix}dt + \begin{bmatrix} 1 & 1 & 1 \\ -B_3(t) & 2B_2(t) & -B_1(t) \end{bmatrix}\begin{bmatrix} dB_1(t) \\ dB_2(t) \\ dB_3(t) \end{bmatrix}.$$

4.2. By Itô's formula we get

$$d\big(\tfrac{1}{3}B_t^3\big) = B_t^2 dB_t + B_t dt .$$

Hence

$$\tfrac{1}{3}B_t^3 = \int\limits_0^t B_s^2 dB_s + \int\limits_0^t B_s ds \ .$$

□

4.3. If we apply the Itô formula with $g(x,y) = x \cdot y$ we get

$$d(X_t Y_t) = d(g(X_t, Y_t)) = \frac{\partial g}{\partial x}(X_t, Y_t)dX_t + \frac{\partial g}{\partial y}(X_t, Y_t)dY_t$$

$$+ \tfrac{1}{2}\frac{\partial^2 g}{\partial x^2}(X_t, Y_t)\cdot(dX_t)^2 + \frac{\partial^2 g}{\partial x \partial y}(X_t, Y_t)dX_t dY_t + \tfrac{1}{2}\frac{\partial^2 g}{\partial y^2}(X_t, Y_t)\cdot(dY_t)^2$$

$$= Y_t dX_t + X_t dY_t + dX_t dY_t \ .$$

From this we obtain

$$X_t Y_t = X_0 Y_0 + \int\limits_0^t Y_s dX_s + \int\limits_0^t X_s dY_s + \int\limits_0^t dX_s dY_s \ .$$

□

4.5. $E[B_t^6] = 15t^3$ if $B_0 = 0$.

4.11. b) By the Itô formula we have

$$d\big(e^{\frac{1}{2}t}\sin B_t\big) = \tfrac{1}{2}e^{\frac{1}{2}t}\sin B_t dt + e^{\frac{1}{2}t}\cos B_t dB_t + \tfrac{1}{2}e^{\frac{1}{2}t}(-\sin B_t)dt$$

$$= e^{\frac{1}{2}t}\cos B_t dB_t \ .$$

Since

$$f(t) := e^{\frac{1}{2}t}\cos B_t \in \mathcal{V}(0,T) \qquad \text{for all } T,$$

we conclude that $X_t = e^{\frac{1}{2}}\sin B_t$ is a martingale.

□

c) By the Itô formula (or Exercise 4.3) we get

$$d\big((B_t + t)\exp(-B_t - \tfrac{1}{2}t)\big) = (B_t + t)\exp(-B_t - \tfrac{1}{2}t)(-1)dB_t$$
$$+ \exp(-B_t - \tfrac{1}{2}t)(dB_t + dt) + \exp(-B_t - \tfrac{1}{2}t)(-1)dt$$
$$= \exp(-B_t - \tfrac{1}{2}t)(1 - t - B_t)dB_t \ ,$$

where we have used that

$$d(\exp(-B_t - \tfrac{1}{2}t)) = -\exp(-B_t - \tfrac{1}{2}t)dB_t \ .$$

Since

$$f(t) := \exp(-B_t - \tfrac{1}{2}t)(1 - t - B_t) \in \mathcal{V}(0,T)$$

for all $T > 0$, we conclude that $X_t = (B_t + t)\exp(-B_t - \tfrac{1}{2}t)$ is a martingale.

□

4.14. a) $B_T(\omega) = \int\limits_0^T 1 \cdot dB_t$.

□

b) By integration by parts we have

$$\int_0^T B_t dt = T B_T - \int_0^T t\, dB_t = \int_0^T (T - t) dB_t \ .$$

□

c) By the Itô formula we have $d(B_t^2) = 2B_t dB_t + dt$. This gives

$$B_T^2 = T + \int_0^T 2B_t dB_t \ .$$

□

d) By the Itô formula we have

$$d(B_t^3) = 3B_t^2 dB_t + 3B_t dt \ .$$

Combined with 4.14 b) this gives

$$B_T^3 = \int_0^T 3B_t^2 dB_t + 3\int_0^T B_t dt = \int_0^T (3B_t^2 + 3(T - t)) dB_t \ .$$

□

e) Since $d\left(e^{B_t - \frac{1}{2}t}\right) = e^{B_t - \frac{1}{2}t} dB_t$ we have

$$e^{B_T - \frac{1}{2}T} = 1 + \int_0^T e^{B_t - \frac{1}{2}t} dB_t$$

or

$$e^{B_T} = e^{\frac{1}{2}T} + \int_0^T e^{B_t + \frac{1}{2}(T-t)} dB_t \ .$$

□

f) By Exercise 4.11 b) we have

$$d\left(e^{\frac{1}{2}t} \sin B_t\right) = e^{\frac{1}{2}t} \cos B_t dB_t$$

or

$$e^{\frac{1}{2}T} \sin B_T = \int_0^T e^{\frac{1}{2}t} \cos B_t dB_t \ .$$

Hence

$$\sin B_T = \int_0^T e^{\frac{1}{2}(t-T)} \cos B_t dB_t \ .$$

□

5.3. $X_t = X_0 \cdot \exp\left(\left(r - \frac{1}{2}\sum_{k=1}^n \alpha_k^2\right)t + \sum_{k=1}^n \alpha_k B_k(t)\right)$ (if $B(0) = 0$).

5.4. (i) $X_1(t) = X_1(0) + t + B_1(t)$,

$X_2(t) = X_2(0) + X_1(0)B_2(t) + \int\limits_0^t s dB_2(s) + \int\limits_0^t B_1(s)dB_2(s),$

assuming (as usual) that $B(0) = 0$.

(ii) $X_t = e^t X_0 + \int\limits_0^t e^{t-s}dB_s$.

(iii) $X_t = e^{-t}X_0 + e^{-t}B_t$ (assuming $B_0 = 0$).

5.6. By the Itô formula we have

$$dF_t = F_t(-\alpha\, dB_t + \tfrac{1}{2}\alpha^2 dt) + \tfrac{1}{2}F_t\alpha^2 dt$$
$$= F_t(-\alpha\, dB_t + \alpha^2 dt) .$$

Therefore, by Exercise 4.3,

$$d(F_t Y_t) = F_t dY_t + Y_t dF_t + dF_t dY_t$$
$$= F_t dY_t + Y_t F_t(-\alpha\, dB_t + \alpha^2 dt) + (-\alpha\, F_t dB_t)(\alpha\, Y_t dB_t)$$
$$= F_t(dY_t - \alpha Y_t dB_t) = F_t r\, dt .$$

Integrating this we get

$$F_t Y_t = F_0 Y_0 + \int\limits_0^t r\, F_s ds$$

or

$$Y_t = Y_0 F_t^{-1} + F_t^{-1} \int\limits_0^t r\, F_s ds$$

$$= Y_0 \exp(\alpha B_t - \tfrac{1}{2}\alpha^2 t) + r\int\limits_0^t \exp(\alpha(B_t - B_s) - \tfrac{1}{2}\alpha^2(t-s))ds .$$

5.7. a) $X_t = m + (X_0 - m)e^{-t} + \sigma\int\limits_0^t e^{s-t}dB_s$.

b) $E[X_t] = m + (X_0 - m)e^{-t}.$
   $\text{Var}[X_t] = \frac{\sigma^2}{2}[1 - e^{-2t}]$.

5.8. $X(t) = \begin{bmatrix} X_1(t) \\ X_2(t) \end{bmatrix} = \exp(tJ)X(0) + \exp(tJ)\int\limits_0^t \exp(-sJ)MdB(s),$

where

$$J = \begin{bmatrix} 0 & 1 \\ -1 & 0 \end{bmatrix}, \quad M = \begin{bmatrix} \alpha & 0 \\ 0 & \beta \end{bmatrix}, \quad dB(s) = \begin{bmatrix} dB_1(s) \\ dB_2(s) \end{bmatrix}$$

and

$$\exp(tJ) = I + tJ + \frac{t^2}{2}J^2 + \cdots + \frac{t^n}{n!}J^n + \cdots \in \mathbf{R}^{2\times 2} .$$

Using that $J^2 = -I$ this can be rewritten as

$$X_1(t) = X_1(0)\cos(t) + X_2(0)\sin(t) + \int_0^t \alpha\cos(t-s)dB_1(s)$$

$$+ \int_0^t \beta\sin(t-s)dB_2(s) \,,$$

$$X_2(t) = -X_1(0)\sin(t) + X_2(0)\cos(t) - \int_0^t \alpha\sin(t-s)dB_1(s)$$

$$+\beta \int_0^t \cos(t-s)dB_2(s) \,.$$

5.11.   Hint: To prove that $\lim_{t\to 1}(1-t)\int_0^t \frac{dB_s}{1-s} = 0$ a.s., put $M_t = \int_0^t \frac{dB_s}{1-s}$ for $0 \le t < 1$ and apply the martingale inequality to prove that

$$P[\sup\{(1-t)|M_t|; t \in [1-2^{-n}, 1-2^{-n-1}]\} > \epsilon] \le 2\epsilon^{-2} \cdot 2^{-n} \,.$$

Hence by the Borel-Cantelli lemma we obtain that for a.a. ω there exists $n(\omega) < \infty$ such that

$$n \ge n(\omega) \Rightarrow \omega \notin A_n \,,$$

where

$$A_n = \{\omega; \sup\{(1-t)|M_t|; t \in [1-2^{-n}, 1-2^{-n-1}]\} > 2^{-\frac{n}{4}}\} \,.$$

5.12.   a) If we introduce

$$x_1(t) = y(t), \quad x_2(t) = y'(t) \quad \text{and} \quad X(t) = \begin{bmatrix} x_1(t) \\ x_2(t) \end{bmatrix}$$

then the equation

$$y''(t) + (1 + \epsilon W_t)y(t) = 0$$

can be written

$$X'(t) := \begin{bmatrix} x_1'(t) \\ x_2'(t) \end{bmatrix} = \begin{bmatrix} y'(t) \\ y''(t) \end{bmatrix} = \begin{bmatrix} x_2(t) \\ -(1 + \epsilon W_t)x_1(t) \end{bmatrix}$$

which is intepreted as the stochastic differential equation

$$dX(t) = \begin{bmatrix} x_2(t)dt \\ -x_1(t)dt - \epsilon x_1(t)dB(t) \end{bmatrix}$$

$$= \begin{bmatrix} 0 & 1 \\ -1 & 0 \end{bmatrix}\begin{bmatrix} x_1(t) \\ x_2(t) \end{bmatrix}dt - \epsilon\begin{bmatrix} 0 & 0 \\ 1 & 0 \end{bmatrix}\begin{bmatrix} x_1(t) \\ x_2(t) \end{bmatrix}dB_t$$

$$= KX(t)dt - \epsilon LX(t)dB_t \,,$$

where

$$K = \begin{bmatrix} 0 & 1 \\ -1 & 0 \end{bmatrix} \quad \text{and} \quad L = \begin{bmatrix} 0 & 0 \\ 1 & 0 \end{bmatrix}.$$

b) By the above interpretation we have

$$y'(t) = y'(0) + \int_0^t y''(s)ds = y'(0) - \int_0^t y(s)ds - \epsilon \int_0^t y(s)dB_s$$

Hence, if we apply a stochastic Fubini theorem,

$$y(t) = y(0) + \int_0^t y'(s)ds$$

$$= y(0) + y'(0)t - \int_0^t \left(\int_0^s y(r)dr \right) ds - \epsilon \int_0^t \left(\int_0^s y(r)dB_r \right) ds$$

$$= y(0) + y'(0)t - \int_0^t \left(\int_r^t y(r)ds \right) dr - \epsilon \int_0^t \left(\int_r^t y(r)ds \right) dB_r$$

$$= y(0) + y'(0)t + \int_0^t (r-t)y(r)dr + \epsilon \int_0^t (r-t)y(r)dB_r .$$

□

5.16.   c) $X_t = \exp(\alpha B_t - \frac{1}{2}\alpha^2 t)\left[x^2 + 2\int_0^t \exp(-2\alpha B_s + \alpha^2 s)ds\right]^{1/2}$.

6.15.   We have

a)
$$X_t = X_0 \exp(\sigma \, dB_t + (\mu - \frac{1}{2}\sigma^2)t) = x \exp(\xi_t - \frac{1}{2}\sigma^2 t) .$$

Therefore

$$\mathcal{M}_t \subseteq \mathcal{N}_t .$$

On the other hand, since

$$\xi_t = \frac{1}{2}\sigma^2 t + \ln \frac{X_t}{x}$$

we see that $\mathcal{N}_t \subseteq \mathcal{M}_t$. Hence $\mathcal{M}_t = \mathcal{N}_t$.

b) Consider the filtering problem

(system) $d\mu = 0, \ \bar{\mu} = E[\mu], \ a^2 = E[(\mu - \bar{\mu})^2] = \theta^{-1}$
(observations) $d\xi_t = \mu \, dt + \sigma \, dB_t; \ \xi_0 = 0$

By Example 6.2.9 the solution is

$$\widehat{\mu} = E[\mu|\mathcal{N}_t] = \frac{\sigma^2 \bar{\mu}}{\sigma^2 + a^2 t} + \frac{a^2}{\sigma^2 + a^2 t}\xi_t$$

$$= (\theta + \sigma^{-2}t)^{-1}(\bar{\mu}\theta + \sigma^{-2}\xi_t) \ .$$

c) The innovation process N_t of the filtering problem in b) is

$$N_t = \xi_t - \int_0^t \widehat{\mu}(s)ds \ .$$

Therefore, by a)

$$\widetilde{B}_t = \int_0^t \sigma^{-1}(\mu - E[\mu|\mathcal{M}_s])ds + B_t$$

$$= \int_0^t \sigma^{-1}(\mu - E[\mu|\mathcal{M}_s])ds + B_t = \sigma^{-1}N_t$$

is a Brownian motion by Lemma 6.2.6.

d) Since $\widetilde{B}_t = \sigma^{-1}(\xi_t - \int_0^t \widehat{\mu}(s)ds)$ we see that $\widetilde{B}_t$ is $\mathcal{N}_t$-measurable and hence $\mathcal{M}_t$-measurable by a).

e) We have

$$\sigma \, d\widetilde{B}_t = d\xi_t - \widehat{\mu}(t)dt = d\xi_t - \frac{\bar{\mu}\theta + \sigma^{-2}\xi_t}{\theta + \sigma^{-2}t}dt \ .$$

Hence

$$d\xi_t - \frac{1}{\sigma^2\theta + t}\xi_t dt = \frac{\bar{\mu}\theta}{\theta + \sigma^{-2}t}dt + \sigma \, d\widetilde{B}_t$$

which gives

$$d\left(\xi_t \exp\left(-\int_0^t \frac{ds}{\sigma^2\theta + s}\right)\right) = \exp\left(-\int_0^t \frac{ds}{\sigma^2\theta + s}\right)\left[\frac{\bar{\mu}\theta}{\theta + \sigma^{-2}t}dt + \sigma \, d\widetilde{B}_t\right]$$

or

$$d\left(\frac{\xi_t}{\sigma^2\theta + t}\right) = \frac{1}{\sigma^2\theta + t}\left[\frac{\bar{\mu}\theta}{\theta + \sigma^{-2}t}dt + \sigma \, d\widetilde{B}_t\right].$$

We conclude that

$$\xi_t = \bar{\mu} - \frac{\bar{\mu}\sigma^2\theta}{\sigma^2\theta + t} + \sigma \int_0^t \frac{d\widetilde{B}_s}{\sigma^2\theta + s} \ ,$$

which shows that ξ_t is $\widetilde{\mathcal{F}}_t$-measurable.

f) By combining (6.3.20), (6.3.24) and a) we have

$$dX_t = X_t(\mu \, dt + \sigma \, dB_t) = X_t d\xi_t$$

$$= X_t(\widehat{\mu}(t)dt + \sigma \, d\widetilde{B}_t)$$
$$= E[\mu|\mathcal{M}_t]X_t dt + \sigma \, X_t \, d\widetilde{B}_t \; .$$

□

7.1. a) $Af(x) = \mu x f'(x) + \frac{1}{2}\sigma^2 f''(x); \quad f \in C_0^2(\mathbf{R}).$

b) $Af(x) = rx f'(x) + \frac{1}{2}\alpha^2 x^2 f''(x); \quad f \in C_0^2(\mathbf{R}).$

c) $Af(y) = r f'(y) + \frac{1}{2}\alpha^2 y^2 f''(y); \quad f \in C_0^2(\mathbf{R}).$

d) $Af(t,x) = \frac{\partial f}{\partial t} + \mu x \frac{\partial f}{\partial x} + \frac{1}{2}\sigma^2 \frac{\partial^2 f}{\partial x^2}; \quad f \in C_0^2(\mathbf{R}^2).$

e) $Af(x_1,x_2) = \frac{\partial f}{\partial x_1} + x_2 \frac{\partial f}{\partial x_2} + \frac{1}{2}e^{2x_1}\frac{\partial^2 f}{\partial x_2^2}; \quad f \in C_0^2(\mathbf{R}^2).$

f) $Af(x_1,x_2) = \frac{\partial f}{\partial x_1} + \frac{1}{2}\frac{\partial^2 f}{\partial x_1^2} + \frac{1}{2}x_1^2\frac{\partial^2 f}{\partial x_2^2}; \quad f \in C_0^2(\mathbf{R}^2).$

g) $Af(x_1,\ldots,x_n) = \sum\limits_{k=1}^{n} r_k x_k \frac{\partial f}{\partial x_k} + \frac{1}{2}\sum\limits_{i,j=1}^{n} x_i x_j \left(\sum\limits_{k=1}^{n} \alpha_{ik}\alpha_{jk} \right) \frac{\partial^2 f}{\partial x_i \partial x_j};$

$f \in C_0^2(\mathbf{R}^n).$

7.2. a) $dX_t = dt + \sqrt{2}\, dB_t \; .$

b) $dX(t) = \begin{bmatrix} dX_1(t) \\ dX_2(t) \end{bmatrix} = \begin{bmatrix} 1 \\ cX_2(t) \end{bmatrix} dt + \begin{bmatrix} 0 \\ \alpha X_2(t) \end{bmatrix} dB_t \; .$

c) $dX(t) = \begin{bmatrix} dX_1(t) \\ dX_2(t) \end{bmatrix} = \begin{bmatrix} 2X_2(t) \\ \ln(1+X_1^2(t)+X_2^2(t)) \end{bmatrix} dt + \begin{bmatrix} X_1(t) & 1 \\ 1 & 0 \end{bmatrix} \begin{bmatrix} dB_1(t) \\ dB_2(t) \end{bmatrix}.$

(Several other diffusion coefficients are possible.)

7.4. a), b). Let $\tau_k = \inf\{t > 0; B_t^x = 0 \text{ or } B_t^x = k\}; \; k > x > 0$ and put

$$p_k = P^x[B_{\tau_k} = k] \; .$$

Then by Dynkin's formula applied to $f(y) = y^2$ for $0 \le y \le k$ we get

$$E^x[\tau_k] = k^2 p_k - x^2 \; . \tag{S1}$$

On the other hand, Dynkin's formula applied to $f(y) = y$ for $0 \le y \le k$ gives

$$kp_k = x \; . \tag{S2}$$

Combining these two identities we get that

$$E^x[\tau] = \lim_{k\to\infty} E^x[\tau_k] = \lim_{k\to\infty} x(k-x)) = \infty \; . \tag{S3}$$

Moreover, from (S2) we get

$$P^x[\exists t < \infty \text{ with } B_t = 0] = \lim_{k\to\infty} P^x[B_{\tau_k} = 0] = \lim_{k\to\infty} (1-p_k) = 1 \; , \tag{S4}$$

so $\tau < \infty$ a.s. P^x.

7.15. By the Markov property (7.2.5) we have

$$E^x[\mathcal{X}_{[K,\infty)}(B_T) \mid \mathcal{F}_t] = E^x[\theta_t \mathcal{X}_{[K,\infty)}(B_{T-t}) \mid \mathcal{F}_t]$$

$$= E^{B_t}[\mathcal{X}_{[K,\infty)}(B_{T-t})] = E^y[f(B_{T-t})]_{y=B_t}$$

$$= \left[\frac{1}{\sqrt{2\pi(T-t)}} \int_{\mathbf{R}} f(z)e^{-\frac{(z-y)^2}{2(T-t)}} dz\right]_{y=B_t}$$

$$= \frac{1}{\sqrt{2\pi(T-t)}} \int_{K}^{\infty} e^{-\frac{(z-B_t)^2}{2(T-t)}} dz \ ,$$

where

$$f(x) = \mathcal{X}_{[K,\infty)}(x) \ .$$

7.18.   a) Define

$$w(x) = \frac{f(x) - f(a)}{f(b) - f(a)} \ ; \qquad x \in [a,b].$$

Then

$$Lw(x) = 0 \qquad \text{in } D := (a,b)$$

and

$$w(a) = 0, \quad w(b) = 1 \ .$$

Hence by the Dynkin formula

$$1 \cdot P^x[X_\tau = b] + 0 \cdot P^x[X_\tau = a] = w(x) + E^x\left[\int_0^\tau Lw(X_t)dt\right] = w(x) \ ,$$

i.e.

$$w = P^x[X_\tau = b] = \frac{f(x) - f(a)}{f(b) - f(a)} \ .$$

c) $p = \left[\exp\left(-\frac{2\mu b}{\sigma^2}\right) - \exp\left(-\frac{2\mu a}{\sigma^2}\right)\right]^{-1}\left(\exp\left(-\frac{2\mu x}{\sigma^2}\right) - \exp\left(-\frac{2\mu a}{\sigma^2}\right)\right).$

8.1.   a) $g(t,x) = E^x[\phi(B_t)]$.

b) $u(x) = E^x[\int_0^\infty e^{-\alpha t}\psi(B_t)dt]$.

8.11.   For $T > 0$ define the measure Q_T on $\mathcal{F}_T$ by

$$dQ_T(\omega) = \exp(-B(T) - \tfrac{1}{2}T)dP(\omega) \qquad \text{on } \mathcal{F}_T \ .$$

Then by the Girsanov theorem $Y(t) := t + B(t); \ 0 \le t \le T$ is a Brownian motion with respect to Q_T. Since

$$Q_T = Q_S \qquad \text{on } \mathcal{F}_t \text{ for all } t \le \min(S,T)$$

there exists a measure Q on $\mathcal{F}_\infty$ such that

$$Q = Q_T \qquad \text{on } \mathcal{F}_T \text{ for all } T < \infty$$

(See Theorem 1.14 in Folland (1984).)

By the law of iterated logarithm (Theorem 5.1.2) we know that

$$P[N] = 1 \,,$$

where

$$N = \{\omega; \lim_{t\to\infty} Y(t) = \infty\}.$$

On the other hand, since $Y(t)$ is a Brownian motion w.r.t. Q we have that

$$Q[N] = 0 \,.$$

Hence P is not absolutely continuous w.r.t. Q.

8.12. Here the equation

$$\sigma(t,\omega)u(t,\omega) = \beta(t,\omega)$$

has the form

$$\begin{bmatrix} 1 & 3 \\ -1 & -2 \end{bmatrix} \begin{bmatrix} u_1 \\ u_2 \end{bmatrix} = \begin{bmatrix} 0 \\ 1 \end{bmatrix}$$

which has the solution

$$u = \begin{bmatrix} u_1 \\ u_2 \end{bmatrix} = \begin{bmatrix} -3 \\ 1 \end{bmatrix}.$$

Hence we define the measure Q on $\mathcal{F}_T^{(2)}$ by

$$dQ(\omega) = \exp(3B_1(T,\omega) - B_2(T,\omega) - 5T)dP(\omega) \,.$$

9.1. a) $dX_t = \begin{bmatrix} \alpha \\ 0 \end{bmatrix} dt + \begin{bmatrix} 0 \\ \beta \end{bmatrix} dB_t$.

b) $dX_t = \begin{bmatrix} a \\ b \end{bmatrix} dt + \begin{bmatrix} 1 & 0 \\ 0 & 1 \end{bmatrix} dB_t$.

c) $dX_t = \alpha X_t dt + \beta dB_t$.

d) $dX_t = \alpha dt + \beta X_t dB_t$.

e) $dX_t = \begin{bmatrix} dX_1(t) \\ dX_2(t) \end{bmatrix} = \begin{bmatrix} \ln(1+X_1^2(t)) \\ X_2(t) \end{bmatrix} dt + \sqrt{2} \begin{bmatrix} X_2(t) & 0 \\ X_1(t) & X_1(t) \end{bmatrix} \begin{bmatrix} dB_1(t) \\ dB_2(t) \end{bmatrix}.$

9.2. (i) In this case we put

$$dX_t = \begin{bmatrix} dt \\ dB_t \end{bmatrix} = \begin{bmatrix} 1 \\ 0 \end{bmatrix} dt + \begin{bmatrix} 0 \\ 1 \end{bmatrix} dB_t \,; \qquad X_0 = \begin{bmatrix} s \\ x \end{bmatrix}$$

and

$$D = \{(s,x) \in \mathbf{R}^2; s < T\} \,.$$

Then

$$\tau_D = \inf\{t > 0; (s+t, B_t) \notin D\} = T - s$$

and we get

$$u(s,x) = E^{s,x}\left[\psi(B_{\tau_D}) + \int_0^{\tau_D} g(X_t)dt\right]$$

$$= E\left[\psi(B_{T-s}^x) - \int_0^{T-s} \phi(s+t, B_t^x)dt\right],$$

where B_t^x is Brownian motion starting at x.

(ii) Define X_t by

$$dX_t = \alpha\, X_t dt + \beta\, X_t dB_t\ ; \qquad X_0 = x > 0\ ,$$

and put

$$D = (0, x_0)\ .$$

If $\alpha \geq \frac{1}{2}\beta^2$ then $\tau_D = \inf\{t > 0; X_t \notin D\} < \infty$ a.s. and

$$X_{\tau_D} = x_0 \quad \text{a.s.} \quad (\text{see Example 5.1.1}).$$

Therefore the only bounded solution is

$$u(x) = E^x[(X_{\tau_D})^2] = x_0^2 \quad (\text{constant}).$$

(iii) If we try a solution of the form

$$u(x) = x^\gamma \qquad \text{for some constant } \gamma$$

we get

$$\alpha\, xu'(x) + \tfrac{1}{2}\beta^2 x^2 u''(x) = (\alpha + \tfrac{1}{2}\beta^2(\gamma - 1))x^\gamma,$$

so $u(x)$ is a solution iff

$$\gamma = 1 - \frac{2\alpha}{\beta^2}\ .$$

With this value of γ we get that the general solution of

$$\alpha\, xu'(x) + \tfrac{1}{2}\beta^2 x^2 u''(x) = 0$$

is

$$u(x) = C_1 + C_2 x^\gamma$$

where C_1, C_2 are arbitrary constants. If $\alpha < \frac{1}{2}\beta^2$ then all these solutions are bounded on $(0, x_0)$ and we need an additional boundary value at $x = 0$ to get uniqueness. In this case $P[\tau_D = \infty] > 0$. If $\alpha > \frac{1}{2}\beta^2$ then $u(x) = C_1$ is the only bounded solution on $(0, x_0)$, in agreement with Theorem 9.1.1 (and (ii) above).

9.3.   a) $u(t, x) = E^x[\phi(B_{T-t})]$.

b) $u(t,x) = E^x[\psi(B_t)]$.

9.8. a) Let $X_t \in \mathbf{R}^2$ be uniform motion to the right, as described in Example 9.2.1. Then each one-point set $\{(x_1, x_2)\}$ is thin (and hence semipolar) but not polar.

b) With X_t as in a) let $H_k = \{(a_k, 1)\}; k = 1, 2, \ldots$ where $\{a_k\}_{k=1}^{\infty}$ is the set of rational numbers. Then each H_k is thin but $Q^{(x_1, 1)}[T_H = 0] = 1$ for all $x_1 \in \mathbf{R}$.

9.10. Let $Y_t = Y_t^{s,x} = (s+t, X_t^x)$ for $t \geq 0$, where $X_t = X_t^x$ satisfies

$$dX_t = \alpha X_t dt + \beta X_t dB_t ; \qquad t \geq 0, \; X_0 = x > 0 .$$

Then the generator $\widehat{A}$ of Y_t is given by

$$\widehat{A}f(s,x) = \frac{\partial f}{\partial s} + \alpha x \frac{\partial f}{\partial x} + \tfrac{1}{2}\beta^2 x^2 \frac{\partial^2 f}{\partial x^2} ; \qquad f \in C_0^2(\mathbf{R}^2) .$$

Moreover, with $D = \{(t,x); x > 0 \text{ and } t < T\}$ we have

$$\tau_D := \inf[t > 0; Y_t \notin D] = \inf\{t > 0; s+t > T\} = T - s .$$

Hence

$$Y_{\tau_D} = (T, X_{T-s}) .$$

Therefore, by Theorem 9.3.3 the solution is

$$f(s,x) = E\left[e^{-\rho T}\phi(X_{T-s}^x) + \int\limits_0^{T-s} e^{-\rho(s+t)}K(X_t^x)dt\right] .$$

9.15. a) If we put $w(s,x) = e^{-\rho s}h(x)$ we get

$$\tfrac{1}{2}\frac{\partial^2 w}{\partial x^2} + \frac{\partial w}{\partial s} = e^{-\rho s}\left(\tfrac{1}{2}h''(x) - \rho h(x)\right),$$

and the boundary value problem reduces to

(1) $\tfrac{1}{2}h''(x) - \rho h(x) = -\theta x^2; \quad a < x < b$

(2) $h(a) = \psi(a), \quad h(b) = \psi(b)$

The general solution of (1) is

$$h(x) = C_1 e^{\sqrt{2\rho}\,x} + C_2 e^{-\sqrt{2\rho}\,x} + \frac{\theta}{\rho}x^2 - \frac{\theta}{\rho^2} ,$$

where C_1, C_2 are arbitrary constants. The boundary values (ii) will determine C_1, C_2 uniquely.

b) To find

$$g(x, \rho) = g(x) := E^x\left[e^{-\rho \tau_D}\right]$$

we apply the above with $\psi = 1$, $\theta = 0$ and get

$$g(x) = C_1 e^{\sqrt{2\rho}\,x} + C_2 e^{-\sqrt{2\rho}\,x},$$

where the constants C_1, C_2 are determined by

$$g(a) = 1, \quad g(b) = 1 .$$

After some simplifications this gives

$$g(x) = \frac{\sinh(\sqrt{2\rho}(b-x)) + \sinh(\sqrt{2\rho}(x-a))}{\sinh(\sqrt{2\rho}(b-a))} ; \qquad a \le x \le b .$$

10.1. a) $g^*(x) = \infty$, τ^* does not exist.

b) $g^*(x) = \infty$, τ^* does not exist.

c) $g^*(x) = 1$, $\tau^* = \inf\{t > 0; B_t = 0\}$.

d) If $\rho < \frac{1}{2}$ then $g^*(s,x) = \infty$ and τ^* does not exist.

If $\rho \ge \frac{1}{2}$ then $g^*(s,x) = g(s,x) = e^{-\rho s} \cosh x$ and $\tau^* = 0$.

10.2. a) Let W be a closed disc centered at y with radius $\rho < |x - y|$. Define

$$\tau_\rho = \inf\{t > 0; B_t^x \in W\} .$$

In $\mathbf{R}^2$ we know that $\tau_\rho < \infty$ a.s. (see Example 7.4.2). Suppose there exists a nonnegative superharmonic function u such that

(1) $u(x) < u(y) .$

Then

(2) $u(x) \ge E^x\big[u(B_{\tau_\rho})\big].$

Since u is lower semicontinuous there exists $\rho > 0$ such that

(3) $\inf\{u(z); z \in W\} > u(x) .$

Combining this with (2) we get the contradiction

$$u(x) \ge E^x\big[u(B_{\tau_\rho})\big] > u(x) . \qquad\qquad \square$$

b) The argument in a) also works for $\mathbf{R}$. Therefore

$$g^*(x) = \sup\{g(x); x \in \mathbf{R}\} = \sup\{x\,e^{-x}; x > 0\} = \frac{1}{e} .$$

c) For $x \ne 0$ we have

$$\Delta(|x|^\gamma) = \sum_{i=1}^{n} \frac{\partial^2}{\partial x_i^2}\left(\Big(\sum_{j=1}^{n} x_j^2\Big)^{\gamma/2}\right) = \sum_{i=1}^{n} \frac{\partial}{\partial x_i}\left(\frac{\gamma}{2}\Big(\sum_{j=1}^{n} x_j^2\Big)^{\frac{\gamma}{2}-1} 2x_i\right)$$

$$= \gamma(\gamma + n - 2)|x|^{\gamma - 2}.$$

So $|x|^\gamma$, and hence $f_\gamma(x) = \min(1, |x|^\gamma)$, is superharmonic iff $\gamma(\gamma + n - 2) \le 0$, i.e.

$$2 - n \le \gamma \le 0 .$$

10.3. $x_0 > 0$ is given implicitly by the equation

$$x_0 = \sqrt{\frac{2}{\rho}} \cdot \frac{e^{2\sqrt{2\rho}\, x_0} + 1}{e^{2\sqrt{2\rho}\, x_0} - 1} ,$$

and $g^*(s, x) = e^{-\rho s} x_0^2 \frac{\cosh(\sqrt{2\rho}\, x)}{\cosh(\sqrt{2\rho}\, x_0)}$ for $-x_0 \le x \le x_0$, where $\cosh \xi = \frac{1}{2}(e^\xi + e^{-\xi})$.

10.9. If $0 < \rho \le 1$ then $\Phi(x) = \frac{1}{\rho} x^2 + \frac{1}{\rho^2}$ but τ^* does not exist. If $\rho > 1$ then

$$\Phi(x) = \begin{cases} \frac{1}{\rho} x^2 + \frac{1}{\rho^2} + C \cosh(\sqrt{2\rho}\, x) & \text{for } |x| \le x^* \\ x^2 & \text{for } |x| > x^* \end{cases}$$

where $C > 0$, $x^* > 0$ are the unique solutions of the equations

$$C \cosh(\sqrt{2\rho}\, x^*) = \left(1 - \frac{1}{\rho}\right)(x^*)^2 - \frac{1}{\rho^2}$$

$$C\sqrt{2\rho} \sinh(\sqrt{2\rho}\, x^*) = 2\left(1 - \frac{1}{\rho}\right) x^* .$$

10.12. If $\rho > r$ then $g^*(s, x) = e^{-\rho s}(x_0 - 1)^+ \left(\frac{x}{x_0}\right)^\gamma$ and $\tau^* = \inf\{t > 0; X_t \ge x_0\}$, where

$$\gamma = \alpha^{-2}\left[\tfrac{1}{2}\alpha^2 - r + \sqrt{(\tfrac{1}{2}\alpha^2 - r)^2 + 2\alpha^2\rho}\,\right]$$

and

$$x_0 = \frac{\gamma}{\gamma - 1} \qquad (\gamma > 1 \Leftrightarrow \rho > r) .$$

10.13. If $\alpha \le \rho$ then $\tau^* = 0$.
If $\rho < \alpha < \rho + \lambda$ then

$$G^*(s, p, q) = \begin{cases} e^{-\rho s} pq & ; & \text{if } 0 < pq < y_0 \\ e^{-\rho s}\left(C_1 (pq)^{\gamma_1} + \frac{\lambda}{\rho + \lambda - \alpha} \cdot pq - \frac{K}{\rho}\right); & \text{if } pq \ge y_0 \end{cases}$$

where

$$\gamma_1 = \beta^{-2}\left[\tfrac{1}{2}\beta^2 + \lambda - \alpha - \sqrt{(\tfrac{1}{2}\beta^2 + \lambda - \alpha)^2 + 2\rho\beta^2}\,\right] < 0 ,$$

$$y_0 = \frac{(-\gamma_1)K(\rho + \lambda - \alpha)}{(1-\gamma_1)\rho(\alpha - \rho)} > 0$$

and

$$C_1 = \frac{(\alpha - \rho)y_0^{1-\gamma_1}}{(-\gamma_1)(\rho + \lambda - \alpha)} .$$

The continuation region is

$$D = \{(s, p, q); pq > y_0\} .$$

If $\rho + \lambda \leq \alpha$ then $G^* = \infty$.

10.14.   First assume that
Case 1: $\rho > \alpha$.
Note that

$$\int_\tau^\infty e^{-\rho(s+t)} P_t dt = e^{-\rho s} \left[\int_0^\infty e^{-\rho t} P_t dt - \int_0^\tau e^{-\rho t} P_t dt \right]$$

and

$$E\left[\int_0^\infty e^{-\rho t} P_t dt \right] = \int_0^\infty e^{-\rho t} E[p \exp(\beta B_t + (\alpha - \tfrac{1}{2}\beta^2)t)]dt$$

$$= \int_0^\infty p\, e^{(\alpha - \rho)t} dt = \frac{p}{\rho - \alpha} .$$

Therefore

$$\Phi(s, p) = \frac{p\, e^{-\rho s}}{\rho - \alpha} + \Psi(s, p) ,$$

where

$$\Psi(s, p) = \sup_\tau E^{(s,p)} \left[\int_0^\tau (-e^{-\rho(s+t)} P_t)dt - C\, e^{-\rho(s+\tau)} \right].$$

This is a problem of the form discussed in Section 10.4, with

$$Y(t) = \begin{bmatrix} s+t \\ P_t \end{bmatrix}, \quad Y(0) = \begin{bmatrix} s \\ p \end{bmatrix} = y \in \mathbf{R}^2$$

and

$$f(y) = f(s, x) = -e^{-\rho s} p , \quad g(s, x) = -C\, e^{-\rho s}.$$

To get an indication of where the continuation region D is situated, we consider the set

$$U = \{y;\, Lg(y) + f(y) > 0\} \qquad (\text{see } (10.3.7)).$$

In this case the generator L is given by

$$L\phi(s,p) = \frac{\partial \phi}{\partial s} + \alpha\, p\frac{\partial \phi}{\partial p} + \tfrac{1}{2}\beta^2 p^2 \frac{\partial^2 \phi}{\partial p^2}$$

and so

$$U = \{(s,p);\, (-\rho)(-C) - p > 0\} = \{(s,p);\, p < \rho\, C\}.$$

In view of this we try a continuation region D of the form

$$D = \{(s,p);\, 0 < p < p^*\}$$

for some $p^* > 0$ (to be determined).

We try a value function candidate of the form

$$\phi(s,p) = e^{-\rho s}\psi(p)$$

where, by Theorem 10.4.1, the function ψ is required to satisfy the following conditions:

(1) $\quad L_0\psi(p) := -\rho\psi(p) + \alpha\, p\, \psi'(p) + \tfrac{1}{2}\beta^2 p^2 \psi''(p) - p = 0;\ 0 < p < p^*$

(2) $\quad L_0\psi(p) \le 0;\quad p > p^*$

(3) $\quad \psi(p) = -C;\quad p > p^*$

(4) $\quad \psi(p) > -C;\quad 0 < p < p^*.$

The general solution of (1) is

$$\psi(p) = K_1\, p^{\gamma_1} + K_2\, p^{\gamma_2} + \frac{p}{\alpha - \rho}$$

where

(5) $\quad \gamma_i = \beta^{-2}\left[\tfrac{1}{2}\beta^2 - \alpha \pm \sqrt{\left(\tfrac{1}{2}\beta^2 - \alpha\right)^2 + 2\rho\,\beta^2}\,\right];\qquad i = 1, 2$

with

$$\gamma_2 < 0 < \gamma_1$$

and K_1, K_2 are arbitrary constants.

Since $\psi(p)$ must be bounded near $p = 0$ we must have $K_2 = 0$. Hence we put

(6) $\qquad \psi(p) = \begin{cases} K_1 p^{\gamma_1} + \frac{p}{\alpha - \rho} & ;\ 0 \le p < p^* \\ -C & ;\ p \ge p^* \end{cases}$

If we require $\psi(p)$ to be continuous at $p = p^*$ we get the equation

(7)
$$K_1(p^*)^{\gamma_1} + \frac{p^*}{\alpha - \rho} = -C$$

If $\psi(p)$ is also C^1 at $p = p^*$ we get

(8)
$$K_1\gamma_1(p^*)^{\gamma_1-1} + \frac{1}{\alpha - \rho} = 0$$

Combining these two equations we get

(9)
$$x^* = \frac{C(\rho - \alpha)\gamma_1}{\gamma_1 - 1}$$

and

(10)
$$K_1 = \frac{(p^*)^{1-\gamma_1}}{\gamma_1(\rho - \alpha)}$$

It is easy to see that
$$\gamma_1 > 1 \Leftrightarrow \rho > \alpha$$

Since we have assumed $\rho > \alpha$ we get by (9) that $p^* > 0$ and by (10) that $K_1 > 0$.

It remains to verify that with these values of p^* and K_1 the function

$$\phi(s, p) = e^{-\rho s}\psi(p) \ ,$$

with ψ given by (6), satisfies all the conditions of Theorem 10.4.1. Many of these verifications are straightforward. However, to avoid false solutions, it is important to check (ii) and (vi), which correspond to (4) and (2) above:

Verification of (4):
Define $h(p) = \psi(p) + C$
Then
$$h(p^*) = h'(p^*) = 0 \qquad \text{by (7) and (8) above.}$$

Moreover, if $0 < p < p^*$ then

$$h''(p) = K_1\gamma_1(\gamma_1 - 1)p^{\gamma_1-2} > 0$$

which implies that $h'(p) < 0$ and hence $h(p) > 0$ for $0 < p < p^*$, as required.

Verification of (2):
For $p > p^*$ we have $\psi(p) = -C$ and hence

$$L_0\psi(p) = (-\rho)(-C) - p = \rho C - p \ ,$$

so

$$L_0\psi(p) \leq 0 \qquad \text{for all } p > p^*$$

if and only if

$$p^* \geq \rho C \,,$$

which holds because $U = \{(s,p); p < \rho C\}$ and $U \subset D$.
The remaining cases,
Case 2: $\rho = \alpha$
Case 3: $\rho < \alpha$
are left to the reader.

11.6. $u^* = \frac{a_1 - a_2 + \sigma_2^2(1-\gamma)}{(\sigma_1^2 + \sigma_2^2)(1-\gamma)}$ (constant),

$$\Phi(s,x) = e^{\lambda(t-t_1)} x^\gamma \qquad \text{for } t < t_1,\, x > 0$$

where

$$\lambda = \tfrac{1}{2}\gamma(1-\gamma)[\sigma_1^2(u^*)^2 + \sigma_2^2(1-u^*)^2] - \gamma[a_1 u^* + a_2(1-u^*)] \,.$$

11.7.   Define

$$dY(t) = \begin{bmatrix} dt \\ dX_t \end{bmatrix} = \begin{bmatrix} 1 \\ \alpha\, u(t) \end{bmatrix} dt + \begin{bmatrix} 0 \\ u(t) \end{bmatrix} dB_t \,; \qquad Y(0) = \begin{bmatrix} s \\ x \end{bmatrix}$$

and

$$G = \{(s,x); x > 0 \text{ and } s < T\}.$$

Then

$$\Phi(s,x) = \Phi(y) = \sup_{u(\cdot)} E^y[g(Y_{\tau_G})],$$

where

$$g(y) = g(s,x) = x^\gamma$$

and

$$\tau_G = \inf\{t > 0; Y(t) \notin G\} = \widehat{\tau} \,.$$

We apply Theorem 11.2.2 and hence look for a function ϕ such that

(1) $\sup_{v \in \mathbf{R}} \{f^v(y) + (L^v\phi)(y)\} = 0$ for all $y \in G$,

where in this case $f^v(y) = 0$ and

$$L^v\phi(y) = L^v\phi(s,x) = \frac{\partial \phi}{\partial s} + av\frac{\partial \phi}{\partial x} + \tfrac{1}{2}v^2\frac{\partial^2 \phi}{\partial x^2} \,.$$

If we guess that $\frac{\partial^2 \phi}{\partial x^2} < 0$ then the maximum of the function $v \to L^v\phi(s,x)$ is attained at

(2) $$v = v^*(s, x) = -\frac{a \frac{\partial \phi}{\partial x}(s, x)}{\frac{\partial^2 \phi}{\partial x^2}(s, x)}.$$

We try a function ϕ of the form

(3) $$\phi(s, x) = f(s)x^\gamma$$

for some function f (to be determined). Substituted in (2) this gives

(4) $$v^*(s, x) = -\frac{af(s)\gamma\, x^{\gamma-1}}{f(s)\gamma(\gamma - 1)x^{\gamma-2}} = \frac{ax}{1 - \gamma}$$

and (1) becomes

$$f'(s)x^\gamma + \frac{a^2 x}{1 - \gamma}f(s)\gamma\, x^{\gamma-1} + \frac{1}{2}\left(\frac{ax}{1 - \gamma}\right)^2 f(s)\gamma(\gamma - 1)x^{\gamma-2} = 0$$

or

(5) $$f'(s) + \frac{a^2\gamma}{2(1 - \gamma)}f(s) = 0.$$

Combined with the terminal condition

$$\phi(y) = g(y) \qquad \text{for } y \in \partial G$$

i.e.

(6) $$f(T) = 1$$

the equation (5) has the solution

(7) $$f(s) = \exp\left(\frac{a^2\gamma}{2(1 - \gamma)}(T - s)\right); \qquad s \leq T.$$

With this value of f it is now easily verified that

$$\phi(s, x) = f(s)x^\gamma$$

satisfies all the conditions of Theorem 11.2.2, and we conclude that the value function is

$$\Phi(s, x) = \phi(s, x) = f(s)x^\gamma$$

and that
$$u^*(s, x) = v^*(s, x) = \frac{ax}{1 - \gamma}$$

is an optimal Markov control.

11.11.   Additional hints:

For the solution of the unconstrained problem try a function $\phi_\lambda(s, x)$ of the form

$$\phi_\lambda(s, x) = a_\lambda(s)x^2 + b_\lambda(s) \, ,$$

for suitable functions $a_\lambda(s), b_\lambda(s)$ with $\lambda \in \mathbf{R}$ fixed. By substituting this into the HJB equation we arrive at the equations

$$a_\lambda'(s) = \frac{1}{\theta}a_\lambda^2(s) - 1 \qquad \text{for } s < t_1$$

$$a_\lambda(t_1) = \lambda$$

and

$$b_\lambda'(s) = -\sigma^2 a_\lambda(s) \qquad \text{for } s < t_1$$

$$b_\lambda(t_1) = 0 \, ,$$

with optimal control $u^*(s, x) = -\frac{1}{\theta}a_\lambda(s)x$.

Now substitute this into the equation for $X_t^{u^*}$ and use the terminal condition to determine λ_0 .

If we put $s = 0$ for simplicity, then $\lambda = \lambda_0$ can be chosen as any solution of the equation

$$A\lambda^3 + B\lambda^2 + C\lambda + D = 0 \, ,$$

where

$$A = m^2(e^{t_1} - e^{-t_1})^2 \, ,$$
$$B = m^2(e^{2t_1} + 2 - 3e^{-2t_1}) - \sigma^2(e^{t_1} - e^{-t_1})^2 \, ,$$
$$C = m^2(-e^{2t_1} + 2 + 3e^{-2t_1}) - 4x^2 - 2\sigma^2(1 - e^{-2t_1})$$
$$D = -m^2(e^{t_1} + e^{-t_1})^2 + 4x^2 + \sigma^2(e^{2t_1} - e^{-2t_1}) \, .$$

11.12.   If we introduce the process

$$dY(t) = \begin{bmatrix} dt \\ dX_t \end{bmatrix} = \begin{bmatrix} 1 \\ u_t \end{bmatrix} dt + \begin{bmatrix} 0 \\ \sigma \end{bmatrix} dB_t \, ; \qquad Y(0) = y = \begin{bmatrix} s \\ x \end{bmatrix}$$

then the problem can be written

$$\Psi(s, x) = \inf_u E^y \left[\int_0^\infty f(Y(t), u(t))dt \right],$$

where

$$f(y, u) = f^u(y) = f^u(s, x) = e^{-\rho s}(x^2 + \theta u^2).$$

In this case we look for a function ψ such that

(1) $\displaystyle\inf_{v \in \mathbf{R}} \left\{ e^{-\rho s}(x^2 + \theta\, v^2) + \frac{\partial \psi}{\partial s} + v \frac{\partial \psi}{\partial x} + \frac{1}{2}\sigma^2 \frac{\partial^2 \psi}{\partial x^2} \right\} = 0; \quad (s,x) \in \mathbf{R}^2.$

If we guess that ψ has the form

$$\psi(s,x) = e^{-\rho s}(ax^2 + b)$$

then (1) gets the form

(2) $\displaystyle\inf_{v \in \mathbf{R}} \{x^2 + \theta\, v^2 - \rho(ax^2 + b) + v\,2ax + \frac{1}{2}\sigma^2 2a\} = 0; \qquad x \in \mathbf{R} .$

The minimum is attained when

(3) $$v = v^*(s,x) = -\frac{ax}{\theta}$$

and substituting this into (2) we get

$$x^2 \left[1 - \rho a - \frac{a^2}{\theta} \right] + \sigma^2 a - \rho b = 0 \qquad \text{for all } x \in \mathbf{R} .$$

This is only possible if

(4) $$a^2 + \rho \theta a - \theta = 0$$

and

(5) $$b = \frac{\sigma^2 a}{\rho}$$

The positive root of (4) is

(6) $$a = \frac{1}{2}\left[-\rho\theta + \sqrt{\rho^2\theta^2 + 4\theta} \right] .$$

We can now verify that with the values of a and b given by (6) and (5), respectively, the function

$$\psi(s,x) = e^{-\rho s}(ax^2 + b)$$

satisfies all the requirements of Theorem 11.2.2 and we conclude that the value function is

$$\Psi(s,x) = \psi(s,x) = e^{-\rho s}(ax^2 + b)$$

and that

$$u^*(s,x) = v^*(s,x) = -\frac{ax}{\theta}$$

is an optimal control.

11.13.   If we introduce

$$dY(t) = \begin{bmatrix} dt \\ dX_t \end{bmatrix} = \begin{bmatrix} 1 \\ 1 - u_t \end{bmatrix} dt + \begin{bmatrix} 0 \\ \sigma \end{bmatrix} dB_t ; \qquad Y(0) = \begin{bmatrix} s \\ x \end{bmatrix}$$

and

$$G = \{(s, x) \in \mathbf{R}^2; x > 0\},$$

then the problem gets the form

$$\Phi(s, x) = \sup_u E^y \left[\int_0^{\tau_G} f(Y(t), u_t) dt \right],$$

where

$$f(y, u) = f(s, x, u) = e^{-\rho s} u .$$

The corresponding HJB equation is

(1) $$\sup_{v \in [0,1]} \left\{ e^{-\rho s} v + \frac{\partial \phi}{\partial s} + (1 - v) \frac{\partial \phi}{\partial x} + \frac{1}{2} \sigma^2 \frac{\partial^2 \phi}{\partial x^2} \right\} = 0 .$$

If we substitute

(2) $$\phi(s, x) = e^{-\rho s} \frac{1}{\rho} \left(1 - \exp\left(-\sqrt{\frac{2}{\sigma^2}} x \right) \right)$$

then (1) gets the form

(3) $$\sup_{v \in [0,1]} \left\{ -1 + \sqrt{\frac{2}{\rho \sigma^2}} \exp\left(-\sqrt{\frac{2\rho}{\sigma^2}} x \right) \right.$$
$$\left. + v \left[1 - \sqrt{\frac{2}{\rho \sigma^2}} \exp\left(-\sqrt{\frac{2\rho}{\sigma^2}} x \right) \right] \right\} = 0 .$$

If $\rho \geq \frac{2}{\sigma^2}$ then $1 - \sqrt{\frac{2}{\rho \sigma^2}} \exp\left(-\sqrt{\frac{2\rho}{\sigma^2}} x \right) \geq 0$ for all x and hence the
supremum in (3) is attained for

$$v = u^*(s, x) = 1 .$$

Moreover, we see that the corresponding supremum in (2) is 0. Hence
by Theorem 11.2.2 we conclude that $\Phi = \phi$ and $u_t^* = 1$.

12.6. a)   no arbitrage
         b)   no arbitrage
         c) $\theta(t) = (a, 1, 1)$ is an arbitrage, where $a \in \mathbf{R}$ is chosen such that
             $V_\theta(0) = 0$
         d)   no arbitrage
         e)   arbitrages exist
         f)   no arbitrage.

12.7. a)   complete
         b)   not complete. For example, the claim

$$F(\omega) = \int_0^T B_3(t)dB_3(t) = \tfrac{1}{2}B_3^2(T) - \tfrac{1}{2}T$$

cannot be hedged.
c)   (arbitrages exist)
d)   not complete
e)   (arbitrages exist)
f)   complete.

12.9.   For the n-dimensional Brownian motion $B(t) = (B_1(t), \ldots, B_n(t))$ the representation formula (12.3.33) gets the form

$$h(B(T)) = E[h(B(T))] + \int_0^T \sum_{j=1}^n \frac{\partial}{\partial z_j} E^z\big[h(B(t-t))\big]_{z=B(t)} dB_j(t).$$

(i)   If $h(x) = x^2 = x_1^2 + \cdots + x_n^2$ we get

$$E^z[h(B(T-t))] = E[(B^z(T-t))^2] = z^2 + n(T-t).$$

Hence

$$\frac{\partial}{\partial z_j} E^z[h(B(T-t))] = 2z_j$$

and we conclude that

$$B^2(T) = x^2 + nT + \int_0^T \sum_{j=1}^n 2B_j(t)dB_j(t), \quad \text{with} \quad B(0) = x.$$

(ii)   If $h(x) = \exp(x_1 + \cdots + x_n)$ we get

$$E^z[h(B(T-t))] = E[\exp(z_1 + B_1(T-t) + \cdots + z_n + B_n(T-t))]$$
$$= \exp\Big(\frac{n}{2}(T-t) + \sum_{i=1}^n z_i\Big).$$

Hence

$$\frac{\partial}{\partial z_j} E^z[h(B(T-t))] = \exp\Big(\frac{n}{2}(T-t) + \sum_{i=1}^n z_i\Big)$$

and we conclude that if $B(0) = x$ then

$$\exp\Big(\sum_{i=1}^n B_i(T)\Big) = \exp\Big(\frac{n}{2}T + \sum_{i=1}^n x_i\Big)$$

$$+ \int_0^T \exp\left(\frac{n}{2}(T-t) + \sum_{i=1}^n B_i(t)\right)(dB_1(t) + \cdots + dB_n(t))$$

12.12. c) $E_Q[\xi(T)F] = \sigma^{-1}x_1(1 - \frac{\alpha}{\rho})(1 - e^{-\rho T})$. The replicating portfolio is $\theta(t) = (\theta_0(t), \theta_1(t))$, where

$$\theta_1(t) = \sigma^{-1}\left[1 - \frac{\alpha}{\rho}(1 - e^{\rho(t-T)})\right]$$

and $\theta_0(t)$ is determined by (12.1.17).

12.15.
$$\Phi(s, x) = \begin{cases} e^{-\rho s}(K - x) & \text{for} \quad x \le x^* \\ e^{-\rho s}(K - x^*)(\frac{x}{x^*})^\gamma & \text{for} \quad x > x^* \end{cases}$$

where

$$\gamma = \beta^{-2}\left[\tfrac{1}{2}\beta^2 - \alpha - \sqrt{\left(\tfrac{1}{2}\beta^2 - \alpha\right)^2 + 2\rho\beta^2}\right] < 0$$

and

$$x^* = \frac{K\gamma}{\gamma - 1} \in (0, K) .$$

Hence it is optimal to stop the first time $X(t) \le x^*$. If $\alpha = \rho$ this simplifies to

$$\gamma = -\frac{2\rho}{\beta^2} \quad \text{and} \quad x^* = \frac{2\rho K}{\beta^2 + 2\rho} .$$

12.16. Recall that (see Theorem 12.3.5)

$$V^\theta(T) = V^\theta(0) + \int_0^T \theta(s)dX(s)$$

iff

$$\bar{V}^\theta(T) = V^\theta(0) + \int_0^T \theta(s)d\bar{X}(s) = V^\theta(0) + \int_0^T \xi(s)\widehat{\theta}(s)\sigma(s)d\widetilde{B}(s) .$$

Therefore, if we seek a portfolio θ such that

$$V^\theta(T) = F \text{ a.s. },$$

we first find ϕ such that

(1) $$\bar{V}^\theta(T) = \xi(T)F = V^\theta(0) + \int_0^T \phi(s)d\widetilde{B}(s)$$

and then put

(2) $$\hat{\theta}(s) = X_0(s)\phi(s)\Lambda(s),$$

where $\Lambda(s)$ is the left inverse of $\sigma(s)$.

a) $F(\omega) = (K - X_1(T,\omega))^+$

In this case (1) gets the form

$$e^{-\rho T}(K - X_1(T,\omega))^+ = E_Q[e^{-\rho T}(K - X_1(T,\omega))^+] + \int_0^T \phi(s)d\tilde{B}(s)$$

or

(3) $$(K - X_1(T,\omega))^+ = E_Q[(K - X_1(T,\omega))^+] + \int_0^T \phi_0(s)d\tilde{B}(s),$$

where

$$\phi_0(s) = e^{\rho T}\phi(s).$$

To find ϕ_0 we use Theorem 12.3.3:
In this case we have

$$dY(t) = dX_1(t) = \alpha X_1(t)dt + \beta X_1(t)dB(t) = \rho X_1(t)dt + \beta X_1(t)d\tilde{B}(t),$$

with

$$d\tilde{B}(t) = dB(t) + \frac{\alpha - \rho}{\beta}dt .$$

Moreover

$$h(y) = (K - y)^+$$

and

$$E_Q^y[h(Y(T-t))] = E_Q^{x_1}[(K - X_1(T-t))^+]$$
$$= E_Q[(K - x_1\exp\{\beta\tilde{B}(T-t) + (\rho - \tfrac{1}{2}\beta^2)(T-t)\})^+].$$

From this we deduce that

(4) $$\frac{d}{dx_1}E_Q^{x_1}[(K - X_1(T-t))^+] = -E_Q[\mathcal{X}_{[0,K]}(x_1\exp\{\beta\tilde{B}(T-t)$$
$$+(\rho - \tfrac{1}{2}\beta^2)(T-t)\}) \cdot X_1^{(1)}(T-t)],$$

where $X_1^{(1)}(T-t) = \exp\{\beta\tilde{B}(T-t) + (\rho - \tfrac{1}{2}\beta^2)(T-t)\}$.
Hence

$$\phi_0(t) = \frac{d}{dx_1}E_Q^{x_1}[(K - X_1(T-t))^+]_{x_1=X_1(t)}\beta X_1(t) = e^{\rho T}\phi(s).$$

Substituting this into (2) we get

$$\hat{\theta}(t) = \theta_1(t)$$

$$= -e^{-\rho(T-t)} E_Q \left[\mathcal{X}_{[0,K]}(X_1(t) \exp\{\beta \widetilde{B}(T-t) \right.$$
$$\left. + (\rho - \tfrac{1}{2}\beta^2)(T-t)\}) X_1^{(1)}(T-t) \right]$$
$$= -(2\pi(T-t))^{-\frac{1}{2}} \int_{\mathbb{R}} \mathcal{X}_{[0,K]}(X_1(t) \exp\{\beta y + (\rho - \tfrac{1}{2}\beta^2)(T-t)\})$$

(5)
$$\cdot \exp\left\{\beta y - \tfrac{1}{2}\beta^2(T-t) - \frac{y^2}{2(T-t)}\right\} dy .$$

Note that $\theta_1(t) < 0$ for all $t \in [0,T]$. Hence it is necessary to *shortsell* at all times in order to replicate the European put option.

References

Aase, K.K. (1982): Stochastic continuous-time model reference adaptive systems with decreasing gain. Advances in Appl. Prop. **14**, 763–788

Aase, K.K. (1984): Optimum portfolio diversification in a general continuous time model. Stoch. Proc. and their Applications **18**, 81–98

Aase, K.K., Øksendal, B., Privault, N. and Ubøe, J. (2000): White noise generalizations of the Clark-Haussmann-Ocone theorem, with application to mathematical finance. Finance & Stochastics **4**, 465–496

Adler, R.J. (1981): The Geometry of Random Fields. Wiley & Sons

Andersen, E.S., Jessen, B. (1948): Some limit theorems on set-functions. Danske Vid. Selsk. Mat.-Fys. Medd. **25**, #5, 1–8

Applebaum, D. (2009): Lévy Processes and Stochastic Calculus. Cambridge University Press

Arnold, L. (1973): Stochastische Differentialgleichungen. Theorie und Anwendung. Oldenbourgh Verlag

Barles, G., Burdeau, J., Romano, M., Samsoen, N. (1995): Critical stock price near expiration. Math. Finance **5**, 77–95

Barndorff-Nielsen, O.E. (1998): Processes of normal inverse Gaussian type. Finance and Stochastics **2**, 41–68

Bather, J.A. (1970): Optimal stopping problems for Brownian motion. Advances in Appl. Prob. **2**, 259–286

Bather, J.A. (1997): Bounds on optimal stopping times for the American put. Preprint, University of Sussex

Beneš, V.E. (1974): Girsanov functionals and optimal bang-bang laws for final-value stochastic control. Stoch. Proc. and Their Appl. **2**, 127–140

Bensoussan, A. (1984): On the theory of option pricing. Acta Appl. Math. **2**, 139–158

Bensoussan, A. (1992): Stochastic Control of Partially Observable Systems. Cambridge Univ. Press

Bensoussan, A., Lions, J.L. (1978): Applications des inéquations variationelles en controle stochastique. Dunod. (Applications of Variational Inequalities in Stochastic Control. North-Holland)

Benth, F.E, (2004): Option Theory with Stochastic Analysis. Springer-Verlag

Bernard, A., Campbell, E.A., Davie, A.M. (1979): Brownian motion and generalized analytic and inner functions. Ann. Inst. Fourier **29**, 207–228

Bers, L., John, F., Schechter, M. (1964): Partial Differential Equations. Interscience

Biagini, F. and Øksendal, B. (2005): A general stochastic calculus approach to insider trading. Appl. Math. Optim. **52**, 167–181

Biagini, F., Hu, Y., Øksendal, B. and Zhang, T. (2007): Stochastic Calculus for Fractional Brownian Motion and Applications. Springer-Verlag (to appear)

Biais, B., Björk, T., Cvitanic, J., El Karoui, N., Jouini, E., Rochet, J.C. (1997): Financial Mathematics. Lecture Notes in Mathematics, Vol. 1656. Springer-Verlag

Björk, T. (2004): Arbitrage Theory in Continuous Time. Second Edition. Oxford University Press

Bingham, N.H., Kiesel, R. (1998): Risk-Neutral Valuation. Springer-Verlag

Black, F., Scholes, M. (1973): The pricing of options and corporate liabilities. J. Political Economy **81**, 637–654

Blumenthal, R.M., Getoor, R.K. (1968): Markov Processes and Potential Theory. Academic Press

Borodin, A.N., Salminen, P. (2002): Handbok of Brownian Motion – Facts and Formulae. Second Edition. Birkhäuser

Breiman, L. (1968): Probability. Addison-Wesley

Brekke, K.A., Øksendal, B. (1991): The high contact principle as a sufficiency condition for optimal stopping. In D. Lund and B. Øksendal (editors): Stochastic Models and Option Values. North-Holland, pp. 187–208

Brown, B.M., Hewitt, J.I. (1975): Asymptotic likelihood theory for diffusion processes. J. Appl. Prob. **12**, 228–238

Bucy, R.S., Joseph, P.D. (1968): Filtering for Stochastic Processes with Applications to Guidance. Interscience

Chow, Y.S., Robbins, H., Siegmund, D. (1971): Great Expectations: The Theory of Optimal Stopping. Houghton Mifflin Co

Chung, K.L. (1974): A Course in Probability Theory. Academic Press

Chung, K.L. (1982): Lectures from Markov Processes to Brownian Motion. Springer-Verlag

Chung, K.L., Williams, R. (1990): Introduction to Stochastic Integration. Second Edition. Birkhäuser

Clark, J.M. (1970, 1971): The representation of functionals of Brownian motion by stochastic integrals. Ann. Math. Stat. **41**, 1282–1291 and **42**, 1778, a correction

Cont, R. and Tankov, P. (2004): Finacial Modelling with Jump Processes. Chapman and Hall 2004

Csink, L., Øksendal, B. (1983): Stochastic harmonic morphisms: Functions mapping the paths of one diffusion into the paths of another. Ann. Inst. Fourier **330**, 219–240

Csink, L., Fitzsimmons, P., Øksendal, B. (1990): A stochastic characterization of harmonic morphisms. Math. Ann. **287**, 1–18

Cutland, N.J., Kopp, P.E., Willinger, W. (1995): Stock price returns and the Joseph effect: A fractional version of the Black-Scholes model. In Bolthausen, Dozzi and Russo (editors): Seminar on Stochastic Analysis, Random Fields and Applications. Birkhäuser, 327–351

Davis, M.H.A. (1977): Linear Estimation and Stochastic Control. Chapman and Hall

Davis, M.H.A. (1980): Functionals of diffusion processes as stochastic integrals. Math. Proc. Cambridge Phil. Soc. **87**, 157–166

Davis, M.H.A. (1984): Lectures on Stochastic Control and Nonlinear Filtering. Tata Institute of Fundamental Research 75. Springer-Verlag

Davis, M.H.A. (1993): Markov Models and Optimization. Chapman & Hall, London

Davis, M.H.A., Vinter, R.B. (1985): Stochastic Modelling and Control. Chapman and Hall

Delbaen, F., Schachermayer, W. (1994): A general version of the fundamental theorem of asset pricing. Math. Ann. **300**, 463–520

Delbaen, F., Schachermayer, W. (1995): The existence of absolutely continuous local martingale measures. Annals of Applied Probability **5**, 926–945

Delbaen, F., Schachermayer, W. (1998): The fundamental theorem of asset pricing for unbounded stochastic processes. Math. Annalen **312**, 215–250

Di Nunno, G., Meyer-Brandis, T., Øksendal, B. and Proske, F. (2006): Optimal portfolio for an insider in a market driven by Lévy processes. Quantitative Finance **6**, 83–94

Di Nunno, G., Øksendal, B. and Proske, F. (2009): Malliavin Calculus for Lévy Processes with Applications to Finance. Springer-Verlag

Dixit, A.K., Pindyck, R.S. (1994): Investment under Uncertainty. Princeton University Press

Doob, J.L. (1984): Classical Potential Theory and Its Probabilistic Counterpart. Springer-Verlag

Duffie, D. (1994): Martingales, arbitrage, and portfolio choice. First European Congress of Mathematics, vol. II, Birkhäuser, 3–21

Duffie, D. (1996): Dynamic Asset Pricing Theory. Second Edition. Princeton University Press

Durrett, R. (1984): Brownian Motion and Martingales in Analysis. Wadsworth Inc

Dynkin, E.B. (1963): The optimum choice of the instant for stopping a Markov process. Soviet Mathematics **4**, 627–629

Dynkin, E.B. (1965 I): Markov Processes, vol. I. Springer-Verlag

Dynkin, E.B. (1965 II): Markov Processes, vol. II. Springer-Verlag

Dynkin, E.B., Yushkevich, A.A. (1979): Controlled Markov Processes. Springer-Verlag

Eberlein, E. and Keller, U. (1995): Hyperbolic distributions in finance. Bernoulli **1**, 281–299

El Karoui (1981): Les aspects probabilistes du contrôl stochastique. Lecture Notes in Math. **876**, 73–238. Springer-Verlag

Elliott, R.J. (1982): Stochastic Calculus and Applications. Springer-Verlag

Elliott, R.J., Kopp, P.E. (2005): Mathematics of Financial Markets. Second Edition. Springer-Verlag

Elliott, R.J. and Van der Hoek, J. (2003): A general fractional white noise theory and applications to finance. Math. Finance **13**, 301–330

Elworthy, K.D. (1982): Stochastic Differential Equations on Manifolds. Cambridge University Press

Emery, M. (1989): Stochastic Calculus in Manifolds. Springer-Verlag

Fakeev, A.G. (1970): Optimal stopping rules for processes with continuous parameter. Theory Probab. Appl. **15**, 324–331

Fleming, W. (1999): Controlled Markov processes and mathematical finance. In Clarke, F.H. and Stern, R.J. (editors): Nonlinear Analysis, Differential Equations and Control. Kluwer, pp. 407–446

Fleming, W.H., Rishel, R.W. (1975): Deterministic and Stochastic Optimal Control. Springer-Verlag

Fleming, W.H., Soner, H.M. (1993): Controlled Markov Processes and Viscosity Solutions. Springer-Verlag

Folland, G.B. (1984): Real Analysis. J. Wiley & Sons

Freidlin, M. (1985): Functional Integration and Partial Differential Equations. Princeton University Press

Friedman, A. (1975): Stochastic Differential Equations and Applications, vol. I. Academic Press

Friedman, A. (1976): Stochastic Differential Equations and Applications, vol. II. Academic Press

Fukushima, M. (1980): Dirichlet Forms and Markov Processes. North-Holland/Kodansha

Gard, T.C. (1988): Introduction to Stochastic Differential Equations. Dekker

Gelb, A., Ed. (1974): Applied Optimal Estimation. MIT

Gihman, I.I., Skorohod, A.V. (1972): Stochastic Differential Equations. Springer-Verlag

Gihman, I.I., Skorohod, A.V. (1974): The Theory of Stochastic Processes, vol. I. Springer-Verlag

Gihman, I.I., Skorohod, A.V. (1975): The Theory of Stochastic Processes, vol. II. Springer-Verlag

Gihman, I.I., Skorohod, A.V. (1979a): The Theory of Stochastic Processes, vol. III. Springer-Verlag

Gihman, I.I., Skorohod, A.V. (1979): Controlled Stochastic Processes. Springer-Verlag

Grue, J. (1989): Wave drift damping of the motions of moored platforms by the method of stochastic differential equations. Manuscript, University of Oslo

Harrison, J.M., Kreps, D. (1979): Martingales and arbitrage in multiperiod securities markets. J. Economic Theory **20**, 381–408

Harrison, J.M., Pliska, S. (1981): Martingales and stochastic integrals in the theory of continuous trading. Stoch. Proc. and Their Applications **11**, 215–260

Harrison, J.M., Pliska, S. (1983): A stochastic calculus model of continuous trading: Complete markets. Stoch. Proc. Appl. **15**, 313–316

Hida, T. (1980): Brownian Motion. Springer-Verlag

Hida, T., Kuo, H.-H,, Potthoff, J., Streit, L. (1993): White Noise. An Infinite Dimensional Approach. Kluwer

Hoel, P.G., Port, S.C., Stone, C.J. (1972): Introduction to Stochastic Processes. Waveland Press, Illinois 60070

Hoffmann, K. (1962): Banach Spaces of Analytic Functions. Prentice Hall

Holden, H., Øksendal, B., Ubøe, J., Zhang, T. (2010): Stochastic Partial Differential Equations. Second Edition. Springer-Verlag

Hu, Y. (1997): Itô-Wiener chaos expansion with exact residual and correlation, variance inequalities. Journal of Theor. Prob. **10**, 835–848

Hu, Y., Øksendal, B. (2003): Fractional white noise calculus and applications to finance. Infinite Dim. Analysis, Quantum Probab. & Related Topics **6**, 1–32

Ikeda, N., Watanabe, S. (1989): Stochastic Differential Equations and Diffusion Processes. Second Edition. North-Holland/Kodansha

Itô, K. (1951): Multiple Wiener integral. J. Math. Soc. Japan **3**, 157–169

Itô, K., McKean, H.P. (1965): Diffusion Processes and Their Sample Paths. Springer-Verlag

Jacka, S. (1991): Optimal stopping and the American put. Mathematical Finance **1**, 1–14

Jacod, J. (1979): Calcul Stochastique et Problemes de Martingales. Springer Lecture Notes in Math. 714

Jacod, J., Shiryaev, A.N. (2003): Limit Theorems for Stochastic Processes. Second Edition. Springer-Verlag

Jaswinski, A.H. (1970): Stochastic Processes and Filtering Theory. Academic Press

Kallianpur, G. (1980): Stochastic Filtering Theory. Springer-Verlag

Kallianpur, G., Karandikar, R.L. (2000): Introduction to Option Pricing Theory. Birkhäuser

Karatzas, I. (1988): On the pricing of American options. Appl. Math. Optimization **17**, 37–60

Karatzas, I. (1997): Lectures on the Mathematics of Finance. American Mathematical Society

Karatzas, I., Lehoczky, J., Shreve, S.E. (1987): Optimal portfolio and consumption decisions for a 'Small Investor' on a finite horizon. SIAM J. Control and Optimization **25**, 1157–1186

Karatzas, I., Ocone, D. (1991): A generalized Clark representation formula, with application to optimal portfolios. Stochastics and Stochastics Reports **34**, 187–220

Karatzas, I., Shreve, S.E. (1991): Brownian Motion and Stochastic Calculus. Second Edition. Springer-Verlag

Karatzas, I., Shreve, S.E. (1997): Methods of Mathematical Finance. Springer-Verlag

Karlin, S., Taylor, H. (1975): A First Course in Stochastic Processes. Second Edition. Academic Press

Kloeden, P.E., Platen, E. (1992): Numerical Solution of Stochastic Differential Equations. Springer-Verlag

Knight, F.B. (1981): Essentials of Brownian Motion. American Math. Soc.

Kopp, P. (1984): Martingales and Stochastic Integrals. Cambridge University Press

Krishnan, V. (1984): Nonlinear Filtering and Smoothing: An Introduction to Martingales, Stochastic Integrals and Estimation. J. Wiley & Sons

Krylov, N.V. (1980): Controlled Diffusion Processes. Springer-Verlag

Krylov, N.V., Zvonkin, A.K. (1981): On strong solutions of stochastic differential equations. Sel. Math. Sov. **I**, 19–61

Kuo, H.H. (1996): White Noise Distribution Theory. CRC Press

Kushner, H.J. (1967): Stochastic Stability and Control. Academic Press

Lamperti, J. (1977): Stochastic Processes. Springer-Verlag

Lamberton, D., Lapeyre, B. (1996): Introduction to Stochastic Calculus Applied to Finance. Chapman & Hall

Levental, S., Skorohod, A.V. (1995): A necessary and sufficient condition for absence of arbitrage with tame portfolios. Ann. Appl. Probability **5**, 906–925

Lin, S.J. (1995): Stochastic analysis of fractional Brownian motions. Stochastics **55**, 121–140

Liptser, R.S., Shiryaev, A.N. (1977): Statistics of Random Processes, vol. I. Springer-Verlag

Liptser, R.S., Shiryaev, A.N. (1978): Statistics of Random Processes, vol. II. Springer-Verlag

Lungu, E.M., Øksendal, B. (1997): Optimal harvesting from a population in a stochastic crowded environment. Math. Biosciences **145**, 47–75

McDonald, R., Siegel, D. (1986): The valueof waiting to invest. Quarterly J. of Economics **101**, 707–727

McGarty, T.P. (1974): Stochastic Systems and State Estimation. J. Wiley & Sons

McKean, H.P. (1965): A free boundary problem for the heat equation arising from a problem of mathematical economics. Industrial Managem. Review **6**, 32–39

McKean, H.P. (1969): Stochastic Integrals. Academic Press

Malliaris, A.G. (1983): Itô's calculus in financial decision making. SIAM Review **25**, 481–496

Malliaris, A.G., Brock, W.A. (1982): Stochastic Methods in Economics and Finance. North-Holland

Malliavin, P. (1997): Stochastic Analysis. Springer-Verlag

Markowitz, H.M. (1976): Portfolio Selection. Efficient Diversification of Investments. Yale University Press

Maybeck, P.S. (1979): Stochastic Models, Estimation, and Control. Vols. 1–3. Academic Press

Merton, R.C. (1971): Optimum consumption and portfolio rules in a continuous-time model. Journal of Economic Theory **3**, 373–413

Merton, R.C. (1990): Continuous-Time Finance. Blackwell Publishers

Metivier, M., Pellaumail, J. (1980): Stochastic Integration. Academic Press

Meyer, P.A. (1966): Probability and Potentials. Blaisdell

Meyer, P.A. (1976): Un cours sur les intégrales stochastiques. Sem. de Prob. X. Lecture Notes in Mathematics, vol. 511. Springer-Verlag, 245–400

Musiela, M., Rutkowski, M. (1997): Martingale Methods in Financial Modelling. Springer-Verlag

Ocone, D. (1984): Malliavin's calculus and stochastic integral: representation of functionals of diffusion processes. Stochastics **12**, 161–185

Øksendal, B. (1984): Finely harmonic morphisms, Brownian path preserving functions and conformal martingales. Inventiones Math. **75**, 179–187

Øksendal, B. (1990): When is a stochastic integral a time change of a diffusion? Journal of Theoretical Probability **3**, 207–226

Øksendal, B. (1996): An Introduction to Malliavin Calculus with Application to Economics. Preprint, Norwegian School of Economics and Business Administration

Øksendal, B. (2005): A universal optimal consumption rate for an insider. Math. Finance **16**, 119–129

Øksendal, B. and Sulem, A. (2007): Applied Stochastic Control of Jump Diffusions. Second Edition. Springer-Verlag

Olsen, T.E., Stensland, G. (1987): A note on the value of waiting to invest. Manuscript CMI, N–5036 Fantoft, Norway

Pardoux, E. (1979): Stochastic partial differential equations and filtering of diffusion processes. Stochastics **3**, 127–167

Pham, H. (2009): Continuous-time Stochastic Control and Optimization with Financial Applications. Springer-Verlag

Platen, E. and Rebolledo, R. (1996): Principles for modelling financial markets. J. Appl. Probab. **33**, 601–630

Privault, N. (2009): Stochastic Analysis in Discrete and Continuous Settings. With Normal Martingales. Springer-Verlag

Protter, P. (2004): Stochastic Integration and Differential Equations. Second Edition. Springer-Verlag

Ramsey, F.P. (1928): A mathematical theory of saving. Economic J. **38**, 543–549

Rao, M. (1977): Brownian Motion and Classical Potential Theory. Aarhus Univ. Lecture Notes in Mathematics 47

Rao, M.M. (1984): Probability Theory with Applications. Academic Press

Revuz, D., Yor, M. (1991): Continuous Martingales and Brownian Motion. Springer-Verlag

Rogers, L.C.G. (1997): Arbitrage with fractional Brownian motion. Math. Finance **7**, 95–105

Rogers, L.C.G., Williams, D. (1987): Diffusions, Markov Processes, and Martingales. Vol. 2. J. Wiley & Sons

Rozanov, Yu.A. (1982): Markov Random Fields. Springer-Verlag

Samuelson, P.A. (1965): Rational theory of warrant pricing. Industrial Managem. Review **6**, 13–32

Schoutens, W. (2003): Lévy Processes in Finance. Wiley

Shiryaev, A.N. (1978): Optimal Stopping Rules. Springer-Verlag

Shiryaev, A.N. (1999): Essentials of Stochastic Finance. World Scientific

Shreve, S.E. (2004): Stochastic Calculus for Finance, vol. I, vol. II. Springer-Verlag

Simon, B. (1979): Functional Integration and Quantum Physics. Academic Press

Snell, J.L. (1952): Applications of martingale system theorems. Trans. Amer. Math. Soc. **73**, 293–312

Stein, J. (2006): Stochastic Optimal Control, International Finance, and Dept Crisis. Oxford University Press

Stratonovich, R.L. (1966): A new representation for stochastic integrals and equations. J. Siam Control **4**, 362–371

Stroock, D.W. (1971): On the growth of stochastic integrals. Z. Wahr. verw. Geb. **18**, 340–344

Stroock, D.W. (1981): Topics in Stochastic Differential Equations. Tata Institute of Fundamental Research. Springer-Verlag

Stroock, D.W. (1993): Probability Theory, An Analytic View. Cambridge University Press

Stroock, D.W., Varadhan, S.R.S. (1979): Multidimensional Diffusion Processes. Springer-Verlag

Sussmann, H.J. (1978): On the gap between deterministic and stochastic ordinary differential equations. The Annals of Prob. **6**, 19–41

Taraskin, A. (1974): On the asymptotic normality of vectorvalued stochastic integrals and estimates of drift parameters of a multidimensional diffusion process. Theory Prob. Math. Statist. **2**, 209–224

The Open University (1981): Mathematical models and methods, unit 11. The Open University Press

Topsøe, F. (1978): An information theoretical game in connection with the maximum entropy principle (Danish). Nordisk Matematisk Tidsskrift **25/26**, 157–172

Turelli, M. (1977): Random environments and stochastic calculus. Theor. Pop. Biology **12**, 140–178

Ubøe, J. (1987): Conformal martingales and analytic functions. Math. Scand. **60**, 292–309

Van Moerbeke, P. (1974): An optimal stopping problem with linear reward. Acta Mathematica **132**, 111–151

Williams, D. (1979): Diffusions, Markov Processes and Martingales. J. Wiley
  & Sons
Williams, D. (1981) (editor): Stochastic Integrals. Lecture Notes in Mathe-
  matics, vol. 851. Springer-Verlag
Williams, D. (1991): Probability with Martingales. Cambridge University
  Press
Wong, E. (1971): Stochastic Processes in Information and Dynamical Sys-
  tems. McGraw-Hill
Wong, E., Zakai, M. (1969): Riemann-Stieltjes approximations of stochastic
  integrals. Z. Wahr. verw. Geb. **12**, 87–97
Yor, M. (1992): Some Aspects of Brownian Motion, Part I. ETH Lectures in
  Math. Birkhäuser
Yor, M. (1997): Some Aspects of Brownian Motion, Part II. ETH Lectures
  in Math. Birkhäuser
Youg, J., Zhou, X.-Y. (1999): Stochastic Cobntrols. Springer-Verlag

List of Frequently Used Notation and Symbols

$\mathbf{R}^n$	n-dimensional Euclidean space
$\mathbf{R}^+$	the non-negative real numbers
$\mathbf{Q}$	the rational numbers
$\mathbf{Z}$	the integers
$\mathbf{Z}^+ = \mathbf{N}$	the natural numbers
$\mathbf{C}$	the complex plane
$\mathbf{R}^{n \times m}$	the $n \times m$ matrices (real entries)
A^T	the transposed of the matrix A
$|C|$	the determinant of the $n \times n$ matrix C
$\mathbf{R}^n \simeq \mathbf{R}^{n \times 1}$	i.e. vectors in $\mathbf{R}^n$ are regarded as $n \times 1$-matrices
$\mathbf{C}^n = \mathbf{C} \times \cdots \times \mathbf{C}$	the n-dimensional complex space
$|x|^2 = x^2$	$\sum\limits_{i=1}^{n} x_i^2$ if $x = (x_1, \ldots, x_n) \in \mathbf{R}^n$
$x \cdot y$	the dot product $\sum\limits_{i=1}^{n} x_i y_i$ if $x = (x_1, \ldots, x_n)$, $y = (y_1, \ldots, y_n)$
x^+	$\max(x, 0)$ if $x \in \mathbf{R}$
x^-	$\max(-x, 0)$ if $x \in \mathbf{R}$
$\operatorname{sign} x$	$\begin{cases} 1 & \text{if } x \geq 0 \\ -1 & \text{if } x < 0 \end{cases}$
$C(U, V)$	the continuous functions from U into V
$C(U)$	the same as $C(U, \mathbf{R})$
$C_0(U)$	the functions in $C(U)$ with compact support
$C^k = C^k(U)$	the functions in $C(U, \mathbf{R})$ with continuous derivatives up to order k
$C_0^k = C_0^k(U)$	the functions in $C^k(U)$ with compact support in U
$C^{k+\alpha}$	the functions in C^k whose k'th derivatives are Lipschitz continuous with exponent α
$C^{1,2}(\mathbf{R} \times \mathbf{R}^n)$	the functions $f(t, x) \colon \mathbf{R} \times \mathbf{R}^n \to \mathbf{R}$ which are C^1 w.r.t. $t \in \mathbf{R}$ and C^2 w.r.t. $x \in \mathbf{R}^n$

$C_b(U)$	the bounded continuous functions on U
$f \vert K$	the restriction of the function f to the set K
$A = A_X$	the generator of an Itô diffusion X
$\mathcal{A} = \mathcal{A}_X$	the characteristic operator of an Itô diffusion X
$L = L_X$	the second order partial differential operator which coincides with A_X on C_0^2 and with $\mathcal{A}_X$ on C^2
B_t (or $(B_t, \mathcal{F}, \Omega, P^x)$)	Brownian motion
$\mathcal{D}_A$	the domain of definition of the operator A
∇	the gradient: $\nabla f = (\frac{\partial f}{\partial x_1}, \ldots, \frac{\partial f}{\partial x_n})$
Δ	the Laplace operator: $\Delta f = \sum_i \frac{\partial^2 f}{\partial x_i^2}$
L	a semielliptic second order partial differential operator of the form $L = \sum_i b_i \frac{\partial}{\partial x_i} + \sum_{i,j} a_{ij} \frac{\partial^2}{\partial x_i \partial x_j}$
R_α	the resolvent operator
iff	if and only if
a.a., a.e., a.s.	almost all, almost everywhere, almost surely
w.r.t.	with respect to
s.t.	such that
$\simeq$	coincides in law with (see Section 8.5)
$E[Y] = E^\mu[Y] = \int Y d\mu$	the expectation of the random variable Y w.r.t. the measure μ
$E[Y \vert \mathcal{N}]$	the conditional expectation of Y w.r.t. $\mathcal{N}$
$\mathcal{F}_\infty$	the σ-algebra generated by $\bigcup_{t>0} \mathcal{F}_t$
$\mathcal{B}$	the Borel σ-algebra
$\mathcal{F}_t, \mathcal{F}_t^{(m)}$	the σ-algebra generated by $\{B_s; s \leq t\}$, B_s is m-dimensional
$\mathcal{F}_\tau$	the σ-algebra generated by $\{B_{s \wedge \tau}; s \geq 0\}$ (τ is a stopping time)
$\perp$	orthogonal to (in a Hilbert space)
$\mathcal{M}_t$	the σ-algebra generated by $\{X_s; s \leq t\}$ (X_t is an Itô diffusion)
$\mathcal{M}_\tau$	the σ-algebra generated by $\{X_{s \wedge \tau}; s \geq 0\}$ (τ is a stopping time)
∂G	the boundary of the set G
$\overline{G}$	the closure of the set G
G^C	the complement of the set G
$G \subset\subset H$	$\overline{G}$ is compact and $\overline{G} \subset H$
$d(y, K)$	the distance from the point $y \in \mathbf{R}^n$ to the set $K \subset \mathbf{R}^n$
τ_G	the first exit time from the set G of a process X_t: $\tau_G = \inf\{t > 0; X_t \notin G\}$
$\mathcal{V}(S,T), \mathcal{V}^n(S,T)$	Definition 3.3.1
$\mathcal{W}, \mathcal{W}^n$	Definition 3.3.2
(H)	Hunt's condition (Chapter 9)

HJB	the Hamilton-Jacobi-Bellman equation (Chapter 11)
I_n	the $n \times n$ identity matrix
$\mathcal{X}_G$	the indicator function of the set G $\mathcal{X}_G(x) = 1$ if $x \in G$, $\mathcal{X}_G(x) = 0$ if $x \notin G$
P^x	the probability law of B_t starting at x
$P = P^0$	the probability law of B_t starting at 0
Q^x	the probability law of X_t starting at x $(X_0 = x)$
$R^{(s,x)}$	the probability law of $Y_t = (s+t), X_t^x)_{t \leq 0}$ with $Y_0 = (s,x)$ (Chapter 10)
$Q^{s,x}$	the probability law of $Y_t = (s+t, X_{s+t}^{s,x})_{t \geq 0}$ with $Y_0 = (s,x)$ (Chapter 11)
$P \ll Q$	the measure P is absolutely continuous w.r.t. the measure Q
$P \sim Q$	P is equivalent to Q, i.e. $P \ll Q$ and $Q \ll P$
E^x, $E^{(s,x)}$, $E^{s,x}$	the expectation operator w.r.t. the measures Q^x, $R^{(s,x)}$ and $Q^{s,x}$, respectively
E_Q	the expectation w.r.t. the measure Q
E	the expectation w.r.t. a measure which is clear from the context (usually P^0)
$s \wedge t$	the minimum of s and t $(= \min(s,t))$
$s \vee t$	the maximum of s and t $(= \max(s,t))$
σ^T	the transposed of the matrix σ
δ_x	the unit point mass at x
δ_{ij}	$\delta_{ij} = 1$ if $i = j$, $\delta_{ij} = 0$ if $i \neq j$
θ_t	the shift operator: $\theta_t(f(X_s)) = f(X_{t+s})$ (Chapter 7)
$\theta(t)$	portfolio (see (12.1.3))
$V^\theta(t)$	$= \theta(t) \cdot X(t)$, the value process (see (12.1.4))
$V_z^\theta(t)$	$= z + \int_0^t \theta(s)dX(s)$, the value generated at time t by the self-financing portfolio θ if the initial value is z (see (12.1.7))
$\overline{X}(t)$	the normalized price vector (see (12.1.8)–(12.1.11))
$\xi(t)$	the discounting factor (see (12.1.9))
$:=$	equal to by definition
$\underline{\lim}$, $\overline{\lim}$	the same as lim inf, lim sup
ess inf f	$\sup\{M \in \mathbf{R}; \quad f \geq M \text{ a.s.}\}$
ess sup f	$\inf\{N \in \mathbf{R}; \quad f \leq N \text{ a.s.}\}$
$\square$	end of proof

"increasing" is used with the same meaning as "nondecreasing", "decreasing" with the same meaning as "nonincreasing". In the strict cases "*strictly increasing/strictly decreasing*" are used.

Index

CPSIA information can be obtained at www.ICGtesting.com
Printed in the USA
LVOW010949030613

336654LV00001B/2/P